Encyclopaedia of Mathematical Sciences

Volume 66

Editor-in-Chief: R. V. Gamkrelidze

Springer
Berlin
Heidelberg
New York
Barcelona
Budapest
Hong Kong
London
Milan
Paris
Tokyo

D. V. Anosov (Ed.)

Dynamical Systems IX

Dynamical Systems with Hyperbolic Behaviour

With 39 Figures

Springer

Consulting Editors of the Series:
A. A. Agrachev, A. A. Gonchar, E. F. Mishchenko,
N. M. Ostianu, V. P. Sakharova, A. B. Zhishchenko

Title of the Russian edition:
Itogi nauki i tekhniki, Sovremennye problemy matematiki,
Fundamental'nye napravleniya, Vol. 66
Dinamicheskie Sistemy 9
Publisher VINITI, Moscow 1991

Library of Congress Cataloging-in-Publication Data

Dynamical systems with hyperbolic behavior / D.V. Anosov (ed.).
p. cm. -- (Dynamical systems ; 9) (Encyclopaedia of mathematical sciences ; v. 66)
Translated from the Russian.
Includes bibliographical references and indexes.
ISBN 3-540-57043-8 (Berlin : acid-free) -- ISBN 0-387-57043-8 (New York : acid-free)
1. Differentiable dynamical systems. 2. Hyperbolic spaces. 3. Chaotic behavior in systems. I. Anosov, D. V. II. Series. III. Series: Dinamicheskie sistemy. English ; 9.
QA805.D5613 1988 vol. 9
[QA614.8]
515.352 s--dc20
[514'.74] 94-47311
CIP

Mathematics Subject Classification (1991):
58F15, 58F09, 58F10, 58F11, 58F13, 58F17, 58F18, 58F19, 34C35, 34D45, 34D30

ISBN 3-540-57043-8 Springer-Verlag Berlin Heidelberg New York
ISBN 0-387-57043-8 Springer-Verlag New York Berlin Heidelberg

Printed in Germany
Typesetting: Camera-ready copy from the translator using a Springer TEX macro package
SPIN: 10046430 41/3142 - 5 4 3 2 1 0 - Printed on acid-free paper

List of Editors, Authors and Translators

Editor-in-Chief

R. V. Gamkrelidze, Russian Academy of Sciences, Steklov Mathematical Institute, ul. Vavilova 42, 117966 Moscow, Institute for Scientific Information (VINITI), ul. Usievicha 20a, 125219 Moscow, Russia, e-mail: gam@ips.ac.msk.su

Consulting Editor

D. V. Anosov, Steklov Mathematical Institute, ul. Vavilova 42, 117966 Moscow GSP-1, Russia

Authors

D. V. Anosov, Steklov Mathematical Institute, ul. Vavilova 42, 117966 Moscow GSP-1, Russia
S. Kh. Aranson, Institute of Applied Mathematics and Cybernetics, Nizhny Novgorod State University, 603187 Nizhny Novgorod, Russia
V. Z. Grines, Institute of Agriculture, 603078 Nizhny Novgorod, Russia
R. V. Plykin, Institute of Atomic Energy, 249020 Obninsk, Russia
A. V. Safonov, Moscow Aviation Institute, 125871, Moscow, Russia
E. A. Sataev, Institute of Atomic Energy, 249020 Obninsk, Russia
S. V. Shlyachkov, Institute of Atomic Energy, 249020 Obninsk, Russia
V. V. Solodov, Institute of Information Transmission Problems of the Russian Academy of Sciences, 101447 Moscow, Russia
A. N. Starkov, Istra Branch of the Electrotechnical Institute, 143500 Istra, Russia
A. M. Stepin, Moscow State University, 117234 Moscow, Russia

Translator

G. G. Gould, School of Mathematics, University of Wales, Senghennydd Road, Cardiff, South Glamorgan CF2 4AG, Great Britain,
e-mail: smaggg@cardiff.ac.uk

Contents

Dynamical Systems with Hyperbolic Behaviour

D.V. Anosov, S.Kh. Aranson, V.Z. Grines, R.V. Plykin, A.V. Safonov, E.A. Sataev, S.V. Shlyachkov, V.V. Solodov, A.N. Starkov, A.M. Stepin

Translated from the Russian
by G.G. Gould

Contents

Preface

This volume is devoted to the "hyperbolic theory" of dynamical systems (DS), that is, the theory of smooth DS's with hyperbolic behaviour of the trajectories (generally speaking, not the individual trajectories, but trajectories filling out more or less "significant" subsets in the phase space. Hyperbolicity of a trajectory consists in the property that under a small displacement of any point of it to one side of the trajectory, the change with time of the relative positions of the original and displaced points resulting from the action of the DS is reminiscent of the motion next to a saddle. If there are "sufficiently many" such trajectories and the phase space is compact, then although they "tend to diverge from one another" as it were, they "have nowhere to go" and their behaviour acquires a complicated intricate character. (In the physical literature one often talks about "chaos" in such situations.) This type of behaviour would appear to be the opposite of the more customary and simple type of behaviour characterized by its own kind of stability and regularity of the motions (these words are for the moment not being used as a strict terminology but rather as descriptive informal terms).[1] The ergodic properties of DS's with hyperbolic behaviour of trajectories (Bunimovich et al. 1985) have already been considered in Volume 2 of this series. In this volume we therefore consider mainly the properties of a topological character (see below for further details).[2]

When at the beginning of the 60's the significance of hyperbolicity was recognized, investigations were first made of DS's in which such behaviour of a trajectory was expressed in the sharpest way — hyperbolicity is, so to speak, "total" (the hyperbolic character, noted above, of the change in time of the relative positions of two phase points, namely the original one and the slightly displaced one, holds under any direction of the displacement to one side of the original trajectory) and "uniform" (uniformity of the inequalities expressing the hyperbolicity with respect to the points, the displacements, and time). The precise formulation of this version of hyperbolicity has led to the so-called hyperbolic sets, whose properties (both topological and metric (ergodic)) have aroused great interest. But in ergodic theory a lot of significant work has been done since then, in connection with the weakening of the conditions of hyperbolicity in various directions. In the broadest terms, this progress

[1] The general definition of hyperbolicity is such that exponential stability turns out to be a particular case of it (cf. hyperbolicity of periodic trajectories in (Anosov et al. 1985, Chap. 1, Sect. 2.4)), while Morse-Smale systems (Anosov et al. 1985, Chap. 2, Sect. 3), which can be regarded as the simplest DS's, are the objects of hyperbolic theory. Thus it deals with both types of behaviour of trajectories, but its novelty is in connection with what occurs when there are "many" hyperbolic trajectories.

[2] In connection with our mention of the ergodic properties of DS's with hyperbolic behaviour of the trajectories it is appropriate to add that there has recently appeared a textbook on the ergodic theory of smooth DS's (Mañé 1987), almost half of which relates to hyperbolicity in one way or another.

is in connection with technical improvements rather than with new ideas of principle at the paradigmatic level. But however that may be, there have been considerable achievements in the metric theory occasioned by weakening the requirements of hyperbolicity. This has not been done in the topological theory (again speaking in the broadest terms). However, this does not mean that since the 60's there have been no significant achievements beyond the framework of total uniform hyperbolicity in the theory of smooth DS's, or that none of these achievements relates to DS's displaying some weaker form of hyperbolicity in their behaviour. Nevertheless, in the topological theory of DS's (so far?) no precise and workable notion of weakened hyperbolicity has been concocted. It would appear that here the gateway that might lead beyond the realm of total uniform hyperbolicity requires new ideas of principle. (At the same time, there remain many unsolved problems even within the bounds of these conditions.)

For all that, something now has emerged beyond the limits of the original version of hyperbolicity, although, as it seems to me, in the two most important cases, namely, DS's with a Lorenz attractor and certain cascades (DS's with discrete time) on surfaces, it is a question not so much of going beyond these limits, but rather of stretching them. (In essence, what have been weakened in these cases are not total and uniform hyperbolicity, but certain other "accompanying" conditions, so to speak. These are the conditions of continuity and smoothness (as ordinarily encountered in analysis); weakening them consists in violating them in some sense at certain points or on certain lines, and the condition that if a hyperbolic set of a flow contains equilibrium points, then they are isolated points of this set). In one way or another, these DS's are considered in the current volume. In this connection, certain information is also given on their metric properties, whereas questions of ergodic theory relating to hyperbolic sets are not touched upon in this volume; here we have nothing to add to the earlier results (given in (Bunimovich et al. 1985)).

The last two articles of this volume may outwardly appear to have no relation to hyperbolicity. In fact, the origin of their subject matter is only partially and fairly tenuously related to hyperbolicity. However in the development of this theme such connections have arisen and have turned out to be essential. These connections are a sufficient motivation for the inclusion of the last two articles in this volume although, of course, their contents deal with other matters. It is therefore worth saying a few words about these articles.

The investigation of cascades on closed surfaces is closely related to the classification of homeomorphisms of surfaces, more precisely, with the classification of isotopy classes of such homeomorphisms. The latter was the concern of topologists back in the 20's. One of the important questions here consisted in choosing "good" representatives of these isotopy classes; it must be recalled that successful representatives have usually turned out to be of great utility in many respects. Naturally, the representatives of some classes possess no hyperbolic properties. Until the 70's it was only such representatives that were known (leaving out the torus). Essential progress was achieved when "good" representatives in the remaining isotopy classes had been successfully chosen. Leaving aside "mixed" (in technical jargon, reducible) cases, when the corresponding homeomorphisms behave differently in different parts of the surface, it can be said that the new representatives generate cascades with "typically hyperbolic" properties. Their difference from objects of ordinary hyperbolic theory is related not to the weakening of hyperbolicity conditions, but to the slight breakdown of smoothness at certain points. In this manner, progress in the theory of homeomorphisms of surfaces in the 70's was in no small degree stimulated by the development of the hyperbolic theory of DS's in the 60's.

What has just been said is only one aspect of the theory of cascades on surfaces, this being the best known and most important. In the article included here, it is considered in the general context of this theory.

The article on certain DS's of algebraic origin, namely, homogeneous flows, occupies a special position in this volume. It is, to some extent, devoted to their ergodic properties. It might appear that this is close to Volume 2, but there is no such article there. The latter is partly accounted for by the fact that there was simply no room, due to the abundance of material in Volume 2. But the real reason is more one of principle. The investigation of the ergodic properties of homogeneous flows has its own specific character. Here an important role is played by the theory of Lie groups and their representations, while those ideas and methods forming the subject matter of Volume 2 recede into the background (although, of course, some of this is essentially used). In a number of cases homogeneous flows possess a certain amount of hyperbolicity (partial but uniform) and related to this are the corresponding geometric properties, as is the case throughout this volume. In essence, the algebra is then required for sorting out this geometry (without always mentioning this explicitly).

In connection with the discussion of the contents of this volume it is appropriate to mention three related sections of the theory of DS's that are not touched upon or only partly dealt with here. The first is certain questions of bifurcation theory where one has to deal with hyperbolicity. Information on this is contained in one of the earlier volumes of this edition (Arnol'd et al. 1986) (it is worth mentioning another new book (Wiggins 1988)); we have given only a few mentions of this. The second is "one-dimensional dynamics", that is, the study of iterations of maps (in general, non-invertible) in a one-dimensional real or complex domain[3] (in the latter case one talks about conformal dynamics whenever the iterated map is conformal). It arose independently of hyperbolic theory (and, if one is talking about conformal dynamics, somewhat earlier) but received an appreciable stimulus from the

[3] The reader should be warned that this terminology is used in another sense, namely, the dynamics of a one-dimensional "chain" (or some other system of particles, etc., on the line). The dimension of the phase space of such a DS is greater than 1 (and increases without bound as the number of particles increases, while if the system has distributed parameters, then it is infinite).

latter when it was realized (around 1970) that there was a "similarity" of behaviour of the trajectories between these two. There is also an influence the other way round: irreversible one-dimensional maps play an important auxiliary or heuristic role in the investigation of certain invertible DS's in higher dimensions that are peculiar to hyperbolic phenomena. It is only in this respect that these maps are referred to in the articles of the present volume. Even if such references had been discussed in greater detail, they would have reflected only part of one-dimensional dynamics and, of course, this arouses interest for other reasons. From a utilitarian point of view, it is simply a question that non-invertible one-dimensional maps arise in various questions of science and technology. Of course, invertible transformations relate more to the original ideas on DS's (as described at the beginning of (Anosov et al. 1985)), but there are nevertheless such problems that directly, or via some indirect route, lead to non-invertible one-dimensional maps.[4] In its conceptual aspects, one-dimensional dynamics admits the almost unique possibility of a fairly full investigation of the complicated behaviour of DS's; that is, the behaviour both in the sense of the qualitative picture in phase space and in the sense of dependence on the parameters.[5] In this connection it is worth mentioning the appropriate literature. In (Bunimovich et al. 1985) there is a small chapter on one-dimensional dynamics (primary attention being given to ergodic questions). In addition to the books and surveys referred to there, one can also mention (Eremenko and Lyubich 1989), (Lyubich 1986), (Sharkovskij et al. 1989), (Sharkovskij, Maistrenko and Romanenko 1986), (de Melo 1989), (Milnor 1990), (Nitecki 1982).

Another contiguous area left completely untouched in this book is hyperbolicity and bifurcations connected with it for infinite-dimensional systems. As is well known, reversion to such systems gives a satisfactory treatment of a number of problems for partial differential equations and for ordinary differential equations with a delay. It is natural that in this connection local questions (or questions of a similar character) should be developed in the first instance; but now hyperbolicity is also brought into play. Apparently, such

[4] With regard to "direct" examples it is customary first of all to refer to the "propagandist" article (May 1976). As for "indirect routes", there are apparently various examples; cf. the use of one-dimensional dynamics in the present volume and at the beginning of (Sharkovskij, Maistrenko and Romanenko 1986). Several bibliographical references to applications (both "direct" and "indirect") can be found in (Nitecki 1982).

[5] Thus, relatively amenable to investigation are those one-dimensional DS's for which opposite types of behaviour of the trajectories are realized in different parts of phase space, see the very beginning of this preface. Even if these actual types are known (albeit partially) for "ordinary" DS's (otherwise this book would not have been written!), what is less understood is the complicated way in which they can combine, or what else can occur. (The theory, of course, describes certain versions of such a combination, but unfortunately, it does not exhaust all the "typical" situations. Even for numerical experiments which, it might appear, ought clearly to demonstrate what is occurring in some DS or other, one does not always succeed in finding a convincing theoretical interpretation.)

non-local entities have not been reflected at the textbook or survey-article level.

Finally, I should like to draw attention to four books of a relatively popular character that bear a relation to our theme. The books (Devaney 1989) and (Ruelle 1989) are introductions to a wide range of questions relating, in particular, to the complex behaviour of trajectories. Further, in such a situation there are so-called "fractal" sets which are quite unlike the usual geometric figures, being so violently "jagged" that it is reasonable to ascribe a fractional dimension to them (whence their name). The book (Falconer 1990) serves as an introduction to this topic. There is an album with coloured diagrams representing the results of numerical experiment giving rise to such objects (three quarters of these diagrams relate to conformal dynamics) and supplied with a certain amount of explanatory text (Pitgen and Richter 1986). The figures are very beautiful — in this respect they produce the same impression both for the specialist and for the complete outsider, mathematician or not.

To read this volume one needs general familiarity with the elements of the theory of DS's, not so much with regard to any advanced theorems, but rather with the general system of concepts, terminology, and so on. All this is contained in the article "Smooth dynamical systems", published in Volume 1 of the present series (Anosov et al. 1985) and to some extent precedes all the articles on DS's included in the other volumes. The requisite material from other branches of mathematics is in the main summarized in the preface to (Anosov et al. 1985), where one can also find the notation of a general mathematical character that is used (which is fairly standard). In certain parts additional material is required which is either recalled or (as the authors hope) is clear from the context. The chapter on homogeneous flows naturally takes up a special position in this respect. The reading of it requires not episodical, but constant (and sufficiently systematic) acquaintance with a number of questions that go beyond the above framework. All this is mentioned at the beginning of that chapter.

D.V. Anosov

References*

Anosov, D.V., Aranson S. Kh., Bronshtein, I.U., Grines, V.Z. (1985): Dynamical Systems I: Smooth Dynamical Systems. Itogi Nauki Tekh., Ser. Sovrem. Probl. Mat., Fundam. Napravleniya **1**, 151–242. [English transl. in: Encycl. Math. Sci. **1**, 149–230. Springer-Verlag, Berlin Heidelberg New York 1988] Zbl. 605.58001

Arnol'd, V.I., Afraimovich, V.S., Il'yashenko, Yu.S., Shil'nikov, L.P. (1986): Dynamical Systems V: Bifurcation Theory. Itogi Nauki Tekh., Ser. Sovrem. Probl. Mat., Fundam. Napravleniya **5**, 5–218. [English transl. in: Encycl. Math. Sci. **5**. Springer-Verlag, Berlin Heidelberg New York 1993]

Bunimovich, L.A., Pesin, Ya.B., Sinai, Ya.G., Yakobson, M.V. (1985): Dynamical Systems II: Ergodic Theory of Smooth Dynamical Systems. Itogi Nauki Tekh., Ser. Sovrem. Probl. Mat., Fundam. Napravleniya **2**, 113–232. [English transl. in: Encycl. Math. Sci. **2**, 99–206. Springer-Verlag, Berlin Heidelberg New York 1989] Zbl. 781.58018

Devaney, R.L. (1989): An Introduction to Chaotic Dynamical Systems. 2nd ed. Addison-Wesley, Redwood City. Zbl. 632.58005

Eremenko, A.Eh., Lyubich, M.Yu. (1989): The dynamics of analytic transformations. Algebra Anal. **1**, 1–70. [English transl.: Leningr. Math. J. **1**, 563–634 (1990)] Zbl. 717.58029

Falconer, K.J. (1990): Fractal Geometry: Mathematical Foundations and Applications. Wiley, New York. Zbl. 689.28003

Lyubich, M.Yu. (1986): The dynamics of rational transformations: the topological picture. Usp. Mat. Nauk **41**, No. 4, 35–95. [English transl.: Russ. Math. Surv. **41**, No. 4, 43–117 (1986)] Zbl. 619.30033

Mañé, R. (1987): Ergodic Theory and Differentiable Dynamics. Springer-Verlag, Berlin Heidelberg New York. Zbl. 616.28007

May, R.M. (1976): Simple Mathematical Models with very Complicated Dynamics. Nature **261**, No. 5560, 459–467

de Melo, W. (1989): Lectures on One-Dimensional Dynamics. Instituto de Matematica Pura et Appl. de CNP-Q

Milnor, J. (1990): Dynamics in One Complex Variable: Introductory Lectures. Preprint Stony Brook: SUNY, Inst. Math. Sci., No. 5

Nitecki, Z. (1982): Topological dynamics on the interval. Ergodic theory and dynamical systems II. Proc. special year, Maryland, 1979–80. Prog. Math. **21**, 1–17. Zbl. 506.54035

Peitgen, H.-O, Richter, P.H. (1986): The Beauty of Fractals. Springer-Verlag, Berlin Heidelberg New York. Zbl. 601.58003

Ruelle, D. (1989): Elements of Differentiable Dynamics and Bifurcation Theory. Academic Press, Boston. Zbl. 684.58001

Sharkovskij, A.N., Kolyada, S.F., Sivak, A.G., Fedorenko, V.V. (1989): Dynamics of One-Dimensional Mappings. Naukova Dumka, Kiev (Russian). Zbl. 759.58001

Sharkovskij, A.N., Maistrenko, Yu.L., Romanenko, E.Yu. (1986): Difference Equations and their Applications. Naukova Dumka, Kiev. [English transl.: Kluwer, Dordrecht 1993] Zbl. 669.39001

Wiggins, S. (1988): Global Bifurcations and Chaos, Analytical Methods. Appl. Mat. Sci. **73**, Springer-Verlag, New York Berlin Heidelberg. Zbl. 661.58001

* For the convenience of the reader, references to reviews in Zentralblatt für Mathematik (Zbl.), compiled using the MATH database, have, as far as possible, been included in this bibliography.

Chapter 1
Hyperbolic Sets

D.V. Anosov, V.V. Solodov[1]

§1. Preliminary Notions

1.1. Definition of a Hyperbolic Set. Hyperbolicity of a compact (two-sided) invariant set A (that is, a set consisting of entire trajectories) of a flow or cascade $\{g^t\}$ ($t \in \mathbb{R}$ or $t \in \mathbb{Z}$) given on a phase manifold M is defined in terms of the restriction over A of the tangent linear extension $\{Tg^t\}$ (see Anosov et al. 1985, Chap. 1, Sect. 2.2), that is, the properties of the DS $\{Tg^t|p^{-1}A\}$, where $p : TM \to M$ is the natural projection. In other words, we are dealing with the behaviour of the solutions of the variational equations along the trajectories of our DS. (Speaking somewhat freely, we also have in mind not only "true" variational equations (and so on) for a flow, but also their analogues for a cascade $\{g^k\}$. In the latter case, the role of the system of variational equations along the trajectory $\{g^k x\}$ is played by the representation of $Tg^k(x)$ in the form of the composite

$$Tg(g^{k-1}x) \circ \ldots \circ Tg(gx) \circ Tg(x),$$

while the role of the solutions of this system (which is more important) is played by the function of discrete time $k \mapsto Tg^k(x)\zeta$ with $\zeta \in T_xM$. The map $Tg^k(x)$ plays the role of the Cauchy matrix of the system of variational equations along the trajectory $\{g^k x\}$.) It is worth recalling, in connection with the term "tangent linear extension", that in (Anosov et al. 1985, Chap. 3, Sect. 5.2) the more general notion of a linear extension of a DS was introduced. In this terminology the tangent linear extension $\{Tg^t\}$ is, in fact, a linear extension of the DS $\{g^t\}$, while its restriction to the closed invariant subset $TM|A = p^{-1}A$, which is a vector bundle over A, is the linear extension of the DS $\{g^t|A\}$.[2]

First we consider the simpler case of discrete time. By a *hyperbolic set* of a cascade $\{g^k\}$ (or diffeomorphism g) we mean a compact invariant set $A \subset M$

---

[1] Sect. 4.2 was written by V.V. Solodov, the rest by D.V. Anosov.

[2] We take the opportunity to correct a somewhat careless formulation in (Anosov et al. 1985). The phase velocity $\mathbf{V}$ of the flow $\{Tg^t\}$ is $JT\mathbf{v}$, where $\mathbf{v} : M \to TM$ is the phase velocity of $\{g^t\}$ and J is the standard involution in TTM. We take local coordinates x^i in M and use the local coordinates $(x^i, \delta x^i)$ in TM and $(x^i, \delta x^i, v^i, \delta v^i)$ in TTM associated with them. (The coordinates of a point $X \in TM$ are written in the following order: first the coordinates of the point $x = pX$, then the coordinates of X as a vector in T_xM; the latter form the "vector part" of the coordinates of X. Similarly for $Y \in TTM$ first we write the coordinates of $X \in TM$ in the fibre over which Y lies, then the "vector part" of the coordinates of Y, that is, the coordinates of Y as a vector in T_XTM.) In this terminology,

such that for each point $x \in A$ the tangent space T_xM decomposes into a direct sum

$$T_xM = E_x^{\mathrm{s}} \oplus E_x^{\mathrm{u}} \tag{1}$$

of two subspaces, namely, a *stable* (*shrinking*) *space* E and an *unstable* (*expanding*) *space* E_x^{u}, with the following properties: for $\xi \in E_x^{\mathrm{s}}$, $\eta \in E_x^{\mathrm{u}}$, $k \geq 0$

a) $|Tg^k(x)\xi| \leq a|\xi|e^{-ck}, \quad |Tg^{-k}(x)\xi| \geq b|\xi|e^{ck},$

b) $|Tg^k(x)\eta| \geq b|\eta|e^{ck}, \quad |Tg^{-k}(x)\eta| \leq a|\eta|e^{-ck},$

where a, b, c are positive constants that are independent of x, ξ, η, k. The norm of the tangent vector is taken with respect to a fixed Riemannian metric on M; if the latter is changed, one merely has to alter the constants a, b, c. (This is because A is compact. It is immaterial whether the phase manifold M is compact, since for the moment we are only interested in what is happening near A.[3]) Clearly the subspaces $E_x^{\mathrm{s}}, E_x^{\mathrm{u}}$ are uniquely defined by their properties a), b) (if $\zeta \in T_xM \backslash (E_x^{\mathrm{s}} \cup E_x^{\mathrm{u}})$, then $|Tg^k(x)\zeta| \to \infty$ as $|k| \to \infty$). It is also easy to prove that their dimensions are locally constant (as functions of $x \in A$), while the subspaces themselves depend continuously on x and

$$Tg(x)E_x^{\mathrm{s}} = E_{gx}^{\mathrm{s}},\ Tg(x)E_x^{\mathrm{u}} = E_{gx}^{\mathrm{u}}. \tag{2}$$

The unions $E^{\mathrm{s}} = \bigcup\{E_x^{\mathrm{s}};\ x \in A\}$, $E^{\mathrm{u}} = \bigcup\{E_x^{\mathrm{u}};\ x \in A\}$ are vector subbundles of the restriction $TM|A$ of the tangent bundle of M to A, and

---

$$(x^i, \delta x^i, v^i, \delta v^i) \overset{J}{\leftrightarrow} (x^i, v^i, \delta x^i, \delta v^i).$$

We verify that $\mathbf{V} = JT\mathbf{v}$. In fact,

$$\mathbf{v} : (x^i) \mapsto (x^i, v^i(x)), \quad T\mathbf{v} : (x^i, \delta x^i) \mapsto (x^i, v^i(x), \delta x^i, \delta v^i(x, \delta x)),$$

where

$$\delta v^i(x, \delta x) = \sum_j \frac{\partial v^i(x)}{\partial x^j} \delta x^j,$$

$\mathbf{V}(x, \delta x) = (x^i, \delta x^i,$ vector part of coordinates of $\mathbf{V}) = (x^i, \delta x^i, (x^i)^{\cdot}, (\delta x^i)^{\cdot}) = (x^i, \delta x^i, v^i(x), \delta v^i(x, \delta x)) = JT\mathbf{v}(x, \delta x)$.

It might be thought that the description of $\mathbf{V}$ should play a significant role. However, this is not the case. Meanwhile, in the hyperbolic theory the explicit form of the variational equations is used principally in the examination of geodesic flows (and flows related to them), while there is another description for these equations that is given in Riemannian geometry.

[3] For the same reason the map g need not be defined outside some neighbourhood of A; similarly, in the case of a flow $\{g^t\}$, to be considered later, it is permissible for $g^t x$ not to be defined for all t if $x \neq A$; it is also possible that the phase velocity field is defined only near A. The branches of the theory of DS's in which one is concerned with the behaviour of a trajectory not in the entire phase space but in a neighbourhood of some invariant set A, which is generally more "extensive" than in the traditional local theory (where one considers a neighbourhood of a periodic trajectory) are sometimes referred to as semilocal. In such theories the DS, properly speaking, may not be a DS in the strict sense of the word (cf. Anosov et al. 1985, Chap. 1, Sect. 1.6).

$TM|A = E^s \oplus E^u$ (Whitney sum). (Here we allow the slight but fairly obvious generalization of the notion of a vector bundle in which the fibres can have different dimensions over different parts of the base A which, of course, are at a positive distance from each other.) These subbundles are invariant with respect to Tg and are called the *stable* and *unstable bundles*, respectively (for A, g and $\{g^k\}$).

A closed invariant subset of a hyperbolic set and unions of a finite number of hyperbolic sets are hyperbolic sets. A hyperbolic set A decomposes into the disjoint closed invariant sets

$$A_j = \{x \in A;\ \dim E^s_x = j\}.$$

When studying the properties of the hyperbolic sets themselves (including the behaviour of trajectories near to them), we can consider the A_j individually, that is, we may suppose at the outset that $\dim E^s_x$ takes the same value throughout A. The situation is different, however, when one says that some set A is hyperbolic in some DS. (This is particularly the case if the definition of A by itself bears no relation to hyperbolicity and A plays a significant role in the global qualitative picture. Thus in what follows we shall give great attention to DS's in which the set of nonwandering points is hyperbolic.) In this case the constancy of E^s_x on A can certainly not be taken for granted; A may quite well consist of several A_j.

The simplest example of a hyperbolic set (in fact one already known to us) is a hyperbolic periodic trajectory l (see Anosov et al. 1985, Chap. 1, Sect. 2.4)[4] or a finite system of such trajectories. If $x \in l$ and the (minimal) period of l is equal to τ, then in accordance with the decomposition (1) we need to take as E^s_x, E^u_x the invariant subspaces of the linear transformation $Tg^\tau(x) : T_xM \to T_xM$ corresponding to the eigenvalues λ with $|\lambda| < 1$ or $|\lambda| > 1$ respectively. It is then clear that inequalities a) and b) hold for some a, b, c. The constants a, b, c can be chosen to be independent of the point x for the simple reason that there are now only a finite number of such points.

In this example the motion is highly regular (as in the Preface, regularity is not a precise term, but rather an appeal to intuition; what could be more regular than a periodic sequence?). Furthermore, they could well be (although they do not have to be) stable, in which case the stability is then exponential; this is the strongest version of stability (in the sense of the nature of the dependence of the trajectory on the starting point). But clearly, the notion of a hyperbolic set has not been introduced for the sake of this example. More complicated examples will be given in Sect. 2. The behaviour of the trajectories in them is quite different in that it is complicated, irregular, unstable and, to some degree, imitating a random process. It is remarkable that such disparate objects fall under the same common definition based on the essentially simple notion of uniform exponential behaviour of the solutions

[4] Hyperbolicity of l means that there are no eigenvalues (that is, eigenvalues of the transformation $Tg^\tau(x)$, where $x \in l$ and τ is the period of l) of modulus 1.

of the variational equations along all the trajectories in the set in question. (Here uniformity means that the constants a, b, c in a) and b) can be chosen the same for all x, ξ, η.)

We now consider the case of a flow (that is, a DS with continuous time) $\{g^t\}$ defined on a manifold M by means of a smooth vector field $\mathbf{v}$. By a *hyperbolic set* of a flow $\{g^t\}$ we mean a compact invariant set $A \subset M$ such that: 1) if a point of A is an equilibrium point, then it is hyperbolic[5] (consequently, there are only a finite number of such points); 2) the set

$$B = A \backslash \{x : \mathbf{v}(x) = 0\}$$

is closed and for each point $x \in B$ the tangent space $T_x M$ decomposes into a direct sum

$$T_x M = E_x^{\mathrm{s}} \oplus E_x^{\mathrm{u}} \oplus \mathbb{R}\mathbf{v}(x) \tag{3}$$

of subspaces, the third of which is spanned by the phase velocity vector, while the properties of the first two are similar to those of $E_x^{\mathrm{s}}, E_x^{\mathrm{u}}$ in the discrete-time case; namely, for $\xi \in E_x^{\mathrm{s}}, \eta \in E_x^{\mathrm{u}},\ t \geq 0$

$$\begin{array}{lll} \text{a)} & |Tg^t(x)\xi| \leq a|\xi|e^{-ct}, & |Tg^{-t}(x)\xi| \geq b|\xi|e^{ct}, \\ \text{b)} & |Tg^t(x)\eta| \geq b|\eta|e^{ct}, & |Tg^{-t}(x)\eta| \leq a|\eta|e^{-ct}, \end{array}$$

where a, b, c are positive constants independent of x, ξ, η, t. Everything that was said earlier (for discrete-time), namely, about the Riemannian metric, the nomenclature and properties of the subspaces $E_x^{\mathrm{s}}, E_x^{\mathrm{u}}$, the bundles formed by them, the finite unions of hyperbolic sets, the closed invariant subsets of hyperbolic sets and the parts of them with constant $\dim E_x^{\mathrm{s}}, \dim E_x^{\mathrm{u}}$, carries over word for word. Only this time there are points ζ in $T_x M$ for which $|Tg^t(x)\zeta|$ stays within certain positive limits for all t; these are the non-zero points of $\mathbb{R}\mathbf{v}(x)$ (we recall that

$$Tg^t(x)\mathbf{v}(x) = \mathbf{v}(g^t x), \tag{4}$$

(see Anosov et al. 1985, Chap. 1, Sect. 2.2) and only these. The subspace $\mathbb{R}\mathbf{v}(x)$ is called the *neutral* or *centre* subspace. The second name is explained by a certain analogy between it (and to an even greater extent the trajectory $\{g^t x\}$) and the centre manifolds in the local qualitative theory (see Il'yashenko 1985, Chap. 3, Sect. 4.2 and Chap. 6, Sect. 2.3), which also is distinguished by the "small" (non-exponential) speed of the motions occurring in it. Accordingly, this subspace can also be denoted by E_x^{n} or E_x^{c}, and sometimes the notation

[5] We recall (see Anosov et al. 1985, Chap. 1, Sect. 2.4) that in the theory of smooth DS's an equilibrium point is said to be hyperbolic if none of its eigenvalues (that is, the eigenvalues of the coefficient matrix corresponding to the linearized system) lies on the imaginary axis. (We do not exclude the possibility that their real parts are all of the same sign; in this regard the terminology differs from that which arose earlier in the qualitative theory of differential equations, where a hyperbolic point refers to a saddle, but not a node or a focus.)

E_x^0 (alluding to the "zero growth" of its elements under the action of $\{Tg^t\}$) is also used.

The one-dimensional vector bundle

$$E^{\mathrm{n}} = \{\mathbb{R}\mathbf{v}(x);\ x \in B\} \tag{5}$$

with base B is called the *neutral bundle* (or *centre bundle*, only then it is denoted by E^{c}; finally, the notation E^0 is used as well). Using this notation we can say that

$$TM|B = E^{\mathrm{s}} \oplus E^{\mathrm{u}} \oplus E^{\mathrm{n}}. \tag{6}$$

If there is an equilibrium point x_j in A, then we can formally append to the bundles $E^{\mathrm{s}}, E^{\mathrm{u}}$ (defined for the time being over B) the stable (or unstable) subspaces $E^{\mathrm{s}}_{x_j}, E^{\mathrm{u}}_{x_j} \subset T_{x_j}M$ (corresponding to the eigenvalues λ with $\operatorname{Re}\lambda < 0$ (or $\operatorname{Re}\lambda > 0$)) by regarding them as the fibres corresponding to the bundles over the isolated points x_j (as isolated points of A). At these points we have

$$T_{x_j}M = E^{\mathrm{s}}_{x_j} \oplus E^{\mathrm{u}}_{x_j}, \tag{7}$$

there being no neutral bundles, or we can say that E^{n} is zero-dimensional over these points. (Incidentally, $E^{\mathrm{s}}_{x_j}$ or $E^{\mathrm{u}}_{x_j}$ may be absent or, speaking more formally, reduce to zero.) These considerations provide the analogue of (6) for $TM|A$.

In the case of a flow the definition obtained would appear to be somewhat inhomogeneous. What is essentially new (by comparison, say, with (Anosov et al. 1985)) is the notion of a hyperbolic set B that does not contain an equilibrium point. Furthermore, we can include among the hyperbolic sets finite collections $\{x_1, \ldots, x_r\}$ of hyperbolic equilibrium points and, finally, unions of sets of these two types (B and $\{x_1, \ldots, x_r\}$) are also called hyperbolic. This addition of equilibrium points has much to recommend it and not merely for the convenience of certain formulations. It is occasioned by general ideas: hyperbolicity is characterized by the property that along any trajectories in A the uniform exponential behaviour is peculiar to the solutions of the variational equations, namely, all the solutions or (when this is impossible in view of (4)) solutions forming a set of codimension 1 in the set of all solutions, in which case the remaining solutions have a neutral component. (That is, in every case, there are as many solutions with purely exponential behaviour as is generally possible.) The inhomogeneity of the definition, noted above, simply indicates that in this regard there are in fact various possibilities for the case of continuous time; one of them is realized for equilibrium points (cf. (7)) and the other (6), already encountered in (Anosov et al. 1985, Chap. 1, Sect. 2.4), for hyperbolic closed trajectories.[6]

A set consisting of a finite number of hyperbolic periodic trajectories of a cascade or flow is a hyperbolic set. But a hyperbolic set may contain a

[6] As in (Anosov et al. 1985, Chap. 1, Sect. 1.3), a closed trajectory is a periodic trajectory that is not an equilibrium point. It is hyperbolic if it has no multipliers on the unit circle, apart from one.

countably infinite family of periodic trajectories (in the case of a flow, this means an infinite family of closed trajectories, since the number of equilibrium points in a hyperbolic set is always finite in view of its compactness and the isolatedness of such points). We note that the family of periodic trajectories contained in a hyperbolic set A is always at most countable and that A cannot reduce to an infinite family of periodic trajectories, for if A contains such a family then it also contains aperiodic trajectories (see Sect. 1.3).

If the dimension $\dim E_x^{\mathrm{u}}$ is constant for a hyperbolic set A, then it is called the (*Morse*) *index* of A. This is in agreement with the terminology adopted in (Anosov et al. 1985, Chap. 2, Sect. 2.5) for a hyperbolic periodic trajectory of a cascade or a hyperbolic equilibrium point of a flow, but for a hyperbolic closed trajectory of a flow, the index in (Anosov et al. 1985) is 1 higher than here. In general, a hyperbolic set consists of several parts of different indices.

The reader should be warned that in the case of a flow for hyperbolic sets of "type B" (that is, containing no equilibrium points) there is also another version of the notation and terminology. The spaces denoted above by $E_x^{\mathrm{s}}, E_x^{\mathrm{u}}$ are denoted by $E_x^{\mathrm{ss}}, E_x^{\mathrm{uu}}$ and are called *strongly stable* and *strongly unstable* spaces, the previous notation and terminology being reserved for their direct sums with the neutral space, so that the *stable* and *unstable* spaces are now defined as

$$E_x^{\mathrm{s}} = E_x^{\mathrm{ss}} \oplus \mathbb{R}\mathbf{v}(x), \quad E_x^{\mathrm{u}} = E_x^{\mathrm{uu}} \oplus \mathbb{R}\mathbf{v}(x). \tag{8}$$

(Similar notation and terminology is, of course, used for the corresponding bundles.) This proves convenient when one is dealing merely with flows, and this is what is done in (Anosov et al. 1985) for hyperbolic closed trajectories. However, when we are giving a parallel discussion of flows and cascades we shall use the version given at the beginning. In this version, instead of (8) we write

$$E_x^{\mathrm{sn}} = E_x^{\mathrm{s}} \oplus \mathbb{R}\mathbf{v}(x), \quad E_x^{\mathrm{un}} = E_x^{\mathrm{u}} \oplus \mathbb{R}\mathbf{v}(x) \tag{9}$$

(the order of the letters in the upper index can be reversed, for example, E_x^{ns}) and talk about *weakly stable* and *weakly unstable* spaces and bundles (this terminology is adopted in (Bunimovich et al. 1985, Chap. 7) for various objects in M itself associated with them, such as manifolds, foliations, laminations (see Sect. 1.3, 1.4 below)). The letter n, of course, refers to the nomenclature and notation for the bundle (5) and its fibres. If the adjective "neutral" is replaced by "centre", then n is replaced by c and one talks about *centre stable* and *centre unstable* subspaces and bundles. Finally, n is sometimes replaced by 0.

Hyperbolic sets were introduced by Smale (see Smale 1967a, 1967b). Earlier I had singled out a class of DS's for which, in modern terminology, the entire phase manifold is a hyperbolic set (now called *Anosov DS's*, see Sect. 1.4, 4.2). Their definition was based at the very outset on the hyperbolicity of the tangent linear extension. Even before that, Smale discovered another class of hyperbolic sets, the "horseshoes" (see Sect. 2.2). Their definition had (and still has) the nature of a prescription: such and such a construction gives a horseshoe.

1.2. Comments

a) As was noted earlier, we do not exclude the possibility that the bundles E^{s} or E^{u} are absent over some part $C \subset A$ (that is, their fibres reduce to zero). This, however, is possible only in the simplest case when C consists of a finite number of hyperbolic periodic trajectories (in the general meaning of the word, including equilibrium points) which are completely unstable when $\dim E_x^{\mathrm{s}} = 0$ over their points and stable when $\dim E_x^{\mathrm{u}} = 0$. This easily follows from the statement, given in Sect. 1.3 concerning a "periodic" (under the action of g^t) unstable or stable manifold. (Here C is at a positive distance from the rest of A, so that, from the point of view of the semilocal theory, the present situation "stands aside" as it were, like equilibrium points of the flow.) Note that when $\dim M = 1$ this case is the only possibility, since we then have $\dim E_x^{\mathrm{s}} + \dim E_x^{\mathrm{u}} \leq 1$, while for a flow with $\dim M = 2$ it follows from $\mathbf{v}(x) \neq 0$ that $\dim E_x^{\mathrm{s}} + \dim E_x^{\mathrm{u}} = 1$, that is,

$$A = C \cup \{\text{equilibrium points}\}.$$

Leaving aside the case $A \supset C$, we can say that close to any fixed trajectory in A the behaviour of the neighbouring trajectories (which are now not necessarily taken in A) in relation to this trajectory (that is, the changes in the relative positions of the points moving along these trajectories) is reminiscent of the behaviour of the trajectories near a saddle.

b) By the *direct product* of two DS's $\{g_1^t\}$, $\{g_2^t\}$ with phase spaces M_1, M_2 we mean the DS $\{h^t\} = \{g_1^t \times g_2^t\}$ acting "coordinatewise" on $M_1 \times M_2$:

$$h^t(x, y) = (g_1^t x, g_2^t y).$$

Suppose that the "factor" DS's are smooth. The direct product $A_1 \times A_2 \subset M_1 \times M_2$ of two hyperbolic sets $A_i \subset M_i$ is a hyperbolic set in precisely two cases: if the $\{g_i^t\}$ are cascades or if they are flows and (at least) one of the sets A_i reduces to equilibrium points. Here, in the obvious notation we have

$$E^{\mathrm{s}}_{(x,y)} = E^{\mathrm{s}}_{1x} \oplus E^{\mathrm{s}}_{2y}, \quad E^{\mathrm{u}}_{(x,y)} = E^{\mathrm{u}}_{1x} \oplus E^{\mathrm{u}}_{2y}.$$

The "components" on the right hand sides of these equalities are subspaces of $T_x M_1$, $T_x M_2$, while the operation of forming the direct sum is the same as the restriction to these subspaces of the direct sum in the natural formula

$$T_{(x,y)}(M_1 \times M_2) = T_x M_1 \oplus T_y M_2.$$

c) A hyperbolic set clearly remains hyperbolic under time reversal (see Anosov et al. 1985, Chap. 1, Sect. 1.4), while the bundles $E^{\mathrm{s}}, E^{\mathrm{u}}$ change roles, the first becoming unstable and the second stable. A hyperbolic set also remains hyperbolic under a smooth change of time in a flow (that is, the phase velocity $\mathbf{v}$ is replaced by $\phi\mathbf{v}$, where ϕ is a positive smooth function). This is obvious for an equilibrium point x_j, since in this case, in terms of local coordinates, the matrix of the linearized system is simply multiplied by the number $\phi(x_j)$. Here $E^{\mathrm{s}}_{x_j}$ and $E^{\mathrm{u}}_{x_j}$ are not altered, as is clear. Our claim is slightly less clear for a hyperbolic set of type B; in this case the previous E^{s}_x, E^{u}_x are no longer suitable, in general. The new E^{s}_x, which we denote by $\bar{E}^{\mathrm{s}}_x$, inclines (with respect to E^{s}_x), so to speak, to the side of the neutral subspace: there exists a continuous linear bundle map $z : E^{\mathrm{s}} \to E^{\mathrm{n}}$ (which is projected to the identity map of the bases, that is, $z(x)E^{\mathrm{s}}_x \subset \mathbb{R}\mathbf{v}(x)$) such that

$$\bar{E}^{\mathrm{s}}_x = \{\xi + z(x)\xi;\ \xi \in E^{\mathrm{s}}_x\}$$

(see (Anosov and Sinai 1967), where the particular case when $B = M$ is considered; the latter does not change anything).

d) Suppose that the flow $\{g^t\}$ in M is a Smale suspension over the diffeomorphism $f : V \to V$ (see Anosov et al. 1985, Chap. 1, Sect. 2.3). In terms of the description given there, we regard V as a submanifold of M by identifying V with $V \times 0$. Then f is the successor map induced by $\{g^t\}$ on the global section of V. If f has a hyperbolic set A then the trajectories of the flow $\{g^t\}$ passing through A form a hyperbolic set B of this flow (it is of exactly the same type as the previous B, that is, it contains no equilibrium points — they do not exist at all in the present case). Conversely, if B is a hyperbolic set of the flow $\{g^t\}$, then $A = B \cap V$ is a hyperbolic set of the cascade $\{f^k\}$ (and B is formed by the trajectories of the flow passing through A). Here it is not so important that the cross-section be global and the recurrence time constant; the latter is not essential at all , while if the former fails to hold, then we simply make the obvious stipulations. Suppose that we are given a smooth flow $\{g^t\}$ on some manifold M, V a cross-section, and f the corresponding successor map (defined, in general, on some open subset of V). If A is a hyperbolic set of the cascade $\{f^k\}$ (implying in particular, that all the f^k are defined on A), then the union B of the trajectories $\{g^t x\}$ of the points $x \in A$ is a hyperbolic set of the flow $\{g^t\}$ (clearly of "type B"). Conversely, if B is a hyperbolic set for $\{g^t\}$ such that all the trajectories in B intersect V infinitely often (so that, in particular, B is of the same type as the previous B) and $A = B \cap V$ is compact, then A is a hyperbolic set of the cascade $\{f^k\}$.

e) In (Anosov et al. 1985, Chap. 3, Sect. 5.2) the notion of a hyperbolic linear extension of a DS was given (there we formally talked about the linear extension of a flow, but in the present instance precisely the same definition is suitable for the case of a cascade). In this terminology one can say that the hyperbolicity of a closed invariant set A of a cascade $\{g^k\}$ is equivalent to that of the linear extension $\{Tg^k|p^{-1}A\}$ of the cascade $\{g^k|A\}$ (we recall that $p : TM \to M$ is the natural projection, so that $p^{-1}A$ is the restriction $TM|A$ of the bundle TM over the set A of the base). If we have a flow $\{g^t\}$ with velocity field $\mathbf{v}$, then it is more complicated to state in the same terms the definition of the hyperbolicity of a closed invariant set B containing no equilibrium points. In this case, hyperbolicity of B means that $TM|B$ can be represented as a Whitney sum $E^{\mathrm{h}} \oplus E^{\mathrm{n}}$ of two invariant bundles with respect to $\{Tg^t|p^{-1}B\}$ the second of which is our previous neutral bundle E^{n}, and the restriction $\{Tg^t|E^{\mathrm{h}}\}$ is hyperbolic (which is why we have used the index h). The following formulation seems to be more elegant. In view of the invariance of E^{n}, the induced maps

$$G^t : (TM|B)/E^{\mathrm{n}} \to (TM|B)/E^{\mathrm{n}},$$
$$TM_x/E^{\mathrm{n}}_x \to TM_{g^tx}/E^{\mathrm{n}}_{g^tx},$$
$$G^t(\xi \bmod E^{\mathrm{n}}_x) = Tg^t(x)\xi \bmod E^{\mathrm{n}}_{g^tx}$$

of the quotient bundle are well defined. They are fibrewise linear and induce a flow in the quotient bundle which is a linear extension of the flow $\{g^t|B\}$. It turns out that the hyperbolicity of B is equivalent to that of its linear extension.

f) If the linear extension $\{G^t\}$ acting in the vector bundle E is hyperbolic, then we can introduce a Euclidean metric in E so as to do away with the multipliers of the exponential functions in the inequalities for $|G^t\xi|$, $|G^t\eta|$ ($\xi \in E^{\mathrm{s}}$, $\eta \in E^{\mathrm{u}}$) expressing hyperbolicity (these are the constants a, b in formulae a), b) in Sect. 1.1). By analogy with the Lyapunov quadratic functions in stability theory (Arnol'd and Il'yashenko 1985, Chap. 1, Sect. 4) this metric was introduced in (Anosov 1967) as the *Lyapunov metric.* In the West it is also called the *Mather metric*, since it was also introduced by Mather in (Mather 1968). (Strictly speaking, (Anosov 1967) and (Mather 1968) were dealing with a particular case, but this is inessential.)

For a linear extension $\{G^k\}$ of a cascade in the base A the inequality referred to above reduces to the property that for $\xi \in E^{\mathrm{s}}$, $\eta \in E^{\mathrm{u}}$

$$|G\xi| \le \lambda|\xi|, \quad |G^{-1}\xi| \ge \frac{1}{\lambda}|\xi|, \quad |G\eta| \ge \frac{1}{\lambda}|\eta|, \quad |G^{-1}\eta| \le \lambda|\eta| \tag{9'}$$

for some $\lambda \in (0, 1)$. If one is dealing with $\{Tg^k|(TM|A)\}$, then the Lyapunov metric in $TM|A$ can be further extended to a continuous Riemannian metric on the whole of M. (It is also called a Lyapunov or Mather metric. Apparently it would be more precise to refer to it as a Riemannian metric that is Lyapunov on A.) As a result of the extension one may obtain a non-smooth metric. But there are arbitrarily smooth Lyapunov metrics on M. In fact, it is clear from (9′) that a metric on M that is sufficiently close to a Lyapunov metric (in the sense of C^0) is itself Lyapunov, so that it suffices to approximate the extended metric on M by a smooth metric.

For a linear extension $\{G^t\}$ of a flow in the base A the above inequality has the form:

$$|G^t\xi| \le e^{-ct}|\xi| \quad (t \ge 0) \tag{10}$$

and so on. However, we can achieve more. There is a Euclidean metric in E such that for any vector $\zeta \in E$ the derivative

$$\frac{d}{dt}\bigg|_{t=0} |G^t \zeta|^2$$

exists and is continuously dependent on $\zeta \in E$ (in terms of local coordinates it is jointly continuous in the fibre and the base), where

$$\begin{aligned} &\frac{d}{dt}\bigg|_0 |G^t \xi|^2 \le -\alpha |\xi|^2, \\ &\frac{d}{dt}\bigg|_0 |G^t \eta|^2 \ge \alpha |\eta|^2 \quad \text{for } \xi \in E^{\mathrm{s}},\ \eta \in E^{\mathrm{u}}, \end{aligned} \tag{11}$$

the constant $\alpha > 0$ being the same for all ξ, η (taken over all possible points of the base). It follows that estimates of the form (10) hold.

Meanwhile, for a hyperbolic set A of a flow $\{g^t\}$ with phase velocity $\mathbf{v}$ we were talking about the metric in $(TM|A)/E^{\mathrm{n}}$ or in E^{h}, rather than in $TM|A$. It can be extended to a metric in $TM|A$ such that $|\mathbf{v}(x)| = 1$ for all $x \in A$. The metric so obtained can then be extended "to the whole of M" (that is, to the whole of TM) so as to obtain a continuous Riemannian metric for which the "derivative due to the flow"

$$D_{\mathbf{v}} |\zeta|^2 = \frac{d}{dt}\bigg|_{t=0} |Tg^t \zeta|^2 \tag{12}$$

exists with the same continuity property with respect to ζ as above, only now on the whole of TM. Furthermore one can ensure that $|\mathbf{v}| = 1$ around A. (Of course, this condition cannot be guaranteed on the whole of M if M has equilibrium points.) Finally, this Riemannian metric can be approximated by a smooth Riemannian metric $\|\cdot\|$ (or even of class C^∞ or C^ω), whose derivative due to our flow is C^0-close to (12): for any $\epsilon > 0$ there exists $\|\cdot\|$ such that

$$|D_{\mathbf{v}} \|\zeta\|^2 - D_{\mathbf{v}} |\zeta|^2| < \epsilon |\zeta|^2. \tag{13}$$

The proof uses the arguments in (Anosov 1967, end of Sect. 6 and Sect. 7). Clearly inequalities of the form (11) hold for this metric on A, as before. But the equality $\|\mathbf{v}\| = 1$ does not hold in general, even on A itself; although in approximation one can obtain the following inequalities on A and around A:

$$1 - \epsilon < \|\mathbf{v}\| < 1 + \epsilon, \quad -\epsilon < D_{\mathbf{v}} \|\mathbf{v}\| < \epsilon. \tag{14}$$

With regard to the condition $\|\mathbf{v}\| = 1$, this too can be secured (with preservation of inequalities of the form (11)) by means of a further (sufficiently obvious) modification of the metric. However, here one must be satisfied that the metric is of the same smoothness class as the field $\mathbf{v}$. If the smoothness of $\mathbf{v}$ is not very high, then it may prove to be more convenient for the Riemannian metric to be, say, of class C^∞, but instead of the equality $\|\mathbf{v}\| = 1$ the weaker condition (14) would hold around A. (As is clear, in the case of a flow the expression "Lyapunov metric" can have some slightly different meanings.)

g) Finally, we look into the various modifications of the definition of hyperbolicity which, in the final reckoning, turn out to be equivalent to the definition in Sect. 1.1 but which, in certain cases, are more convenient to use. Properly speaking, the majority of these refer not to hyperbolic sets, but rather to hyperbolic linear extensions. Several of these modifications were indicated in (Anosov et al. 1985, Chap. 3, Sect. 5.2); this question is considered more fully in (Bronshtein 1984). We recall just two versions which so far have turned out to be the most useful. For the sake of brevity we shall refer only to cascades, although the first version is used also for hyperbolic sets of flows. We keep the earlier notation $p : E \to A$, $\{G^i\}$, $\{g^i\}$, E_x (the time is discrete, so that we write i instead of t).

1) In the first version one is dealing with families of cones (or, in another terminology, sectors) with specified properties. In essence, this is also a natural transfer to the given domain of the geometrically formulated notion of Lyapunov quadratic functions; this technique has been resorted to more than once. It was already used in (Anosov 1967), although the corresponding formulations were not explicitly provided there. Here is one version of the explicit statement.

Let L be a vector space with the Euclidean metric, and L_1 a vector subspace of it. By the *cone K of angle $\alpha \in (0, \pi/2)$ in L with axial space L_1* we mean the set of non-zero $\zeta \in L$ forming an angle less than α with L_1. (Cones in this sense are non-convex, in contrast to the situation when the other meaning of this term is applied. We do not exclude the cases $L_1 = \{0\}$ or $L_1 = L$, when $K = \emptyset$ or $K = L \backslash \{0\}$, although it is the case $0 < \dim L_1 < \dim L$ that is needed in practice.)

Suppose that for each point $x \in A$ in E_x we are given two non-intersecting cones $K_x^{\mathrm{s}}, K_x^{\mathrm{u}}$ with axial spaces $F_x^{\mathrm{s}}, F_x^{\mathrm{u}}$ such that

a) $E_x = F_x^{\mathrm{s}} \oplus F_x^{\mathrm{u}}$ for all $x \in A$;

b) $G^{-1}(K_x^{\mathrm{s}}) \subset K_{g^{-1}x}^{\mathrm{s}}$, $GK_x^{\mathrm{u}} \subset K_{gx}^{\mathrm{u}}$ for all $x \in A$;

c) there exist $a > 0$ and $\lambda > 1$ such that for all $x \in A$, $\xi \in K_x^{\mathrm{s}}$, $\eta \in K_x^{\mathrm{u}}$, $i \geq 0$

$$|G^{-i}\xi| \geq a\lambda^i |\xi|, \quad |G^i \eta| \geq a\lambda^i |\eta|.$$

(The angle of the cone may depend on x; it is not required that this dependence be continuous.) Then the linear extension is hyperbolic. Conversely if it is hyperbolic, then there exist cones with the indicated properties (and even of constant angle).

A somewhat different version is given in (Newhouse 1980) with a reference to the work of Newhouse and Palis. For hyperbolic sets of flows one can apply to both versions the formulation in which we are dealing with cones not in the bundle $\{T_x M / \mathbb{R}\mathbf{v}(x);\ x \in A\}$ but directly in $\{TM|A\}$. This departs from the general scheme in which the hyperbolicity of a set is a special case of the hyperbolicity of a linear extension but is more convenient.

Using Lyapunov metrics and cones it is easy to prove the following statement. Let A be a hyperbolic set of a DS $\{g^t\}$, and B a closed invariant set of the DS $\{h^t\}$ sufficiently close to $\{g^t\}$ (in the sense of C^1), the set B being entirely contained in a sufficiently small neighbourhood of A.[7] Then B is also hyperbolic (as an invariant set of the DS $\{h^t\}$).

2) We denote by $\Gamma(E)$ the vector space of continuous sections of the bundle E (that is, continuous maps $\phi : A \to E$ for which $p\phi = 1_A$) and turn it into a Banach space by providing it with the norm $\|\phi\| = \max\{|\phi(x)|;\ x \in A\}$. The following bounded linear operator G_* is induced in $\Gamma(E)$:

$$(G_*\phi)(x) = G(\phi(g^{-1}x)).$$

It turns out that the hyperbolicity of $\{G^k\}$ is equivalent to the property that the spectrum of G_* contains no points on the unit circle $|\lambda| = 1$. (As usual, when one is talking about the spectrum, the operator is understood to be extended in standard fashion to the complexified space $\Gamma_{\mathbb{C}}(E)$. The spectrum is the complement in $\mathbb{C}$ of the resolvent set, that is, of the set of those λ for which the operator $G_* - \lambda 1_{\Gamma_{\mathbb{C}}(E)}$ has an everywhere-defined bounded two-sided inverse.) Furthermore if the aperiodic points are everywhere dense in A, then the spectrum of G_* consists of the circles $|\lambda| =$ const. The hyperbolicity condition then reduces to the requirement that 1 does not belong to the spectrum.

This characterization of hyperbolicity was proposed (formally, in a particular case) by Mather (see (Mather 1968)).

The set of numbers $\{\ln|\lambda|;\ \lambda \in$ the spectrum of $G_*\}$ is sometimes called the *dynamical spectrum* of the linear extension. In this terminology Mather's hyperbolicity condition reads as follows: the dynamical spectrum of the tangent linear extension does not contain zero.

1.3. Stable and Unstable Manifolds. The *stable* and *unstable sets* $W^s(x)$, $W^u(x)$ of a point x (in a cascade or flow $\{g^t\}$) are defined as

$$W^s(x) = \{y; \rho(g^t x, g^t y) \to 0 \text{ as } t \to \infty\},$$
$$W^u(x) = \{y; \rho(g^t x, g^t y) \to 0 \text{ as } t \to -\infty\}.$$

---

[7] For flows the C^r-topology is the C^r-topology in the space of vector fields (phase velocity) of class C^r, while for cascades it is the topology in the space of C^r-diffeomorphisms generating them. It is convenient in our statements to use the C^r-metric ρ_{C^r} in the space of the DS, in spite of the arbitrariness of its choice. Probably the simplest of all is to embed M in some $\mathbb{R}^N$ and regard the vector fields or diffeomorphisms as maps $M \to \mathbb{R}^N$ (here it is not important that not all these maps correspond to vector fields or diffeomorphisms). One can then use the C^r-metric in the space of all such maps. This reduces to the C^r-norm in the space $C^r(M)$ of ordinary (scalar) functions of class C^r. However, the C^0-metric is understood in the usual "intrinsic" sense: for maps $M \to M$

$$\rho_C(f,g) = \sup\{\rho(fx, gx);\ x \in M\},$$

where the distance is taken in the sense of the Riemannian metric on M.

Here ρ is the distance on M in the sense of some Riemannian metric, the specific choice of which is immaterial in view of the compactness of M. (The term "unstable" is somewhat conventional in the general case; rather it signifies "stability under time reversal". Notwithstanding the etymology, this "unstability" does not exclude "stability"; it can even happen that $W^{\mathrm{s}}(x) = W^{\mathrm{u}}(x) \neq \{x\}$. See Fig. 1, where x is an equilibrium point.)

If L is the trajectory of a cascade, then its *stable* and *unstable sets* are, by definition,

$$W^{\mathrm{s}}(L) = \bigcup\{W^{\mathrm{s}}(x);\ x \in L\},\ W^{\mathrm{u}}(L) = \bigcup\{W^{\mathrm{u}}(x);\ x \in L\}.$$

The situation is more complicated if $L = \{g^t x\}$ is the trajectory of a flow. Then, by definition, $y \in W^{\mathrm{s}}(L)$ (or $y \in W^{\mathrm{u}}(L)$) if and only if there exists a surjective homeomorphism h of the positive (or negative) semi-axis such that $\rho(g^{h(t)}x, g^t y) \to 0$ as $t \to \infty$ (or as $t \to -\infty$). Thus $W^{\mathrm{s}}(L)$ or $(W^{\mathrm{u}}(L))$ consists of trajectories approaching L infinitesimally closely as $t \to \infty$ (or as $t \to -\infty$). Here it needs to be made clear that this approach is to be understood in the natural sense as applied to oriented curves. (Hence, of couse, it follows that $g^t y$ approaches the set L infinitesimally closely, but the latter property is not equivalent, in general, to the approach of trajectories in our sense, although for sufficiently simple (say, periodic) L the two notions are equivalent.)

Clearly we have $x \in W^{\mathrm{s}}(x) \cap W^{\mathrm{u}}(x)$, $L \subset W^{\mathrm{s}}(L) \cap W^{\mathrm{u}}(L)$ and if the point y lies in $W^{\mathrm{s}}(x)$ (or in $W^{\mathrm{u}}(x)$), then $x \in W^{\mathrm{s}}(y)$ (or $x \in W^{\mathrm{u}}(y)$). Similarly, if the trajectory L_1 lies in $W^{\mathrm{s}}(L)$ (or in $W^{\mathrm{u}}(L)$), then $L \subset W^{\mathrm{s}}(L_1)$ (or $L \subset W^{\mathrm{u}}(L_1)$). In other words, the sets $W^{\mathrm{s}}(x), W^{\mathrm{u}}(x), W^{\mathrm{s}}(L), W^{\mathrm{u}}(L)$ form partitions of the phase space M. Clearly, $g^t W^{\mathrm{s}}(x) = W^{\mathrm{s}}(g^t x)$, $g^t W^{\mathrm{u}}(x) = W^{\mathrm{u}}(g^t x)$, $g^t W^{\mathrm{s}}(L) = W^{\mathrm{s}}(L)$, $g^t W^{\mathrm{u}}(L) = W^{\mathrm{u}}(L)$.

In the case that we need, when the point x or the trajectory L lie in a hyperbolic set A, the sets $W^{\mathrm{s}}(x), W^{\mathrm{u}}(x), W^{\mathrm{s}}(L), W^{\mathrm{u}}(L)$ are submanifolds of the phase manifold M (only immersed, in general) having the same smoothness class as the DS under consideration. That is why in this case one talks about *stable* and *unstable manifolds* of the point and the trajectory. Furthermore, it turns out that in the case of a flow we also have

$$W^{\mathrm{s}}(L) = \bigcup\{W^{\mathrm{s}}(x);\ x \in L\}, \quad W^{\mathrm{u}}(L) = \bigcup\{W^{\mathrm{u}}(x);\ x \in L\}. \tag{15}$$

This is connected with the fact that if $g^t y$ approaches the trajectory $L = \{g^t x\}$ infinitesimally closely (lying in a hyperbolic set, as before), then $\rho(g^t y, g^t g^\tau x) \to 0$ for some τ (called the "*asymptotic phase*" of the motion $\{g^t y\}$; when required, one must say, more precisely: "asymptotic phase on the trajectory of L with initial point x"). Then the tangent spaces to these submanifolds at the point x are given by

$$T_x W^{\mathrm{s}}(x) = E^{\mathrm{s}}_x, \quad T_x W^{\mathrm{u}}(x) = E^{\mathrm{u}}_x,$$

where $W^{\mathrm{s}}(x)$ and $W^{\mathrm{u}}(x)$ are diffeomorphic to E^{s}_x and E^{u}_x. In the case of a cascade $W^{\mathrm{s}}(L)$ and $W^{\mathrm{u}}(L)$ are disconnected (and have $W^{\mathrm{s}}(x)$ (or $W^{\mathrm{u}}(x)$) as

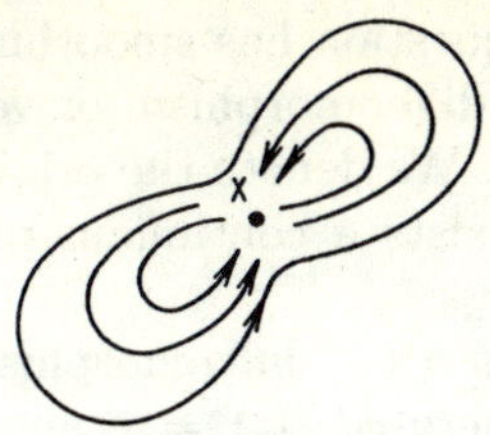

Fig. 1

their connected components) provided that L does not reduce to a fixed point. For the case of a flow, if L is not an equilibrium point, then

$$T_xW^{\mathrm{s}}(L) = E_x^{\mathrm{sn}}, \quad T_xW^{\mathrm{u}}(L) = E_x^{\mathrm{un}},$$

where $W^{\mathrm{s}}(L)$ and $W^{\mathrm{u}}(L)$ are diffeomorphic to E_x^{sn} and E_x^{un}, except for the case when the map

$$W^{\mathrm{s}}(x) \times \mathbb{R} \to W^{\mathrm{s}}(L) \quad (y,t) \mapsto g^t y$$

or the analogous map for W^{u} is not injective. Under this map $W^{\mathrm{s}}(x) \times t$ is taken to $g^tW^{\mathrm{s}}(x) = W^{\mathrm{s}}(g^tx)$, therefore the injectiveness fails only in the case when $W^{\mathrm{s}}(g^{t_1}x) = W^{\mathrm{s}}(g^{t_2}x)$ for some $t_1 \neq t_2$, that is, $W^{\mathrm{s}}(x)$ is "periodic" (under the action of $\{g^t\}$):

$$g^\tau W^{\mathrm{s}}(x) = W^{\mathrm{s}}(g^\tau x) = W^{\mathrm{s}}(x) \text{ for some } \tau > 0. \tag{16}$$

It turns out that, both in the case of a flow and a cascade, (16) can hold only when $W^{\mathrm{s}}(x)$ contains a periodic point y of period τ that also belongs to A. The minimal period of y may be some kth submultiple of τ, in which case we clearly have

$$g^{\tau/k}W^{\mathrm{s}}(x) = g^{\tau/k}W^{\mathrm{s}}(y) = W^{\mathrm{s}}(y) = W^{\mathrm{s}}(x).$$

Thus $W^{\mathrm{s}}(L)$ fails to be diffeomorphic to E_x^{sn} if and only if $W^{\mathrm{s}}(L)$ contains a closed trajectory L_1 (if $L_1 \neq L$, then L is "coiled" on L_1) with $L_1 \subset A$, so that it makes sense to talk about the vector bundle $E^{\mathrm{s}}|L_1$. Then $W^{\mathrm{s}}(L) = W^{\mathrm{s}}(L_1)$ is diffeomorphic to the space of this bundle. All this is also true for W^{u}.

We note that in the general case (when no conditions, such as lying in a hyperbolic set, are imposed on x or L) W^{s} and W^{u} may not be manifolds (see, for example, Fig. 1; this example is still very "tame"), while for a trajectory of a flow (even a closed trajectory) (15) may not hold (when g^ty approaches L slowly there may not exist an asymptotic phase).

So far we have been dealing with $W^{\mathrm{s}}(x)$ for an individual point x in a hyperbolic set A or for points of an individual trajectory $L \subset A$. There arises the question of the dependence of $W^{\mathrm{s}}(x)$ on x. Each $W^{\mathrm{s}}(x)$ is tangent to E_x^{s} at x and is diffeomorphic to E_x^{s}. This means that $\bigcup\{W^{\mathrm{s}}(x);\ x \in A\}$ can be naturally "equated" to the entire bundle E^{s}.

Suppose that the DS in question has smoothness class C^r (that is, this is the smoothness class of the diffeomorphism or vector field generating it; the case $r = \infty$ is not excluded). We denote the origin of E^s_x by 0_x.

It turns out that there exists a continuous map $\phi^s : E^s \to M$ with the following properties:

1) $\phi^s_x = \phi^s | E^s_x$ establishes a C^r-diffeomorphism between E^s_x and the immersed manifold $W^s(x)$, where $\phi^s_x(0_x) = x$ and $T\phi^s_x(0_x) = 1_{E^s_x}$ under the natural identification $T_{0_x}E^s_x = E^s_x$.

2) For $r = \infty$ all the fibrewise jets ϕ^s, and for $r < \infty$ all the fibrewise jets ϕ^s of order $\leq r$ are, so to speak, jointly continuous with respect to the fibre and the base.

We recall that the jth order jets of a smooth map $F : N^k \to M^m$ at the point $y \in N^k$ are represented in terms of the local coordinates $y_1, \ldots, y_k$ in N and $w_1, \ldots, w_m$ in M as the set of all partial derivatives

$$\partial^h w_i / \partial y_1^{h_1} \ldots \partial y_k^{h_k}, \quad h_1, \ldots, h_k \geq 0, \ h_1 + \ldots + h_k = h$$

with $h = 0, \ldots, j$, calculated at the point y (that is, at the values of the local coordinates corresponding to this point). The fibrewise jets are, of course, the jets of the restrictions of ϕ^s to the fibres. In our case the latter are vector spaces and we can use ordinary cartesian coordinates in them. If E^s were a direct product $A \times \mathbb{R}^k$, then "joint continuity of the fibrewise jets with respect to the base and fibre" would mean continuity of the corresponding partial derivatives

$$\partial^h w_i(x, y) / \partial y_1^{h_1} \ldots \partial y_k^{h_k}$$

with respect to $(x, y) \in A \times \mathbb{R}^k$. However, in our case, we need to use local representations of the vector bundle in the form of a direct product.

Everything that has been said concerning $\{W^s(x)\}$ and E^s is also true for $\{W^u(x)\}$ and E^u, that is, the unstable manifolds $W^u(x)$ are images of E^u_x under some continuous map $\phi^u : E^u \times M$ whose properties are analogous to those of ϕ^s. The assertions concerning the existence of suitable ϕ^s, ϕ^u are obtained from each other via time reversal.

The determination of the maps is by no means unique. Along with some fixed ϕ^s any $\phi^s h$ is suitable, where h is a fibrewise homeomorphism $E^s \to E^s$ that is a sufficiently smooth diffeomorphism of the fibres with fibrewise jets that are jointly continuous in the base and the fibre, and where $h(0_x) = 0_x$ and $T_{0_x}(h | E^s_x) = 1_{E^s_x}$. It would appear that there are no reasonable grounds for considering some specific choice of ϕ^s, ϕ^u to be preferable. The proof of the existence of ϕ^s, ϕ^u is carried out by some construction of them; but we will refrain from mentioning that it is not clear why the ϕ^s, ϕ^u so constructed are "better" than others; this very construction contains an arbitrariness. The latter is entirely due to the arbitrariness in carrying out certain preliminary steps (say, in the choice of the Lyapunov Riemannian metric), but it is an arbitrariness just the same.

Generally speaking, the map ϕ^s is not injective, since it is entirely possible that there are points $x \neq y$ in A such that $W^s(x) = W^s(y)$. (Then $\phi^s E^s_x = \phi^s E^s_y$, although $E^s_x \cap E^s_y = \emptyset$). Furthermore if ϕ^s is injective, that is, if $u, v \in A$, $u \neq v$ implies that $W^s(u) \neq W^s(v)$, then it can be proved that A reduces to a finite collection of periodic trajectories. (From this the assertion in Sect. 1.1 would follow, namely, that if A consists entirely of periodic trajectories then there are a finite number of them. This is because two periodic points cannot lie in one stable manifold.)

In many cases it is not the manifolds $W^s(x)$ in their entirety that are required, but only small "pieces" of them around x, the so-called *local stable manifolds.* They can be denoted by $W^s_{\text{loc}}(x)$ or (if one needs to point out that they are "of size ϵ") by $W^s_\epsilon(x)$. This expression and notation can be given a precise meaning in various ways. The situation here is reminiscent of the situation with the word "neighbourhood of a point x": it is true that it has a completely precise "wide" meaning (open set containing x or any superset of such an open set), but one must not forget that one often uses neighbourhoods of various special types, depending on the context. In the wide sense $W^s_{\text{loc}}(x)$ is a neighbourhood of x on $W^s(x)$ in the topology of the immersed manifold. This too is a precise notion but too wide; one ordinarily uses local manifolds of several specialized types. Here are some versions of the meaning of the notation $W^s_\epsilon(x)$.

1) $W^s_\epsilon(x)$ is the ϵ-neighbourhood of the point x on the immersed manifold $W^s(x)$ in the Riemannian metric induced on $W^s(x)$ by some Riemannian metric on M.

2) $W^s_\epsilon(x)$ is the connected component containing x of the intersection of $W^s(x)$ with the ϵ-neighbourhood $U_\epsilon(x)$ of x on M.

3) $W^s_\epsilon(x) = \{y; g^t y \in U_\epsilon(g^t x) \text{ for all } t \geq 0, \lim\limits_{t\to\infty} \rho(g^t y, g^t x) = 0\}$.
In the case of a cascade the limit condition can be omitted.

4) For sufficiently small $r > 0$ the exponential geodesic map $\exp_x : T_x M \to M$ (associating the vector $\zeta \in T_x M$ with the point lying on the geodesic line emanating from x in the direction of ζ and at a distance $|\zeta|$ along this line from x) diffeomorphically maps the region

$$\{\xi + \eta;\ \xi \in E^s_x,\ \eta \in E^u_x,\ |\xi| < r,\ |\eta| < r\} \quad \text{in the case of a cascade,}$$
$$\{\xi + \eta;\ \xi \in E^s_x,\ \eta \in E^{un}_x,\ |\xi| < r,\ |\eta| < r\} \quad \text{in the case of a flow}$$

to a neighbourhood $D_r(x)$ of x in M; in this way, the "coordinates" (ξ, η) are introduced into $D_r(x)$. In these terms $W^s_\epsilon(x)$ is a neighbourhood of x on $W^s(x)$ which is the graph of some function $\eta = \eta_x(\xi)$, $|\xi| \leq \epsilon$.

If we agree to use the Lyapunov metric and replace $U_\epsilon(x)$ in 2) and 3) by $D_r(x)$ in 4), then for sufficiently small ϵ versions 2), 3) and 4) coincide. But this stipulation itself also involves an act of arbitrariness; however, we have already spoken about the arbitrariness in the choice of the Lyapunov metric.

With appropriate modifications, everything mentioned above applies to W^u as well.

The construction of the stable manifolds begins with the construction of $W^s_\epsilon(x)$ of type 4); here we usually use the Lyapunov metric. In fact the map ϕ^s is constructed on some ϵ-neighbourhood of the zero section of the bundle E^s. En route one finds that if $\lim_{t\to\infty} \rho(g^t x, g^t y) = 0$, then $g^t y \in W^s_\epsilon(g^t x)$ for some $t > 0$. The latter means that the set

$$W^s = \bigcup \{g^{-t} W^s_\epsilon(g^t x);\ t > 0\}$$

is stable, after which it is a straightforward matter to deduce that it is an immersed submanifold with the properties indicated above. In this connection, the extension of ϕ^s to a map $E^s \to M$ is realized, in the main, by means of the DS itself; only we still need some "transition zone" where a smooth transition occurs from the already constructed ϕ^s to this extension.

The construction of $W^s_\epsilon(x)$ is a modification of a similar construction for equilibrium points and fixed points. Historically, the latter was realized by four methods (the first of which is applicable only in the analytic case): the power-series expansion of the function whose graph is $W^s_\epsilon(x)$; the geometric considerations of Bohl (see Bohl 1974 and Bohl 1904); the methods of Hadamard and Perron (see Anosov 1967 for references to these and other historical information). The last two methods are used in modern expositions and are suitable for various modifications as required, for example, for the purposes of the hyperbolic theory (where one needs to obtain more specific information on the dependence of $W^s_\epsilon(x)$ on x; this means that the proof itself needs to be modified). These methods exist in various versions; see (Anosov 1967) (and my earlier work cited there), (Hirsch and Pugh 1970), (Hirsch, Pugh, and Shub 1970), (Hirsch, Pugh, and Shub 1977) (taking into account the corrections in (Foster 1976)), (Irwin 1970) and (Irwin 1972). The corresponding results are called the Hadamard-Perron theorem in view of the immense significance of the work of these two mathematicians. Accounts of this theorem for hyperbolic sets also appear in the text books (Nitecki 1971), (Shub 1978) and more schematically in (Alekseev, Katok, and Kushnirenko 1972).

In problems relating one way or another to exponential conditional stability, other invariant families of manifolds or invariant manifolds can feature. For example, one can consider manifolds consisting of points y for which $g^t y$ not only approaches some trajectory, but approaches it at least as rapidly as $e^{-\alpha t}$ for fixed $\alpha > 0$, whereas it can approach in other directions as well, but more slowly. This circle of ideas is also involved in centre manifolds in the local qualitative theory (see Arnol'd and Il'yashenko 1985, Chap. 3, Sect. 4.2 and Chap. 6, Sect. 2.3), although this is not so transparent. All these and objects related to them play an important role in various questions of the theory of differential equations and DS's, which is reflected in a number of earlier volumes of this series (see Anosov et al. 1985, Arnol'd et al. 1986, Arnol'd and Il'yashchenko 1985, Arnol'd, Kozlov, and Neishtadt 1985 and Bunimovich et al. 1985). In connection with those aspects of the theory of invariant families of manifolds etc. that are close to the present volume, see (Fenichel 1971), (Fenichel 1974), (Hartman 1971), (Hirsch and Pugh 1970), (Hirsch, Pugh, and Shub 1970), (Hirsch, Pugh, and Shub 1977).

1.4. Stable and Unstable Manifolds in Certain Special Cases. As is clear from Fig. 5 given below (see Sect. 2.1), $W^{u}(x)$ may contain points y that are distant from x in the intrinsic metric of the manifold $W^{u}(x)$ (induced on it by the Riemannian metric on M) but close to the original hyperbolic set in the sense of the metric on the whole of M. Here the direction of the tangent space $T_yW^{u}(x)$ may differ appreciably from the directions of E^{u}_{z} at points $z \in A$ close to y. Furthermore, the direction of $T_yW^{u}(x)$ can change abruptly at a small distance (in the intrinsic metric of $W^{u}(x)$) — the further from x, the smaller this distance. If we consider $\{W^{u}_{r}(x)\}$ with fixed r, then such phenomena are not observed. In this connection the family $W^{u}_{r}(x)$ has a "more regular" structure than $\{W^{u}(x)\}$: as one moves further away from A (along the manifolds $W^{u}(x)$) the properties of the latter family can worsen. Similar considerations apply to W^{s}.

Those situations when such a "worsening" cannot arise merit special attention.

A DS (cascade or flow) is called an *Anosov system* (*cascade* or *flow*) if the entire phase manifold M is a hyperbolic set. A diffeomorphism generating an Anosov cascade is called an *Anosov diffeomorphism.*[8] For an Anosov DS no "deterioration of the properties of the family $\{W^{s}(x)\}$ or $\{W^{u}(x)\}$ as one moves away from the hyperbolic set $A = M$" can arise for the simple reason that no such "moving away" is possible — the whole of $W^{s}(x)$ or $W^{u}(x)$ lies entirely in the hyperbolic set. It turns out that for an Anosov system the families $\{W^{s}(x)\}$ and $\{W^{u}(x)\}$ and, in the case of a flow, the families $\{W^{s}(L)\}$ and $\{W^{u}(L)\}$ as well (L being trajectories) form so-called foliations (the definition of a foliation is repeated below). It should be noted that the verification of the hyperbolicity of M is often carried out by means of cones (Sect. 1.2).

There is also a criterion in which hyperbolicity must be verified not on the whole of M, but only on the set $\mathcal{R}$ of chain-recurrent points (see Anosov et al. 1985, Chap. 3, Sect. 2.2), but then there are two further conditions. Clearly such a criterion can be helpful when $\mathcal{R} \neq M$; indeed, it is used in the construction of an example of an Anosov flow with the latter property (see Franks and

[8] The conditions defining these objects were originally called "У conditions" for the sake of brevity (see Anosov 1962 and Anosov 1967). Consequently, the terms *У-system*, *У-flow*, *У-cascade*, *У-diffeomorphism* were applied. (Here the Cyrillic letter У stands for the Russian word *uslovie*, meaning "condition".) But this terminology did not take root since it called to mind the word *ustoichivyi* ("stable"), whereas the individual trajectories in these DS's are unstable. (True, one can associate "У" with *usy* ("whiskers") or *uslovnaya ustoichivost'* ("conditional stability") if one wants; and some (Russian) authors changed the У to the Roman "U", which clearly stands for "unstable".) In the Western literature one sometimes encounters "C" (for "condition").

Williams 1980). Here, in fact, the following additional conditions are verified: for $L_1, L_2 \subset \mathcal{R}$ the manifolds $W^s(L_1), W^u(L_2)$ have only transversal intersections; for $L \subset \mathcal{R}$ the dimension $\dim W^s(L)$ is independent of L. (Formally the extra conditions in (Franks and Williams 1980) look different.)

The above criterion emerged in the course of the following question. Suppose that a hyperbolic set of the DS $\{g^t\}$ is a closed manifold N (of class C^1 or even C^∞); then is $\{g^t|N\}$ an Anosov DS? It turns out that this is not always the case (see Franks and Robinson 1976 and Mañé 1977). (An example is constructed with discrete time, but using the Smale suspension one can immediately obtain an example of a flow as well. In the presence of the condition that $\{g_t\}$ be an Anosov DS the question remains open.

Sect. 4.2 and the survey (Solodov 1991) are devoted to Anosov DS's.

One also encounters objects that are something like foliations, except that the leaves do not fill the whole of M but only some closed subset A. Here is the precise definition. We set $I = (-1, 1)$, and in $I^m = I^k \times I^{m-k}$ we use the coordinates (v, w), $v \in I^k$, $w \in I^{m-k}$.

By a *k-dimensional lamination* defined on A (in short, a lamination on A) we mean a partition of A into connected immersed smooth (of class C^1) k-dimensional submanifolds, called *laminas*, with the following property. For each point $x \in A$ there exist a neighbourhood U of this point in A, a subset $W \subset I^{m-k}$ and a homeomorphism $I^k \times W \to U$ such that ϕ maps each k-dimensional cube $I^k \times w$, $w \in W$, diffeomorphically onto a connected component of the intersection of some lamina with U, where

$$\phi_v(v, w) : \mathbb{R}^k \to T_{\phi(v,w)}M$$

is continuous in (v, w) (in terms of local coordinates, this map is described by the matrix $(\partial\phi_i(v, w)/\partial v_j)$).

By the *tangent field* of the lamination we mean the field of k-dimensional tangent subspaces E^k_x that associates the tangent space at the point x to the lamina passing through x with the point x. It is defined and continuous in A.

By requiring that the laminas and diffeomorphisms $\phi|I^k \times w$ be of class C^r and, in terms of the local coordinates, that the partial derivatives

$$\partial^j \phi_i(v, w)/\partial v_1^{j_1} \dots \partial v_k^{j_k}$$

with $j_1 + \ldots + j_k = j \le r$ (or, when $r = \infty$, with any j) be continuous in (v, w), we obtain the definition of C^r-laminations. (Thus the definition above is that of a C^1-lamination.) An equivalent but outwardly different definition is given in (Hirsch, Pugh, and Shub 1970). We note that if $A = M$, then the lamination is a foliation with smooth leaves and a continuous tangent field. (In the above definition, $W = I^{m-k}$.) But a C^r-lamination on M is not a C^r-foliation: for the latter it is required that $\phi : I^m \to U$ be a C^r-diffeomorphism. For an Anosov DS of class C^r the foliations referred to above are C^r-laminations on M but not necesarily C^r-foliations.

Let A be a closed invariant set of a flow of class C^r containing no equilibrium points. Then the trajectories of the flow that lie in A form a C^r-lamination in A.

We now recall the notion of an *attractor*. This can be defined by several equivalent methods. (One uses the compactness of M in the verification of the equivalence. Meanwhile it would be premature to judge which definition would be best in the non-compact and certainly the infinite-dimensional case.)

1) An attractor is a closed invariant set A that is Lyapunov stable and asymptotically stable (see Anosov et al. 1985, Chap. 3, Sect. 2.1). In other words: a) for any neighbourhood U of A there exists a neighbourhood V of it such that $g^tV \subset U$ for all $t \geq 0$; b) there exists a neighbourhood W of A such that for any point x of it the ω-limit set[9] $\omega(x) \subset A$ (in other words, $\rho(g^tx, A) \to 0$ as $t \to \infty$).

2) An attractor is a closed invariant set A for which there is a neighbourhood U such that $\bigcap\{g^tU;\ t \geq 0\} = A$.

3) An attractor is a closed A for which there exists a neighbourhood V such that the closure $\overline{g^tV} \subset V$ for $t > 0$ and $\bigcap\{g^tV;\ t \geq 0\} = A$. (The invariance of A follows from this.) Such a neighbourhood is said to be *absorbing*.

It must be pointed out that the W in 1) does not necessarily enjoy the property of U in 2), and the U in 2) does not necessarily enjoy the property of W in 1); on the other hand the V in 3) does enjoy the property of W in 1). All the points of V, apart from those of A itself, "come into V from somewhere outside", that is, for $x \in V\backslash A$ there exists τ such that $g^{-1}x \notin V$ for $t > \tau$. Hence it follows that A contains the entire unstable set $W^u(x)$ of any point x of it. The zone of attraction $W^s(A) = \{x; \omega(x) \subset A\}$ (see Anosov et al. 1985, Chap. 3, Sect. 2.1) is an open set. We therefore usually talk about the *domain of attraction* for attractors. Clearly it contains the whole of $W^s(L)$ with $L \subset A$ and the whole of $W^s(x)$ with $x \in A$, but it may not be filled out by them. The points of $W^s(A)\backslash A$ are wandering points (see Anosov et al. 1985, Chap. 3, Sect. 4.1 for the definition).

A set A is called a *repeller* if it is an attractor for the DS obtained from the DS under consideration under time reversal. It is not difficult to rephrase statements about attractors in terms of repellers.

In Fig. 2 the horizontal segment is an attractor although most of the trajectories converge to two points, namely, the end points. For this reason various extra conditions are stated that ensure that the trajectories are "attracted" to the attractor as a whole and not to some parts of it. We give one such condition in Chap. 2. A second condition (stochastic attractor in the sense of Sinai) is given in (Bunimovich et al. 1985, Chap. 8, Sect. 2.1). A third condition, proposed by Sharkovskij, is given in the books written by him and collaborators and cited in the Bibliography at the end of the Preface of this volume. (On the

[9] The definition of ω-limit and α-limit sets $\omega(x) = \omega(L)$, $\alpha(x) = \alpha(L)$ of the trajectory L of a point x is exactly the same as in (Arnol'd and Il'yashenko 1985, Chap. 1, Sect. 5.5). Points (or trajectories) lying in $\omega(L)$ or $\alpha(L)$ are themselves called ω- or α-limit points (or trajectories) for L.

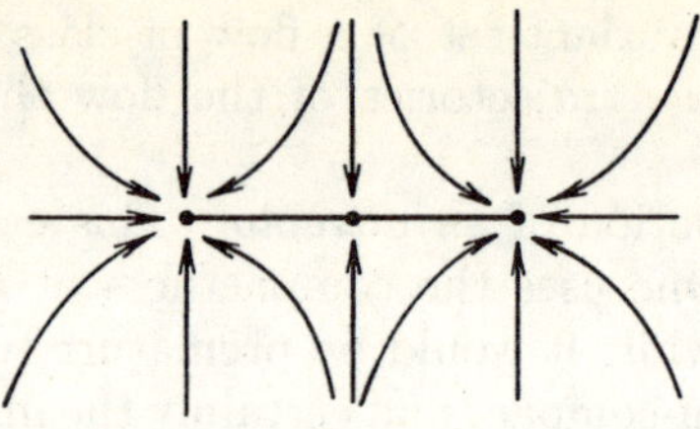

Fig. 2

other hand, apparently there are other cases when some set "attracts many trajectories", but even so, it does not attract all the trajectories of a neighbourhood of it, no matter how small the latter is taken. Such a set should also deserve a special name, if not "attractor", then "quasi-attractor" or whatever. However this type of situation has so far not been described with complete precision; and if such a description were presented, then one would need to have some reasons for considering situations corresponding precisely to this description to be sufficiently widespread.) It should be pointed out that in any case the above definition is still of value, because the appearance of "an attractor in an attractor" compromising the interval in Fig. 2 is of interest in its own right from another point of view (see (Anosov et al. 1985, Chap. 3, Sect. 2)).

We are interested in hyperbolic attractors (and repellers), that is, hyperbolic sets that are attractors (or repellers). They satisfy all the proposed extra conditions (although occasionally with slight reservations which do not essentially change matters), so that the previous questions do not arise for them.

The unstable manifolds of points or trajectories of a hyperbolic attractor A are contained in it and form a lamination on A (a C^r-lamination if the DS is of class C^r). With regard to the stable manifolds, for the hyperbolic attractor

$$W^s(A) = \bigcup \{W^s(L);\ L \subset A\} = \bigcup \{W^s(x);\ x \in A\}.$$

$W^s(A)$ is open but not closed, except for the case when $A = M$. Therefore the stable manifolds, with the above exception, do not form laminations, although, of course, the fibres fill out a closed set. But in other respects the partitioning of $W^s(A)$ into stable manifolds of points or trajectories has the properties of a lamination indicated in its definition. In particular, the tangent spaces to the fibres form a continuous field in $W^s(A)$.

Sect. 1 of Chap. 2 is devoted to certain types of hyperbolic attractors.

We must warn the reader that in the literature laminations are sometimes referred to as foliations and sometimes one calls a family of manifolds (of type $\{W^s(x);\ x \in A\}$) a lamination (or even foliation), although is not necessarily a lamination in the precise sense defined above. This, of course, is not an error, but merely an abuse of language.

§2. Some Examples

2.1. Heteroclinic and Homoclinic Points. These were already introduced in (Anosov et al. 1985, Chap. 1, Sect. 3.1) for Morse-Smale DS's, but the definition of them by itself relates to a more general situation; here it is necessary to include certain explicit indications regarding the properties of the type of hyperbolicity and transversality (for Morse-Smale DS's, their fulfilment is guaranteed by the very definition of the latter). Thus let L_1, L_2 be two different periodic hyperbolic trajectories of a smooth cascade $\{g^k\}$ or two different hyperbolic closed trajectories of a smooth flow $\{g^t\}$ with identical Morse indices (that is, the dimensions $\dim W^{\mathrm{u}}(L_i)$ of the corresponding unstable manifolds). A point x of transversal intersection of the manifolds $W^{\mathrm{u}}(L_1)$ and $W^{\mathrm{s}}(L_2)$ is called a (transversal) *heteroclinic point*, and its trajectory L a (transversal) *heteroclinic trajectory.*[10] Clearly all the points of L are also points of transversal intersection of $W^{\mathrm{u}}(L_1)$ and $W^{\mathrm{s}}(L_2)$, that is, they are (transversal) heteroclinic points.

Let L_1 be a periodic hyperbolic trajectory of a smooth DS. A point x of transversal intersection of $W^{\mathrm{u}}(L_1)$ and $W^{\mathrm{s}}(L_1)$ is called a (transversal) *homoclinic point* and its trajectory L is called a (transversal) *homoclinic trajectory*; all the points of L are (transversal) homoclinic. This definition differs from the previous one in that now we have $L_1 = L_2$ (here, naturally, there is no need to make special reference to the fact that the indices are the same. Furthermore, in the case of a flow the trajectory L_1 in the conditions of the above definition must be closed: in Footnote 10 one must put $L_1 = L_2$).

In the heteroclinic case the manifold $W^{\mathrm{u}}(L_1)$ as it approaches L_2 "along" L_1 starts to "nestle up" to $W^{\mathrm{u}}(L_2)$; larger and larger "pieces" of $W^{\mathrm{u}}(L_1)$ go along $W^{\mathrm{u}}(L_2)$ and closer and closer to it (see Fig. 3 in which L_1, L_2 are fixed points, $W^{\mathrm{u}}(L_i)$ is depicted by the continuous line and $W^{\mathrm{s}}(L_i)$ by the dashed line). Even so, one cannot say how the whole of $W^{\mathrm{u}}(L_1)$ behaves; the situation merely applies to some of its "pieces". (In Fig. 3 it "oscillates" along $W^{\mathrm{u}}(L_2)$, and there is one further heteroclinic trajectory apart from the heteroclinic

[10] In (Anosov et al. 1985) the definition of a heteroclinic point is somewhat more general. A priori it is allowed that the L_i can be equilibrium points (the rest of the definition is the same). Then in the case of a flow, in view of the transversality of the intersection and the fact that $L \subset W^{\mathrm{u}}(L_1) \cap W^{\mathrm{s}}(L_2)$, we have

$$\dim W^{\mathrm{u}}(L_1) + \dim W^{\mathrm{s}}(L_2) - \dim M \geq \dim L = 1,$$

and since $\dim W^{\mathrm{u}}(L_1) = \dim W^{\mathrm{u}}(L_2)$, it follows that

$$\dim W^{\mathrm{u}}(L_2) + \dim W^{\mathrm{s}}(L_2) \geq \dim M + 1.$$

Hence L_2 is a closed trajectory. However, the trajectory L_1 may be an equilibrium point. But all the same, the case when L_1 is also a closed trajectory is selected as the main case (or at any rate the important case) (for flows); in other words, this is what we usually have in mind when we speak of heteroclinic points of flows. As this is exactly what is required by us, this is indicated in the main text.

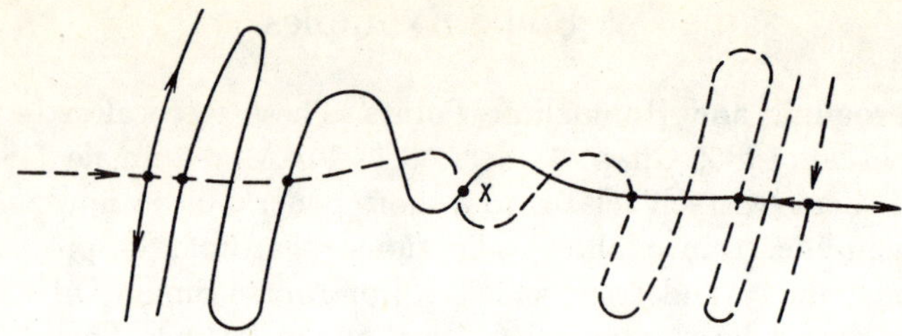

Fig. 3

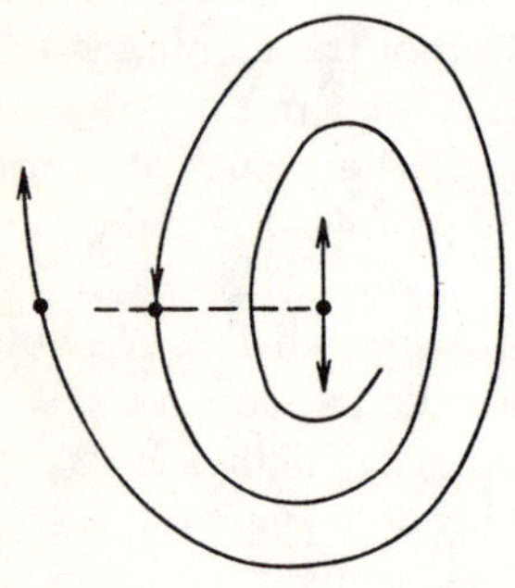

Fig. 4

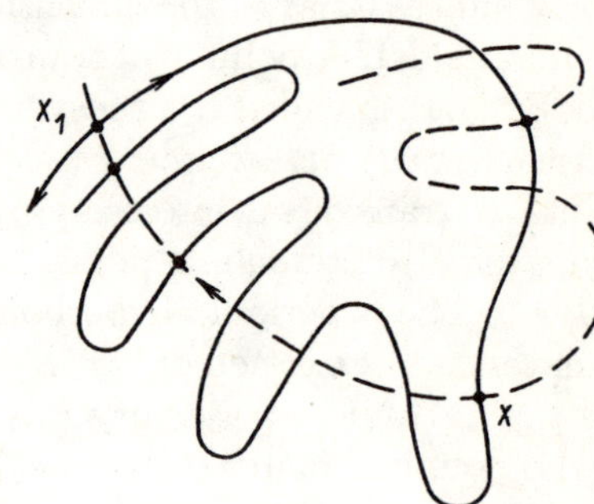

Fig. 5

trajectory indicated by the bold-face points. But even in a compact domain on the plane this is not bound to happen, see Fig. 4).

What has been said concerning $W^u(L_1)$ is also true when $L_1 = L_2$, that is, in the homoclinic case (Fig. 5). In this case, as they approach L_1 along L certain pieces of $W^u(L_1)$ nestle up to $W^u(L_1)$ itself, first to a locally unstable manifold (a neighbourhood of L_1 on $W^u(L_1)$) and second, (as a consequence), to larger and larger parts of $W^u(L_1)$ obtained as the locally unstable manifold broadens under the action of the flow. As a consequence of this, a large piece of $W^u(L_1)$ must also intersect $W^s(L_1)$ after it has approached L_1 sufficiently closely, and so on. In particular, one sees that if indeed there are homoclinic trajectories (doubly asymptotic to L_1), then there must be an infinite number of them. By thinking about Fig. 5, one can realize that among the trajectories there are some that are arbitrarily close to $\bar{L}$ (in the sense that for any neighbourhood U of $\bar{L}$ there is a homoclinic trajectory $L' \neq L$ lying entirely in U. On the other hand, the motions $t \mapsto g^t x'$ of points $x' \in L'$ are by no means close to the motions $t \mapsto g^t x$ of points $x \in L$. In Fig. 5 the latter makes, so to speak, one anticlockwise rotation, while $g^t x'$ makes several rotations). As we have seen, in the presence of heteroclinic trajectories the picture can turn out to be simpler, although not necessarily so. One can merely say that by itself a heteroclinic trajectory does not create such a complex situation as a homoclinic one.

But some heteroclinic trajectories inevitably create a complicated picture if they form a *heteroclinic cycle*. This consists of $n \geq 2$ pairwise disjoint periodic hyperbolic trajectories $L_1, \ldots, L_n$ and n heteroclinic trajectories $L'_1, \ldots, L'_n$ joining them in cyclic order; this means that (with the appropriate numbering) L'_1 goes from L_1 to L_2 (that is, $L'_1 \subset W^{\mathrm{u}}(L_1) \bigcap W^{\mathrm{s}}(L_2)$), L'_2 from L_2 to $L_3, \ldots,$ and L'_n from L_n to L_1. For example, let $n = 2$. The portions of $W^{\mathrm{u}}(L_1)$ that are sufficiently close to a sufficiently large portion of $W^{\mathrm{u}}(L_2)$ coming from L_2 intersect $W^{\mathrm{s}}(L_1)$ together with the latter, so that there are not only heteroclinic but also homoclinic trajectories (and, moreover, as is easily seen, in any neighbourhood $U \supset \overline{L'_1 \bigcup L'_2}$). The same conclusion holds for large n as well.

In view of (15), in the heteroclinic case there exist $x_1 \in L_1, x_2 \in L_2$ such that $x \in W^{\mathrm{u}}(x_1) \cap W^{\mathrm{s}}(x_2)$. This is also true of the homoclinic case, only $L_1 = L_2$ (this certainly does not imply that $x_1 = x_2$ as well). As $t \to \infty$ the "portions" $W^{\mathrm{u}}(g^t x_1)$ with "centre" at $g^t x$ "nestle up" not only to $W^{\mathrm{u}}(L_2)$ but more precisely to $W^{\mathrm{u}}(g^t x_2)$ "stretching out" along it. (For cascades there is essentially nothing to add to this; but for flows one needs to bear in mind that alongside L_2 the relative displacement of two points in the direction of the trajectories of the flow under the action of g^t does not change appreciably, while in the direction of the manifolds $W^{\mathrm{u}}(y)$, $y \in L_2$, a dilation occurs.) In particular, the tangent subspace $T_{g^t x} W^{\mathrm{u}}(g^t x_1)$ approaches $E^{\mathrm{u}}_{g^t x_2} = T_{g^t x_2} W^{\mathrm{u}}(g^t x_2)$.

It turns out that the closure $\bar{L} = L \cup L_1 \cup L_2$ (where the homoclinic case $L_1 = L_2$ is also allowed) is a hyperbolic set, and for $w \in L_1 \cup L_2$ the corresponding decompositions (1) or (3) of E^{s}_w and E^{u}_w involve those subspaces that have already been so designated for the periodic trajectories L_1 and L_2; and for $w \in L$, $w \in W^{\mathrm{u}}(w_1) \cap W^{\mathrm{s}}(w_2)$, $w_1 \in L_1, w_2 \in L_2$ we have

$$E^{\mathrm{s}}_w = T_w W^{\mathrm{s}}(w_2), \; E^{\mathrm{u}}_w = T_x W^{\mathrm{u}}(w_1).$$

After we have said this, this claim would seem to be at least plausible. A rigorous proof in somewhat different terminology can be found in (Palmer 1984).

Having established the hyperbolicity of $\bar{L}$ one can apply the general result in Sect 1.3 concerning the map $\phi^{\mathrm{u}} : E^{\mathrm{u}} \to M$, which proves the above claim on the behaviour of $W^{\mathrm{u}}(g^t x_1)$ around L_2, which was first made on the basis of the pictures. We know, of course, that the portions of $W^{\mathrm{u}}(g^t x_1)$ around $g^t x$ of interest to us are also local unstable manifolds $W^{\mathrm{u}}_{\mathrm{loc}}(g^t x)$.

As long as we speak only about $\bar{L}$, there is no great difference between homoclinic and heteroclinic trajectories. In particular, $\bar{L}$ is a hyperbolic set in both cases and consists of a finite number of trajectories. (The reader should be warned that there also exist hyperbolic sets consisting of a finite number of trajectories including some that are neither periodic, homoclinic nor heteroclinic.) The difference emerges as soon as we turn our attention to the motions of phase points along $\bar{L}$ but not necessarily on $\bar{L}$, and we have already partially spoken about this.

In (Anosov et al. 1985, Chap. 3, Sect. 3.1) the notion of an *isolated* or *locally maximal* invariant set was introduced. This is an invariant (as almost always, compact) set A such that some neighbourhood U of it (called an *isolating neighbourhood*) contains no larger invariant sets (compact or not, here this is not important). The latter is equivalent to the property that any trajectory lying entirely in U lies in A. It turns out that a hyperbolic set consisting of a finite number of trajectories is isolated if and only if the trajectories occurring in it are either periodic or heteroclinic and this set contains no heteroclinic cycles.

We make a final comment. In this subsection (as in Sect. 2.3, where we continue our study of homoclinic trajectories) $\{g^n\}$ may not be a smooth cascade in the strict sense of the word, that is, the domain of definition of g, g^{-1} is not necessarily a closed manifold and the images of some points may be outside this domain. (Thus, it can happen that in Fig. 5 the maps g and g^{-1} are defined only in an annular domain G containing the arcs of $W^{\mathrm{s}}(x_1)$ and $W^{\mathrm{u}}(x_1)$ from x_1 to x (Fig. 6); this domain is mapped by g and g^{-1} into a larger annular domain whose boundary is indicated in Fig. 6 by the dotted line. The images of some of the arcs of $W^{\mathrm{u}}(x_1), W^{\mathrm{s}}(x_1)$ may emerge from G; the further mappings g, g^{-1} on these arcs are not defined, as a consequence of which $W^{\mathrm{u}}(x_1), W^{\mathrm{s}}(x_1)$ themselves "tear" when being extended beyond x.) This remark is fairly important, because it clearly implies that in the present case there is no need for a global cross-section in passing from a flow to a cascade; it suffices merely that there be a local cross-section with the appropriate properties. (We will not give the full details, but their nature is clear from what has been said.) Such a cross-section always exists.

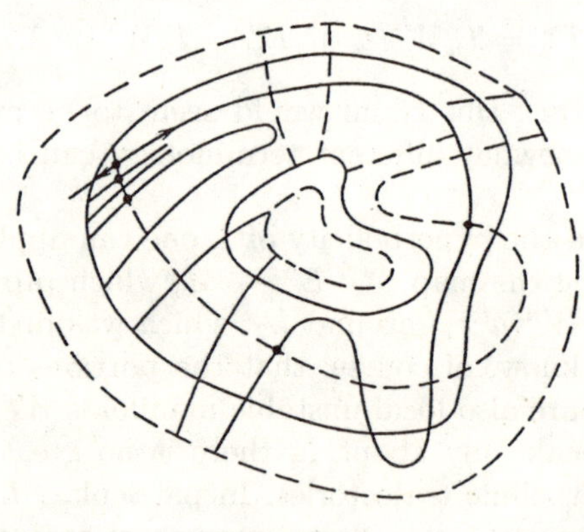

Fig. 6

2.2. The Smale Horseshoe. Hyperbolic sets consisting of several periodic, heteroclinic and homoclinic trajectories are too simple for stimulating or motivating the introduction of the general notion of a hyperbolic set. (This is by contrast with what occurs in a neighbourhood of a homoclinic trajectory, which for a long time was too difficult to sort out.) In this subsection we describe a "genuine" (that is, not so trivial) example of a hyperbolic set which at the same time is sufficiently transparent.

We take the unit square $K = [0,1]^2$ and first we act on it by the homogeneous linear map with matrix $A = \left(\begin{smallmatrix}\lambda & 0\\ 0 & \mu\end{smallmatrix}\right)$, where $\lambda \in (0, 1/2)$, $\mu > 2$ are parameters of the construction. The narrow vertical strip AK so obtained is then acted upon by a second map h which bends its central portion $K_2 = [0,\lambda] \times [\nu, \mu - \nu]$ (where $\nu \in (1, \mu/2)$ is a parameter of the construction) so as to obtain a horseshoe. The latter is laid over the original square K as indicated in Fig. 7, where we see that the action of h on the lower portion $K_1 = [0,\lambda] \times [0,\nu]$ of the vertical strip amounts to displacing it by a constant (vertical) vector u_1, while it causes the upper portion $K_3 = [0,\lambda] \times [\mu - \nu, \mu]$ to be rotated by 180° and then displaced by a constant (vertical) vector u_2.

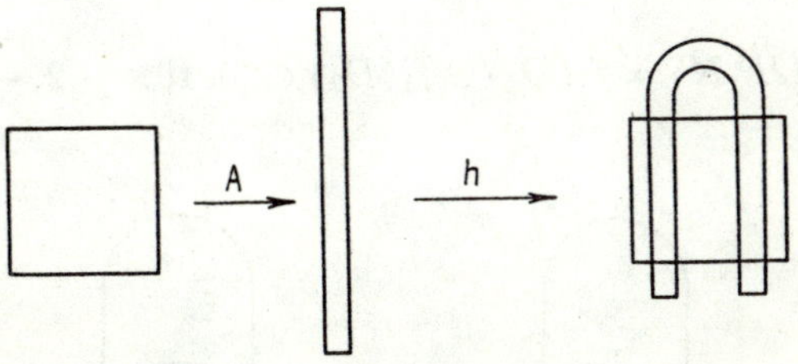

Fig. 7

We suppose for definiteness that the centre of K_1 is taken to the centre of the left half $[0,1/2] \times [0,1]$ of K, while the centre of K_3 is taken to the centre of the right half. What is essential here is that hK_1 should be to the left of hK_3, the images of the lower side of the rectangle K_1 and the upper side of the rectangle K_3 should be below K, and the images of the upper side of K_1 and the lower side of K_3 above K. As a result, $hK_1 \cap K$ and $hK_3 \cap K$ are rectangles B_1 and B_2 of the form $[a, a + \lambda] \times [0,1]$.

At present, the map $f = hA : K \to \mathbb{R}^2$ is defined only on K. But by identifying $\mathbb{R}^2$ with a "punctured" two-dimensional sphere (a sphere from which a point has been removed) and regarding K as a subset of the "standard" sphere $\mathbf{S}^2$ and f as a map $K \to \mathbf{S}^2$, it is not difficult to extend f to a diffeomorphism of this sphere. We describe one such extension which we denote by F.

We suppose that the "sole" of the horseshoe fK (that is, the image of the lower and upper sides of K) is at a height $x_2 = -\alpha$, and the whole of fK lies in $[0,1] \times [-\alpha, 1 + \beta]$. We can suppose that $\beta \geq \alpha$.

We take an infinitely smooth non-self-intersecting closed convex line situated in the rectangle $[0,1] \times [-3\alpha, 1+2\beta]$ and containing the straight-line segments $0 \times [-2\alpha, 1+\beta]$, $1 \times [-2\alpha, 1+\beta]$. It encloses a region H, where $H \backslash K$ has two connected components D_1 (lower) and D_2. We also take a function $\gamma : [-3\alpha, 1+2\beta] \to \mathbb{R}$ with the following properties: $\gamma \in C^\infty$; the derivative $\gamma' > 0$ everywhere; $\gamma(x) = \mu x$ for $x \in [0,1]$; $\gamma(x) = -\alpha + (x+3\alpha)/4$ for $x \in [-3\alpha, -\alpha]$; $\gamma(x) = \mu + \alpha - \alpha(1+2\beta-x)/2\beta$ for $x \in [1+\beta, 1+2\beta]$. Such a γ clearly exists since the values of γ prescribed by our conditions at the ends of the intervals $[-\alpha, 0]$ and $[1, 1+\beta]$ are $-\alpha/2$, 0 and $\mu, \mu+\alpha/2$, that is, they are smaller at the left hand end than at the right hand end. We further note that the map $h : AK \to \mathbb{R}^2$ can be extended to the entire vertical strip $[0, \lambda] \times \mathbb{R}$ by supposing that everywhere below K_2 the extended map h_1 reduces to a displacement by u_1 and above K_2 it reduces to a rotation by 180° and a displacement by u_2. For $x = (x_1, x_2) \in H$ we set

$$A_1(x) = (\lambda x_1, \gamma(x_2)), \quad f_1(x) = h_1 A_1(x).$$

Then (see Fig. 8)

$$f_1(H \backslash K) = f_1(D_1) \cup f_1(D_2) \subset [0,1] \times [-2, -\alpha]. \tag{17}$$

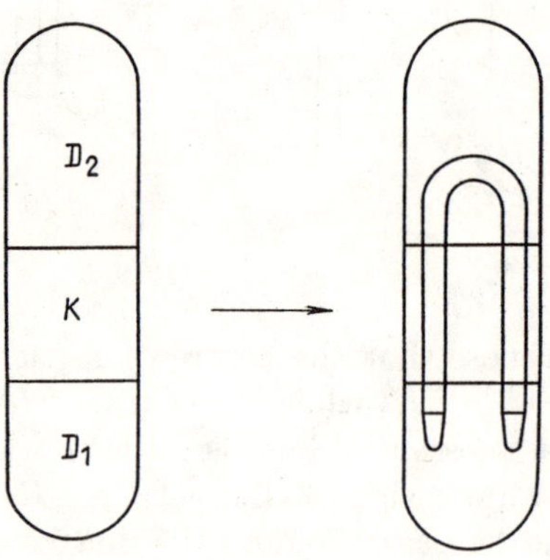

Fig. 8

On the set $H \backslash ([0,1] \times [-\alpha, 1+\beta])$ the map A_1 contracts horizontally λ^{-1} times, and vertically 4 times on the lower connected component of this set and $2\beta/\alpha \geq 2$ times on the upper component, where the image of this set lies in the two parts of the strip $0 \leq x_1 \leq \lambda$ and h_1 is isometric. Therefore

$$f_1([0,1] \times [-2\alpha, -\alpha]) \subset f_1(H \backslash K) \subset [0,1] \times [-2\alpha, -\alpha]$$

and $f_1 | [0,1] \times [-2\alpha, -\alpha]$ is the contracting map. Such a map has an exponentially stable fixed point b and

$$[0,1] \times [-2\alpha, -\alpha] \subset W^{\mathrm{s}}(b).$$

Then, in view of (17), $H \backslash K \subset W^{\mathrm{s}}(b)$. Finally, we identify H with the lower hemisphere

$$\{z = (z_1, z_2, z_3);\ z \in \mathbf{S}^2, z_3 \leq 0\},$$

and some disk $|x| \leq R$ containing H in its strict interior with the spherical segment H' defined by the condition $z_3 \leq 1/2$. The map $f_1 : H \to H$ can then be extended to a C^∞-diffeomorphism $F : \mathbf{S}^2 \to \mathbf{S}^2$, for which the north pole a is an exponentially unstable (in all directions) fixed point and which takes H' to H so that under the action of F all points of the "upper" spherical segment $\mathbf{S}^2 \backslash H'$ (going over to the northern hemisphere) recede from a. For the cascade $\{F^n\}$ so constructed we have

$$\begin{aligned} F(H\backslash K) &\subset H\backslash K \subset W^{\mathrm{s}}(b), \\ F^{-1}(\mathbf{S}^2\backslash H) &\subset \mathbf{S}^2\backslash H \subset W^{\mathrm{u}}(a). \end{aligned} \tag{18}$$

If $x \in K$ but $F(x) \notin K$, then by construction, $F(x) = f_1(x) \in H\backslash K$ and all the subsequent images $F^n(x) \notin K$, $F^n(x) \to b$. If $x \in K$ but $F^{-1}(x) \notin K$, then $F^{-1}(x) \in \mathbf{S}^2\backslash H$ (we recall that if $F^{-1}(x) \in H\backslash K$, then, in view of (18), we would have $x \in H\backslash K$), therefore all the subsequent images $F^{-n}(x) \in \mathbf{S}^2\backslash H$, $F^{-n}(x) \to a$. Therefore $\mathbf{S}^2\backslash(W^{\mathrm{u}}(a) \cup W^{\mathrm{s}}(b))$ consists of those trajectories of the cascade $\{F^n\}$ that lie entirely in K and form there a maximal invariant subset $C = \bigcap_{n=-\infty}^{\infty} F^n K$ of this square.

We now deal with this C, while for a trajectory of points not belonging to C we consider only those that stay close to C and only as long as they stay close to C. Such questions can be qualified as "local". Now it may be recalled that the "local theory" can be understood to be more specific and narrower, namely, referring only to neighbourhoods of periodic trajectories. For this reason, as was mentioned in (Anosov et al. 1985, Chap 1, Sect. 1.6), it is appropriate to use the terminology "semilocal". In any event, we can now deal with just one original f, its inverse map f^{-1} and their iterations, although this f is defined only on K, while the other maps are defined on even smaller sets. Formally this sometimes requires a certain amount of care, for example, $f^n K$ $(n > 0)$ is, strictly speaking, the image under the action of f^n not of the whole of K, but only that part of it where f^n is defined; but such refinements are fairly obvious and we usually omit them (especially since, if necessary, we can nevertheless regard f as an extension to a diffeomorphism of the sphere).

Ignoring all such reservations, it remains to see what $f^{-1}(K)$ (Fig. 9) or $f^2 K$ (Fig. 10) can look like. It is clear from Fig. 9 that $f^{-1}K$ also has the form of a horseshoe, which can therefore be obtained not as has been done in Fig. 9, but by analogy with the construction of fK, except that the roles of the coordinates of x_1 and x_2 are interchanged: K is "flattened" under the action of A^{-1}, the strip $A^{-1}K$ is bent along its central portion and the horseshoe so obtained is laid over K so that its ends protrude slightly to the left. (In fact, it

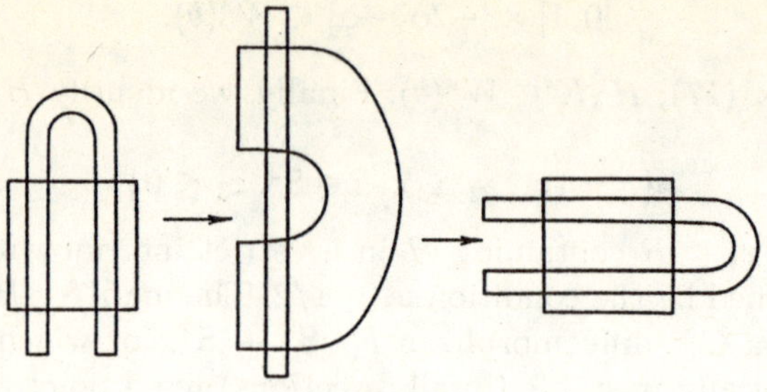

Fig. 9

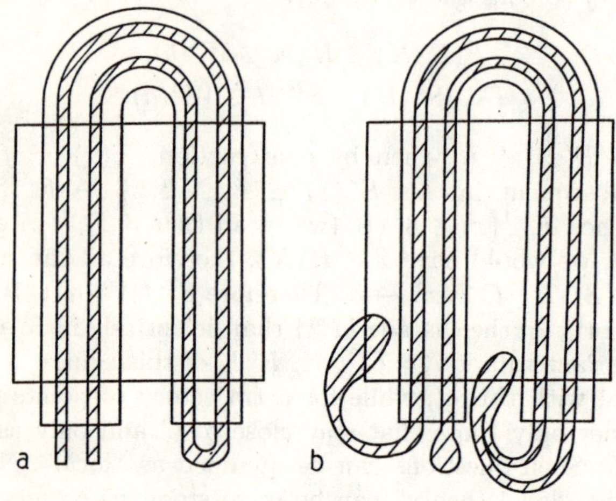

Fig. 10

does not follow from the definition of f that outside $f^{-1}(K \cap fK)$ the inverse map f^{-1} actually contracts in one direction and dilates in another, so that the part of $f^{-1}K$ lying outside K may look different.) The intersection $K \cap f^{-1}K$ consists of two horizontal strips $B^i = f^{-1}B_i$ of the form $[0, 1] \times [a + \mu^{-1}]$, with B^1 below B^2. The form of f^2K also partly depends on what f is like outside K (see, for example, Fig. 10, a and b. The first is obtained for F). But $K \cap fK \cap f^2K$ is the same in both figures. It consists of four narrow vertical rectangles $B_{ij} = B_i \cap fB_j$ $(i, j = 1, 2)$, the lower and upper sides of which are situated on the lower and upper sides of K.

In general, independently of how f is defined outside K, the set $\bigcap_{i=0}^{k} f^i K$ consists of 2^k vertical strips

$$B_{j_0 \dots j_{k-1}} = B_{j_0} \cap fB_{j_1} \cap \dots \cap f^{k-1}B_{j_{k-1}}$$

of the form $[a, a + \lambda^k] \times [0, 1]$, while $\bigcap_{i=0}^{n} f^{-i}K$ consists of 2^n horizontal strips

$$B^{i_1\dots i_n} = B^{i_1} \cap f^{-1}B^{i_2} \cap \dots \cap f^{-n+1}B^{i_n} = f^{-1}B_{i_1} \cap f^{-2}B_{i_2} \cap \dots \cap f^{-n}B_{i_n}$$

of the form $[0,1] \times [a, a+\mu^{-n}]$. When k or n is increased by 1, two narrower strips stay inside each of the corresponding strips. Finally, we have $C = \bigcap_{-\infty}^{\infty} f^i K = C_1 \times C_2$, where the C_i are sets of Cantor type. This C is called the *Smale horseshoe* (although the only thing that resembles a horseshoe is the set fK featuring in its construction).

Let us verify the hyperbolicity of C. We note that $AB^1 \subset K_1$ and $AB^2 \subset K_3$, which means that f has the form $x \mapsto Ax + \text{const}$ on B^1, while on B^2 it has the form $x \mapsto -Ax + \text{const}$. Hence $T_x f = A$ for $x \in B^1$ and $T_x f = -A$ for $x \in B^2$. Clearly $E^{\mathrm{s}}_x = \mathbb{R} \times 0$, $E^{\mathrm{u}}_x = 0 \times \mathbb{R}$ enjoy the required properties: both A and $-A$ preserve these directions, contracting $1/\lambda$ times in the first direction and expanding μ times in the second.

If $x = (x_1, x_2) \in C$, then on the entire vertical segment $[0,1] \times x_2$ passing through x all the iterations f^n, $n \geq 0$, that do not take it outside K are defined. After n iterations the segment is uniformly contracted $1/\lambda^n$ times. Hence it is $W^{\mathrm{s}}_{\mathrm{loc}}(x)$. But there are many other points of $W^{\mathrm{s}}(x)$ in K; since we are now dealing with the semilocal theory, we shall write $W^{\mathrm{s}}_{\mathrm{sl}}(x)$. In fact, the horizontal segment of unit length $W^{\mathrm{s}}_{\mathrm{loc}}(f^n x)$ intersects the 2^n rectangles $B_{j_0\dots j_{n-1}}$ and only one of these intersections is $f^n W^{\mathrm{s}}_{\mathrm{loc}}(x)$, while all the others are images under the map f^n of certain other horizontal segments of unit length. On the other hand, if $y \in W^{\mathrm{s}}_{\mathrm{sl}}(x)$, then, starting from some $n = k$, $f^n y$ always remains in a small neighbourhood of the point $f^n x$. This means that $f^k y \in W^{\mathrm{s}}_{\mathrm{sl}}(g^k x)$, because for all other points z close to $f^k x$, the vertical distance between $f^n z$ and $f^{k+n} x$ increases for a while until they eventually diverge at some finite distance. One finds that $W^{\mathrm{s}}_{\mathrm{sl}}(x)$ consists of a system of horizontal segments of unit length. Similarly, the vertical segment $x_1 \times [0,1]$ is $W^{\mathrm{u}}_{\mathrm{loc}}(x)$, while all the $W^{\mathrm{u}}_{\mathrm{sl}}(x)$ consist of a countable system of similar vertical segments.

By extending to a diffeomorphism of the sphere $\mathbf{S}^2$, one can talk about the manifolds $W^{\mathrm{s}}(x), W^{\mathrm{u}}(x)$, as in Sect. 1.3. In the present case they are curves. They contain $W^{\mathrm{s}}_{\mathrm{sl}}(x)$ and $W^{\mathrm{u}}_{\mathrm{sl}}(x)$, but they may also contain other curves "passing over" K; this depends on the method of extending f. (These "other curves" do not intersect C.) In the extension F described above, there are no such curves.

Independently of the choice of the extension, $W^{\mathrm{s}}(C) = \bigcup\{W^{\mathrm{s}}(x);\ x \in C\}$. This set can be extremely complicated. If everyone is now used to the Cantor set then even for a "good" extension F, the set $\overline{W^{\mathrm{s}}(C)}$ turns out to be an object that still appears "exotic" (although this too is not new). Namely, in this case, the set $\overline{W^{\mathrm{s}}(C)} = W^{\mathrm{s}}(C) \cup \{b\}$ is an indecomposable one-dimensional Brouwer-Janiszewski continuum, which is homeomorphic to the first of the examples of such continua given in (Kuratowski 1969, Sect. 48, V). Apparently this fact was first noted in print by V.M. Alekseev (in a footnote on p. 167

of the Russian translation of (Nitecki 1971)). A generalization of Alekseev's remark appears in (Kamaev 1977).[11]

A clear representation of the motions in C can be obtained using symbolic dynamics (see Anosov et al. 1985, Chap. 1, Sect. 1.2). We associate with a point $x \in C$ a two-sided infinite sequence of symbols $\omega = \{\omega_n\}$ equal to 1 or 2, in accordance with the following rule: $\omega_n = 1$ if $f^n x \in B_1$, and $\omega_n = 2$ if $f^n x \in B_2$. We obtain a map

$$\phi : C \to \Omega_2 \quad x \mapsto \omega,$$

where Ω_2 is the space of all two-sided infinite sequences of two symbols endowed with a suitable topology. This map turns out to be a homeomorphism "onto". For fixed $i_1, \ldots, i_n$ (equal to 1 or 2)

$$\begin{aligned} C \cap B^{i_1 \ldots i_n} &= \{x \in C;\ fx \in B_{i_1}, \ldots, f^n x \in B_{i_n}\} = \\ &= \{x; (\phi x)_1 = i_1, \ldots,\ (\phi x)_n = i_n\}, \end{aligned}$$

that is, $C \cap B^{i_1 \ldots i_n}$ is the inverse image under ϕ of the corresponding cylinder subset of Ω_2. Similarly, for fixed $j_0, \ldots, j_{k-1}$

$$C \cap B_{j_0 \ldots j_{k-1}} = \phi^{-1}\{\omega; \omega_0 = j_0,\ \omega_{-1} = j_1, \ldots,\ \omega_{-k+1} = j_{k-1}\}.$$

Finally,

$$\phi^{-1}\{\omega; \omega_{-k+1} = j_{k-1}, \ldots, \omega_0 = j_0,\ \omega_1 = i_1, \ldots,\ \omega_n = i_n\}.$$

The latter is the part of C lying in the rectangle formed by the intersection of the horizontal and vertical strips $B^{i_1 \ldots i_n}$, $B_{j_0 \ldots j_{k-1}}$. The rectangle decreases in size as n and k increase. From this it is easy to deduce that ϕ is bijective (that is, it is a one-to-one "onto" map) and that ϕ and ϕ^{-1} are continuous.

Furthermore, a Bernoulli topological automorphism (shift) σ acts in Ω_2. If we imagine the elements ω_n of the sequence ω as being written out in the same order in which their indices are arranged on the number axis:

$$\omega = \ldots \omega_{-2}\omega_{-1}\omega_0,\ \omega_1\omega_2 \ldots,$$

where the position of the zeroth element is indicated by a comma to the right of it, then σ shifts this sequence by one step to the left (but the comma is not shifted):[12]

$$\sigma\omega = \ldots \omega_{-1}\omega_0\omega_1,\ \omega_2\omega_3 \ldots.$$

The map ϕ establishes an isomorphism between $f|C$ and σ (that is, an isomorphism of the cascades $\{(f|C)^n\}$ and $\{\sigma^n\}$).

---

[11] Digressing a little, I recall another paper (Barge 1987) where the behaviour of indecomposable continua in the theory of DS's was also noted. In the general case the conditions in (Barge 1987) are scarcely verifiable, but they hold for certain non-transversal homoclinic points of a two-dimensional cascade.

[12] In some works, by a Bernoulli automorphism is meant a shift by one step to the right; but left shifts are more common.

In passing from C to Ω_2 it is customary to speak of a "coding" (of the points of C by means of the sequence of symbols and of the map $f|C$ by means of the shift). This expression is informal and is used in the descriptive presentation; but in psychological terms it is fairly expressive, calling to mind the association with the coding of continuous signals in communications technology. Besides, the expression "coding by means of the sets $B_1, \dots, B_k$" (forming a partition of the phase space) has a precise meaning: as above, the sequence ω such that $f^i x \in B_{\omega_i}$ is associated with the phase point x.

The cascade $\{\sigma^n\}$ (and hence $\{(f|C)^n\}$ as well) are nothing like the simple DS's chiefly considered in (Anosov et al. 1985). The motions (under the action of σ^n) on its points are extremely diverse, as is clear from the following properties of it (apart from the last one, which is Krieger's theorem, they are all very easily proved). 1) The periodic points are everywhere dense in the phase space Ω_2. 2) $W^{\mathrm{s}}(\omega)$ and $W^{\mathrm{u}}(\omega)$ are dense in Ω_2. This follows from the property that

$$W^{\mathrm{s}}(\omega) = \{\xi; \xi_i = \omega_i \text{ for sufficiently large } i\},$$
$$W^{\mathrm{u}}(\omega) = \{\xi; \xi_{-i} = \omega_{-i} \text{ for sufficiently large } i\},$$

as is easy to see. (How large the i must be depends on ξ.) 3) The cascade $\{\sigma^n\}$ is topologically transitive and has the property of mixing of regions.[13] 4) It has minimal sets (Anosov et al. 1985, Chap. 3, Sect. 4.3) that are distinct from periodic trajectories. 5) Its topological entropy is positive. (For the definition of topological entropy see (Alekseev 1976), (Alekseev and Yakobson 1979), and (Martin and England 1981). It gives a certain numerical characterization for the diversity of the motions occurring in a DS. Somewhat remarkably, it also characterizes the complexity of these motions — again in a certain precisely formalized sense, see (Alekseev 1976), (Alekseev and Yakobson 1979), and (Brudno 1983). 6) It has a continuum of invariant normalized measures with respect to which it is ergodic. Certain measures of probability origin are recalled in (Anosov et al. 1985, Chap. 1, Sect. 1.2). It is, in fact, considerably larger than the latter, as is evident from the following theorem due to Krieger, see (Krieger 1972), and (Denker, Grillenberger, and Sigmund 1976). Any ergodic automorphism of a Lebesgue space with a continuous invariant measure and with a (metric) entropy (see (Bunimovich et al. 1985) and (Martin and England 1981) concerning this) $h < \log n$[14] is metrically isomorphic to the restriction of the Bernoulli shift σ in Ω_n (the space of two-sided sequences of n symbols) to a certain closed invariant subset A on which there is a unique invariant normalized measure μ (with respect to which the metric isomorphism

[13] See (Anosov et al. 1985, Chap. 3, Sect. 4.3) concerning topological transitivity. A DS $\{g^t\}$ possesses the property of *mixing of regions* if for any two open sets U, V there exists T such that $g^t U \cap V \neq \emptyset$ for all $t > T$.

[14] Here we mean the logarithm with respect to the same base as in the definition of topological entropy. As is well known, it is often taken to the base 2, but sometimes the natural logarithm is used; in going over from one base to the other, all the entropies are changed the same number of times.

is taken), where $\mu(V) > 0$ for each (relatively) open subset $V \subset A$. (The minimality of A follows from this; hence there are a large number of minimal sets.)

What changes if instead of f one takes a map g that is C^1-close to f? As before, K is first compressed to a tall narrow strip, but now it is not strictly rectangular. It is then bent and laid over K, more or less in the same way as was done before. Clearly, as before, $K \cap gK$ consists of two connected components B_i stretching from the lower side of K to the upper side and close to the previous B_i, only their lateral sides are slightly curved. Similarly our other diagrams and constructions are, so to speak, slightly deformed, but only just. Therefore g has a maximal invariant set D in K that is homeomorphic to Ω_2 via the homeomorphism $\psi : D \to \Omega_2$ combining $g|D$ with σ. The set D is also called a Smale horseshoe. The closer g is to f, the closer ψ is to ϕ (but this time only in the C^0-sense), that is, the homeomorphism $\psi\phi^{-1} : C \to D$ conjugating $f|C$ with $g|D$ displaces the point C by a smaller amount. (This is in connection with the fact that for g the intersections of the vertical and horizontal strips

$$B_{j_0 \dots j_{k-1}} \cap B^{i_1 \dots i_n}$$

are close to the rectangles that analogously are the intersections of the corresponding strips for f.) In this sense one can say that C is preserved under small perturbations and, furthermore, the motions in C are also preserved.

D is a hyperbolic set (see Sect. 1.2, g)). The local stable (or unstable) manifolds for $x \in D$ are almost horizontal (or almost vertical) arcs joining the opposite sides of K.

A compact invariant set $A \subset M$ of a given DS is said to be *locally structurally stable* if it has neighbourhoods $U \supset V \supset A$ and for any $\epsilon > 0$ there exists $\delta > 0$ such that for any smooth perturbation of the original DS in U that alters it in the C^1-metric by an amount less than δ, there exists a homeomorphic embedding $\chi : V \to U$ that shifts the point by an amount less than ϵ and realizes an equivalence between the original DS on V and the perturbed DS on χV. (For cascades $\{f^n\}$ and $\{g^n\}$, equivalence simply means the following: if $x \in V$ and $fx \in V$, then $g\chi x \in \chi V$ and $\chi f x = g \chi x$. For flows, equivalence means that χ takes arcs of trajectories of the original DS lying in V to arcs of the trajectory of the perturbed system lying in χV, the orientation on each trajectory being preserved. The positive orientation of a trajectory corresponds to the direction along it as time increases.) Preservation of the velocity of the motion along the trajectory is not required. However, the specific choice of the metric referred to in the definition is immaterial.

Alongside this "semilocal" concept there is also the "global" concept of a structurally stable system (Anosov et al. 1985, Chap. 2, Sect. 1.2); see also (Anosov 1985). There is also the related concept of *Ω-structural stability*, which is weaker. The name refers to the often applied notation Ω for the set of nonwandering points of a DS (Anosov et al. 1985, Chap. 3, Sect. 4.1). A DS is said to be Ω-structurally stable if for any $\epsilon > 0$ there exists $\delta > 0$

such that for any smooth perturbation of the original DS that alters it in the C^1-metric by an amount less than δ, there exists a homeomorphic embedding $\chi : \Omega \to M$ of the set of nonwandering points of the original system into M that shifts the points by an amount less than ϵ and maps Ω onto the set of nonwandering points of the perturbed system, where the trajectories of the original system lying in Ω are taken to trajectories of the perturbed system with preservation of orientation.

It can be proved that the Smale horseshoe is locally structurally stable and the cascade $\{F^n\}$ is structurally stable. To give further clarification on this point, we note that the Ω-structural stability of $\{F^n\}$ is obvious since in the present instance $\Omega = C \cup \{a\} \cup \{b\}$. In fact, the topological transitivity of C ensures that $C \subset \Omega$, and since a is a repeller and b an attractor, it follows that the nonwandering points in $\mathbf{S}^2 \backslash C = W^{\mathrm{u}}(a) \cup W^{\mathrm{s}}(b)$ are merely a and b. The perturbed cascade $\{G^n\}$ has equilibrium points a' (source) and b' (sink) close to a and b as well as a topologically transitive invariant set $D = \bigcap_{-\infty}^{\infty} G^n K$, where the trajectories emerging from K lie in $W^{\mathrm{u}}(a') \cup W^{\mathrm{s}}(b')$; therefore the set of nonwandering points for $\{G^n\}$ is $D \cup \{a'\} \cup \{b'\}$. We have already discussed the construction of $\chi|C : C \to D$; also a and b must be taken to a' and b'.

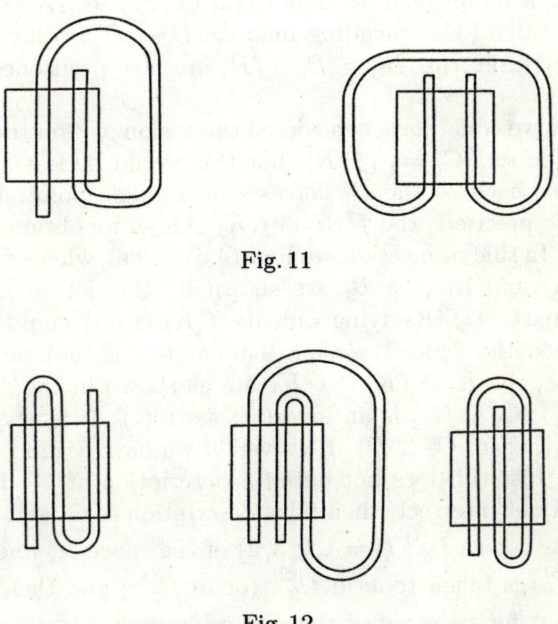

Fig. 11

Fig. 12

The construction of the horseshoe admits certain obvious modifications. Fig. 11 and Fig. 7 exhaust all the possible cases for the plane when $K \cap fK$

consists of two connected components. In these cases the horseshoe is encoded by means of Ω_2 (the two symbols corresponding to the two components). In Fig. 12, $K \cap fK$ consists of three components and in the coding one needs to use three symbols. Fig. 13 refers to the three-dimensional situation. Generalizations to n dimensions are also possible.

In Fig. 14 a) and b) the domain of definition of f consists of two squares K_1 (left) and K_2, and $(K_1 \cup K_2) \cap f(K_1 \cup K_2)$ has four connected components B_i (numbered from left to right), where the images of the lower sides of the K_i are indicated by the thick lines (so that the $f|K_1$ are the same in these two diagrams, while the $f|K_2$ are not). Under the encoding, the symbol i at the nth position of the sequence $\omega = \phi x$ corresponds to $f^n x$ hitting B_i. Here we encounter a new situation: not all the elements of Ω_4 correspond to points of the maximal invariant set $D = \bigcap_{-\infty}^{\infty} f^n(K_1 \cup K_2)$; rather they are the elements of a certain closed invariant subset Ω'. We call the pair of symbols (i, j) admissible if $fB_i \cap B_j \neq \emptyset$. In the preceding examples all such intersections were non-empty, but now this is not so. In both cases a) and b) the admissible pairs are the same. In the oriented *transition graph* in Fig. 14 c) each admissible pair (i, j) corresponds to an edge going from i to j while there are no such edges for inadmissible pairs. As is not difficult to verify, Ω' consists of those sequences $\{\omega_i\}$ for which all the pairs (ω_i, ω_{i+1}) are admissible, so that $\{\sigma^n|\Omega'\}$ is a topological Markov chain (TMC; see (Anosov et al. 1985, Chap. 1, Sect. 1.2)). (The encoding map $\phi : D \to \Omega'$ is different in cases a) and b). The vertical strips $B_{ij} = B_i \cap fB_j$ are now positioned differently in these cases.)

We note that we could have considered the action of f on the K_i separately and obtained the sets $C_i = \bigcap f^n K_i$, but this would be less meaningful—in the present case each of the C_i consists of a single point. By considering the $f^2|K_i$, more precisely, the $f^2|K_i \cap f(K_1 \cup K_2)$, we obtain the two "true" horseshoes C_i'. (In this connection see Fig. 14 d) and e), where $f^2 K_1$ is depicted for examples a) and b) (the B_i are shown by the dotted lines). Properly speaking, the part of $f^2 K_1$ lying outside $f(K_1 \cup K_2)$ could look different; I have illustrated the "good" version (similar to Fig. 9a) rather than 9b)). But in any case, the $K_i \cap fK_j \cap f^2 K_1$ are as shown in the diagram.) Then $C' = C_1' \cup fC_1' \cup C_2' \cup fC_2'$ is an invariant set for f. It is easy to show that $C' \subset K_1 \cup K_2$ (hence, $C' \subset D$). However, if we have a good description for C_i' and $f^2|C_i'$, this still does not entail a description of C'. If the four sets C_i' and fC_i' did not intersect, then for a description of C' and $f|C'$ we would simply take four copies $\Omega_2^{(i)}$ $(i = 1, 2, 3, 4)$ of the space Ω_2 and say that ω in $\Omega_2^{(1)}$ (or in $\Omega_2^{(3)}$) is taken to ω in $\Omega_2^{(2)}$ (or in $\Omega_2^{(4)}$) and then to $\sigma\omega$ in $\Omega_2^{(1)}$ (or in $\Omega_2^{(3)}$). But for us some of these sets intersect and these intersections must be described, say, by providing some identification between some of the points of the corresponding $\Omega_2^{(i)}$. But what is more important is that D is still larger than C'. The sequences corresponding to the points of C' in our encoding of D satisfy one of the four conditions: all the $\omega_{2i} \in \{1, 2\}$; all the

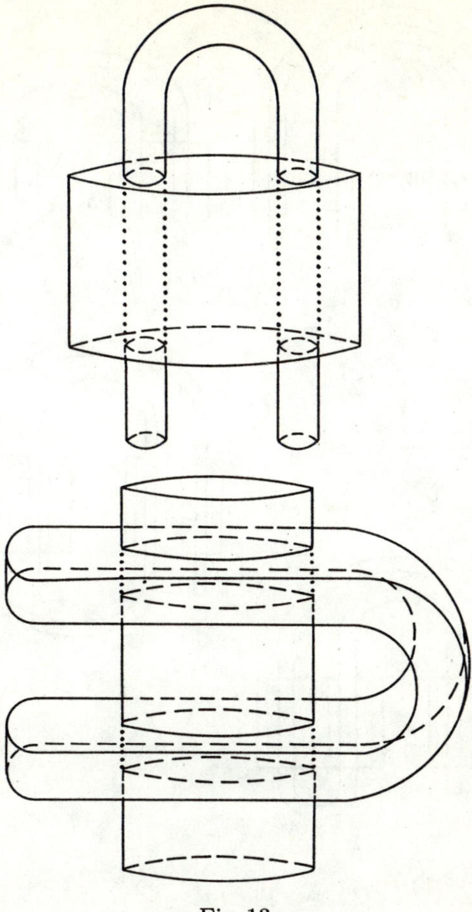

Fig. 13

$\omega_{2i+1} \in \{1,2\}$; all the $\omega_{2i} \in \{3,4\}$; all the $\omega_{2i+1} \in \{3,4\}$. (Thus, if $x \in C_1'$, then all the $f^{2i}x \in K_1$, that is, $f^{2i}x \in B_1 \cup B_2$.) But D contains sequences not satisfying any of these conditions.

In the present instance the cascade $\{f^n|D\}$ is nevertheless equivalent to $\{\sigma^n\}$ in Ω_2. To see this it suffices to realize an encoding using the sets $B_1' = B_1 \cup B_4$, $B_2' = B_2 \cup B_3$. We note that each of them is situated in both K_i; for this reason they do not come to mind all that readily. The horseshoe depicted in Fig. 15 is also encoded by a TMC, where the latter is no longer equivalent to some topological Bernoulli cascade. This is proved by means of topological entropy. For $\{\sigma^n\}$ in Ω_k it is equal to $\log k$, while for our TMC it is equal to $\log(\sqrt{2}+1)$.

As is clear, these slightly modified horseshoes are quite satisfactorily described by the same ideas as in the "standard" horseshoe that we started with,

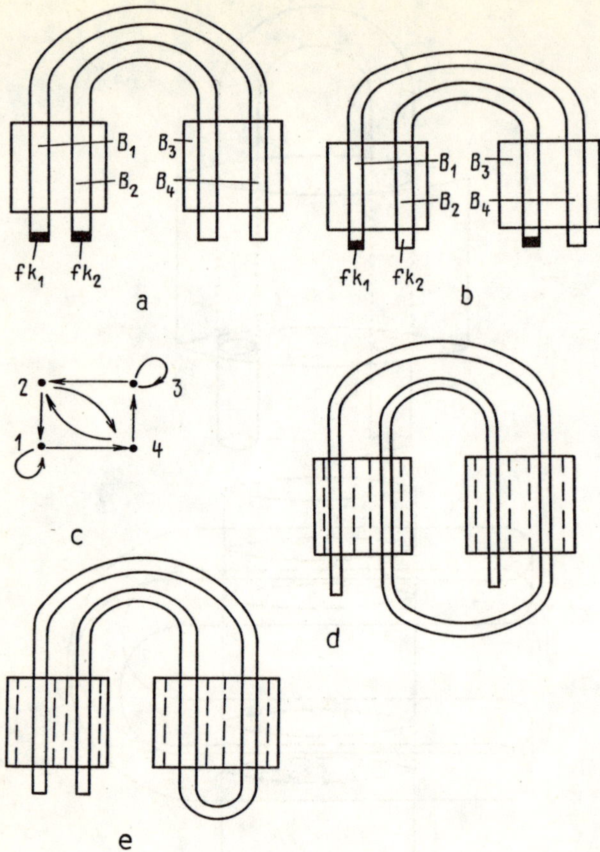

Fig. 14

although this description differs somewhat from that of the standard horseshoe. It is not worth attempting to reduce everything to standard horseshoes.

The horseshoe was invented by Smale in 1961 (Smale 1963), (Smale 1965). At that time several phenomena and objects of a "hyperbolic" character were known and even this character itself had come to be recognized to a considerable extent (especially in ergodic theory); however, the development of the "hyperbolic" theory of DS's, properly speaking, begins with the horseshoe (see Smale's memoir in (Smale 1980)).

2.3. Motions in a Neighbourhood of a Homoclinic Trajectory. ϵ-Trajectories. In Fig. 16 we have depicted a transversal homoclinic trajectory $L = \{g^i x\}$ of a cascade on the plane (we are again dealing with the "semilocal" situation). The point x is a point of transversal intersection of the stable and unstable manifolds of a fixed point x_1. Let K be a curvilinear quadrangle

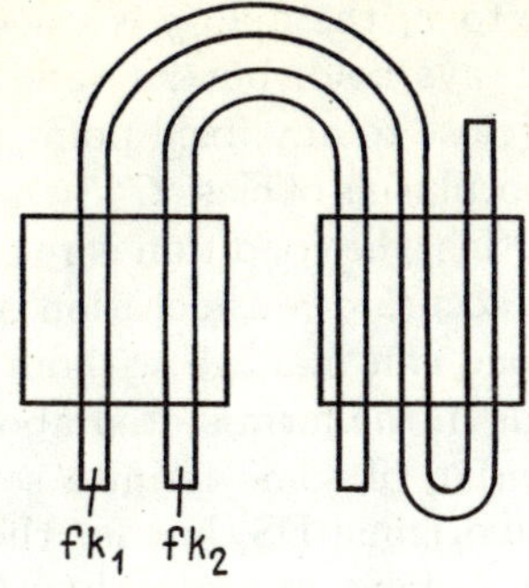

Fig. 15

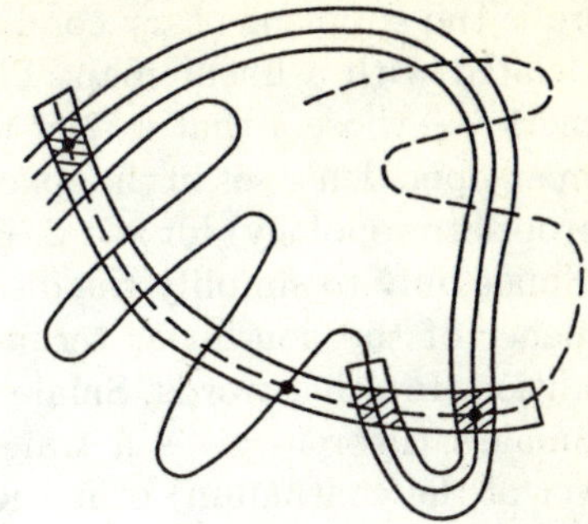

Fig. 16

enclosing the arc $W^s(x_1)$ from x_1 to x (in the figure its boundary is shown by the thick line). As long as we are considering its image gK, there is nothing of interest to be seen (in the case illustrated in Fig. 16). But some iterate g^n stretches K so much along $W^u(x_1)$ (at the same time compressing K in a direction transversal to $W^u(x_1)$) that g^nK, "following along $W^u(x_1)$", as it were, intersects K close to x and, perhaps, several times farther from x_1 (whereas, of course, K and g^nK intersect alongside x_1 for all n). The intersection occurs exactly in the direction of expansion, that is, each connected component of $K \cap g^nK$ (in Fig. 16 they are shaded) is a narrow zone stretched in the direction of expansion (along $W^u(x_1)$) and joining the opposite sides of K. This is the same situation as was dealt with in Sect. 2.2. Thus in Fig. 16 K and g^nK intersect in the same way as K and fK in the right hand diagram of Fig. 12, only now it is not a square but a curvilinear quadrangle. Naturally, in other cases the number of connected components of $K \cap g^nK$ may be different. But we are interested only in those that contain points of $\bar{L}$. In Figure 16 there are two of them and it is easy to see that this can always be achieved by a suitable choice of K.

Let B_1, B_2 be these components. The multiplicity n of the iterate g^n under which g^nK starts to intersect K close to x may be different in different cases. We turn our attention to trajectories of the cascade $\{g^{ni};\ i \in \mathbb{Z}\}$ lying entirely in $B_1 \cup B_2$. In the light of Sect. 2.2 it is clear how these can be described. As a result we conclude that for some n the map g^n has an invariant hyperbolic set C in a neighbourhood of the closure $\bar{L}$ of the homoclinic trajectory L such that $g^n|C$ is equivalent to a Bernoulli topological automorphism σ in the space Ω_2 of sequences of two symbols.

This result is due to Smale (Smale 1965). Although Fig. 16 relates to a cascade on the plane, Smale's arguments are free of restrictions on the phase manifold M, including its dimension. In Fig. 16 the trajectory L is doubly asymptotic to the fixed point x_1, but in Smale's theorem it can be doubly asymptotic to a periodic trajectory; as a matter of fact, this case can be reduced to the previous one by passing to some power of g. In (Smale 1965) there is the supplementary condition that close to x_1 the map g is smoothly associated with a linear map. This does not always hold, but its validity is "generic" — those g that satisfy this condition (close to any fixed point of it) form an open dense set in the space of all diffeomorphisms of class C^r (with the appropriate topology) for $r \geq 2$. But apart from this, the condition is required by Smale only to simplify the discussion. By an accurate consideration of the influence of the non-linear terms close to x_1, one can manage without this condition. In other words, Smale's theorem holds in the form stated above.

Smale's theorem gives a transparent description of some infinite set (of power of the continuum) of motions, not for the original DS, but for the DS $\{g^{ni}; i \in \mathbb{Z}\}$ in a neighbourhood of $\bar{L}$ (properly speaking, in a neighbourhood of the two points of $\bar{L}$: x_1 and x). It is clear from this that one can draw certain conclusions concerning the original DS. Thus, one immediately obtains the assertion stated by Birkhoff in the two-dimensional case: there are periodic points arbitrarily close to a transversal homoclinic point. (The number n does not participate in this formulation, but this is because here we are satisfied with any period including multiples of any n.) Here is another consequence of Smale's theorem (and the well known properties of topological entropy): in the presence of a transversal homoclinic point the topological entropy of a DS is positive (again, n drops out of the statement). According to (Katok 1980), in the two-dimensional case the converse is true for a cascade $\{g^i\}$ of class $C^{1+\epsilon}$: if its topological entropy $h_{\text{top}}(\{g^i\}) > 0$, then it has a hyperbolic set A such that

$$t_{\text{top}}(\{g^i|A\}) \geq h_{\text{top}}(\{g^i\}) - \delta,$$

where $\delta > 0$ is any pre-assigned number, and there are transversal homoclinic points in this set. Judging from all this, one should be able to obtain a similar result for flows with $\dim M = 3$ without any essential change in the proof.

If $n = 1$, then Smale's theorem provides such a simple and complete description of the motion of the original cascade in K that one cannot hope to improve on it. But this is a happy exception, furthermore, if we wish to describe the motions in a sufficiently small neighbourhood of $\bar{L}$, then we must take a small K and then $n > 1$. But if $n > 1$, then we do not have such a satisfactory situation for the description of the motions of $\{g^i\}$ next to $\bar{L}$. The map $\{g^i\}$ has an invariant set $C' = C \cup gC \cup \ldots \cup g^{n-1}C$ and, although this is not entirely obvious, for a suitable choice of K the set C' can be made to lie entirely inside a given neighbourhood of $\bar{L}$. However this C' has drawbacks similar to those pointed out at the end of Sect. 2.2 for C'. (With regard to the second of these, if $y \in C'$, then for some i all the $g^{i+nj}y$, $y \in \mathbb{Z}$, are either close to x or close to x_1. It will be evident from what follows that for trajectories lying in a small neighbourhood of $\bar{L}$ the time intervals between successive occurrences of g^iy in a neighbourhood of x are not necessarily multiples of some $n > 1$.)

As far back as 1935, Birkhoff in Chap. IV of his memoir (Birkhoff 1935) outlined another method of coding which is devoid of these drawbacks. This was a remarkable achievement, even though there are a number of limitations or even defects inherent in this work, see the remarks by Alekseev on p.143 of the Russian translation of (Nitecki 1971). I add that the aim of (Birkhoff 1935) is the investigation of Hamiltonian or related systems; therefore the maps under consideration are assumed to preserve area or a measure with a smooth positive density. Birkhoff does not underline the fact that for the results of Chap. IV this is not important (Birkhoff 1935). It may be that (along with the known inadequacies of Birkhoff's style and the scientific tendencies of that era) this was instrumental in Birkhoff's memoir remaining almost unknown for a long time and failing to exert the influence that was its due.

The approach in accordance with Birkhoff's project was rigorously and fully realized by Shil'nikov (Shil'nikov 1967). In that paper it is flows that are considered, but as before, I shall speak about a cascade $\{g^i\}$ and a transversal homoclinic point x that is doubly asymptotic to a fixed point x_1 (from which one can easily obtain a description of the other cases as well). The description of motions close to $\bar{L}$ given below is formally different from (Shil'nikov 1967) and corresponds to the account given by Alekseev in (Alekseev 1976) and (Alekseev, Katok, and Kushnirenko 1972), where, in addition, the relationship between these two descriptions is pointed out.

The point x_1 has an open neighbourhood U such that outside its closure $\overline{U}$ there lies a "segment" $\{g^i x; -j < i < h\}$ of the trajectory L (we denote the number of points of it, which is $j+h-1$, by k for short), while for $i \leq -j$ and $i \geq h$ all the $g^i x \in U$. A trajectory entering U close to $g^h x$ approaches x_1 for some time, although (in contrast to L) it can then start to go away from x_1 and leave U, staying there for at least time n. Speaking more formally, if y is sufficiently close to $g^h x$, then $g^i y \in U$ for $i = 0, 1, \ldots, n-1$ and the closer the first two points are to each other, the larger n is. If, on the other hand, y is sufficiently close to $g^{-j}x$, then $g^i y \notin U$ for $i = 1, \ldots, k$, and $g^{k+1}y \in U$. Points of a trajectory L' sufficiently close to L lie on segments situated outside $\overline{U}$ (the length of each of them being equal to k), and on the segments lying inside U (the length of each of them being $\geq n$; these lengths can be distinct and possibly infinite). We set

$$\omega_0(y) = 0, \ \text{if } g^{-i}y \in U \text{ for } i = 0, 1, \ldots, n-1,$$

$$\omega_0(y) = \max\{i : g^{-i}y \notin U, \ 0 \leq i < n+k\} + 1 \text{ otherwise}.$$

In other words, as soon as $g^i y$ leaves U (as i increases), say, $g^p y \in U$, $g^{p+1}y \notin U$, $\omega_0(g^i y)$ starts to count off the number of steps made ($\omega_0(g^{p+i}y) = i$) until it reaches $k+n-1$ (at this time $g^{p+k+n-1}y \in U$). After this the indicator of our step counter falls to zero and stays at zero until $g^i y$ again leaves U. If, on the other hand, the entire negative half-trajectory of the point y lies in U, then $\omega_0(y) = 0$. We denote by D the set of points whose trajectories lie entirely in a small neighbourhood V of $\bar{L}$, and we associate with a point $y \in D$ the sequence of symbols $\phi y = \omega \in \Omega_{n+k}$ according to the rule: $\omega_i = \omega_0(g^i y)$. Clearly, $\sigma\phi = \phi(g|D)$. We note that if $0 < \omega_i < n+k-1$, then $\omega_{i+1} = \omega_i + 1$, if $\omega_i = n+k-1$, then $\omega_{i+1} = 0$, while if $\omega_i = 0$, then ω_{i+1} is

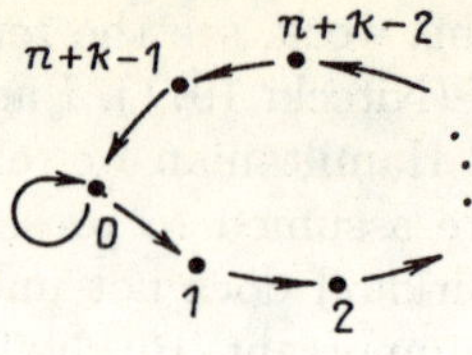

Fig. 17

equal to 0 or 1. We consider in Ω_{n+k} the TMC $\{\sigma|\Omega'\}$ with transition graph corresponding to the above remark (Fig. 17). In accordance with what has been said, $\phi(D) \subset \Omega'$. The function $\omega_0(y)$ is locally constant; this seems to be clear without any special explanation. It is then clear that ϕ is continuous. It is not obvious *a priori* that any sequence $\omega \in \Omega'$ corresponds to some point $y \in D$ and only one such point. This is nevertheless the case if n is sufficiently large and the trajectories L' under consideration lie in some neighbourhood V_n of $\bar{L}$, where the latter can be made arbitrarily small by increasing n. At the time, Shil'nikov had to give a special proof of all this, but now it is easily obtained by applying the general theory of Sect. 3.1.

What has been said above clearly carries over to the case of a heteroclinic cycle. Suppose then that there are two transversal homoclinic trajectories L and L' that are doubly asymptotic to L_1. Points of a trajectory the whole of which is located close to $\bar{L} \cup \bar{L}'$ can lie on segments of three types: those contained in some neighbourhood U of L_1 (their "duration" $\geq n$ and can be infinite); those close to a segment of L that is entirely located outside U (their lengths are the same as the length of this segment, say, k); those close to a segment of L' located entirely outside U (their lengths are also the same, say, k'). Starting from here, one can obtain a description of the motion close to $\bar{L} \cup \bar{L}'$ by means of a TMC in $\Omega_{n+k+k'}$. Similarly, one can consider more complicated combinations of homoclinic and heteroclinic trajectories.

As was shown by Franks and Williams (Franks and Williams 1985), if the flow on a three-dimensional sphere $\mathbf{S}^3$ has a transversal homoclinic trajectory, then it has infinitely many knotted closed trajectories that as knots are pairwise inequivalent. In the proof one uses Birkhoff's description of the invariant set of the first-return map on some local cross-section. Franks and Williams first proved that closed trajectories near to $\bar{L}$ can, via an isotopy deformation, be arranged on some branching surface of roughly the same type as in Fig. 16 of Chap. 2 (although not necessarily exactly as shown there), and then so that these deformed curves (which are no longer trajectories of the original flow) become closed trajectories of a semiflow acting on this surface roughly in the same way as indicated in this diagram. This surface with this semiflow is called a *Lorenz template* in (Franks and Williams 1985).

It remains to solve the following geometrical problem: prove that for any embedding of the Lorenz template into S^3, among its periodic trajectories there are infinitely many knotted ones which are pairwise inequivalent (as knots). This is done in the second part of (Franks and Williams 1985).

As is clear, homoclinic trajectories enable one to make fairly substantial conclusions concerning the properties of DS's. We note, in addition, that in view of the closeness to $\bar{L}$ of the invariant set D constructed above, this set is hyperbolic (Sect. 1.2, g)). The majority of hyperbolic sets in DS's relating to various problems of mechanics, physics, and so on, have in fact been discovered in this way; that is, their existence follows from the fact that there are transversal homoclinic trajectories in these DS's. Such trajectories are quite often found using a computer. In a number of cases homoclinic trajectories arise under a small periodic perturbation of an autonomous Hamiltonian system with one degree of freedom having a closed loop of the separatrix. There are analytical methods due to Mel'nikov and Palmer (Mel'nikov 1963), (Palmer 1984) for investigating this problem, see also (Wiggins 1988). Palmer's method was immediately proposed for the more general n-dimensional case. With regard to the elaboration of the first method, see (Meyer and Sell 1989), (Robinson 1988) and the literature cited therein. (It appeared earlier and there are numerous papers related to it.) In the analytical approach one needs to calculate certain curvilinear integrals which may also require calculations on a computer, but these are simpler.

2.4. Markov Partitioning for a Hyperbolic Automorphism of a Two-Dimensional Torus . We consider a concrete example: the automorphism g of the torus $T^2 = \mathbb{R}^2/\mathbb{Z}^2$ induced by the linear map of the plane with matrix $A = \binom{2\,1}{1\,1}$. It has eigenvalues $\lambda > 1$, $\mu \in (0,1)$. We denote by $W^{\mathrm{u}}(x,y)$ and $W^{\mathrm{s}}(x,y)$ the line in the plane passing through (x,y) and parallel to the eigenvector corresponding to the eigenvalue λ and μ respectively. Under the natural projection $p : \mathbb{R}^2 \to T^2$ these lines become the unstable and stable manifolds of the point $p(x,y)$ for the cascade $\{g^i\}$ (which, incidentally, "winds round" the torus everywhere densely). We denote by T_{hk} the translation of the plane $T_{hk}(x,y) = (x+h, y+k)$.

The lines $W^{\mathrm{u}}(0,0), W^{\mathrm{s}}(0,0), W^{\mathrm{u}}(1,0)$ and $W^{\mathrm{s}}(0,1)$ enclose a square $\widetilde{K}_1$, while the last two lines together with $W^{\mathrm{u}}(1,1)$ and $W^{\mathrm{s}}(1,1)$ enclose a second square $\widetilde{K}_2$. It is clear from Fig. 18a that $W^{\mathrm{u}}(1,1)$ and $W^{\mathrm{u}}(-1,0)$ (the latter is indicated by a dotted line) partition $\widetilde{K}_i$ into smaller rectangles $\widetilde{B}_j$. All these squares and rectangles are to be understood as closed sets. The closed regions $K_i = p\widetilde{K}_i$ cover the entire torus and intersect only on certain arcs (cf. Fig. 18b). The same is true for the five regions $B_j = p\widetilde{B}_j$. We somewhat loosely refer to $\mathcal{B} = \{B_1, \ldots, B_5\}$ as a partitioning of the torus.

The regions $A\widetilde{K}_i$ are depicted in Fig. 18c. In addition we have shown the rectangle $T_{-1,-1}A\widetilde{K}_2$ congruent to $A\widetilde{K}_2$ with respect to the translation. It is bounded by the line segments $W^{\mathrm{u}}(0,0)$, $W^{\mathrm{s}}(2,1)$, $W^{\mathrm{u}}(2,1)$ and $W^{\mathrm{s}}(1,0)$ (the last two being indicated by dotted lines). Comparing Fig. 18a with 18b it is easy to observe that $W^{\mathrm{s}}(1,0)$ partitions $A\widetilde{K}_1$ into two parts $\widetilde{B}_1$ and

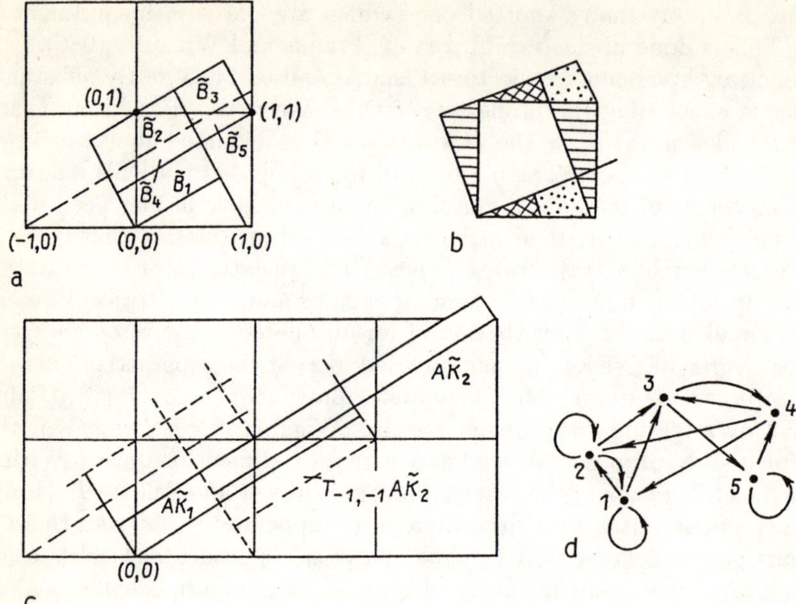

Fig. 18

$T_{1,0}(\widetilde{B}_2 \cup \widetilde{B}_3)$, and that $T_{1,0}(\widetilde{B}_4 \cup \widetilde{B}_5) = T_{-1,-1}A\widetilde{K}_2$. But now it is clear that

$$\begin{gathered} A\widetilde{K}_1 \cap \widetilde{K}_1 = \widetilde{B}_1, \quad T_{-1,0}A\widetilde{K}_1 \cap \widetilde{K}_1 = \widetilde{B}_2, \quad T_{-1,0}A\widetilde{K}_1 \cap \widetilde{K}_2 = \widetilde{B}_3, \\ T_{-2,-1}\widetilde{K}_2 \cap \widetilde{K}_1 = \widetilde{B}_4, \quad T_{-2,-1}\widetilde{K}_2 \cap \widetilde{K}_2 = \widetilde{B}_5. \end{gathered} \tag{19}$$

Furthermore, each $\widetilde{B}_i$ lies in some $\widetilde{K}_j$ joining its two "contracting" sides (which are line segments of the form $W^s(h,k)$) parallel to the "stretched" sides (lying on lines of the form $W^u(h,k)$). Therefore under the action of A the $\widetilde{B}_i$ are contracted in one direction and expanded in another, so that one obtains a band joining the "contracting" sides of $A\widetilde{K}_j$ parallel to its "stretched" sides. Hence and from (19) it is clear that on the torus the intersections $g\mathrm{int}B_i \cap \mathrm{int}B_j$ (where int denotes the interior) are described by the transition graph depicted in Fig. 18d; moreover, the general picture with a combination of contractions and expansions is similar to that in Sect. 2.2. (We do not consider intersections of the type $gB_1 \cap B_5$ which merely reduce to boundary arcs.) The situation with g^{-1} has, of course, yet to be considered. This is left to the reader. Here one finds that corresponding to any finite "path" $(\omega_{-k}, \dots, \omega_l)$ on the graph there is a rectangle, namely, the non-empty intersection of the sets $g^i\mathrm{int}B_{\omega_i}$. This is an open rectangle; in order to go over to infinite paths, we first close it:

$$B(\omega_{-k}, \dots, \omega_l) = \overline{g^{-k}\mathrm{int}B_{\omega_{-k}} \cap \dots \cap g^l\mathrm{int}B_{\omega_l}}.$$

(The arguments of B are not simply the elements $1,\dots,5$; they are endowed with the numbers from $-k$ to l. With this reservation the notation is correct.) For a path $\omega = (\omega_i)$ that is infinite in both directions

$$\bigcap\{B(\omega_{-k},\dots,\omega_l);\ k,l>0\}$$

reduces to a single point which we denote by $\psi\omega$. It is easy to prove that the map $\Omega' \to T^2$ that this gives rise to, where Ω' is the corresponding TMC, is a surjective homomorphism of the corresponding DS's (that is, ψ is continuous and $\psi\sigma = g\psi$).

In contrast with the preceding subsections, we speak of the map $\psi : \Omega' \to T^2$ and not of the coding map $T^2 \to \Omega'$. In the present case ψ is not injective (the torus is not homeomorphic to the zero-dimensional compact set Ω'!). Outside $\bigcup g^i\partial\mathcal{B}$, where $\partial\mathcal{B}$ is the union of the boundaries of the regions B_j, the inverse map ψ^{-1} is single-valued and is obtained by the usual encoding rule:

$$x \mapsto (\omega_i),\ \text{where } \omega_i = j,\ \text{if } g^i x \in B_j. \tag{20}$$

However, at the discarded points ψ^{-1} not only fails to be single-valued in general, but it does not even contain all those points of ω that can be associated with a point x according to the rule (20). Thus, the fixed point $p(0,0) \in B_1 \cap B_2 \cap B_5$, and (20) associates with it any sequence of symbols 1, 2, 5. Of these, the graph 18d can be applied only to sequences for which all the $\omega_i \in \{1,2\}$ and sequences for which all the $\omega_i = 5$. And in fact, $\psi^{-1}p(0,0)$ consists of three sequences in all: in one of them all the $\omega_i = 1$, in the second all the $\omega_i = 2$, in the third all the $\omega_i = 5$. (Corresponding to the sequences ω in which both symbols 1, 2 feature are the points $\psi\omega \neq p(0,0)$.)

In the given example the set $\bigcup g^i\partial\mathcal{B} = W^{\mathrm{u}}(p(0,0)) \cup W^{\mathrm{s}}(p(0,0))$ is an exceptional set. It is everywhere dense but of the first category. It is clear from the nature of the action of g on $W^{\mathrm{u}}, W^{\mathrm{s}}$ that the only finite invariant measure concentrated on this set is the measure concentrated at the fixed point $p(0,0)$. The only periodic trajectory intersecting the exceptional set, and even the only minimal set intersecting it, consists of this same fixed point. It follows that in questions touching upon ergodic theory, periodic points or minimal sets, the exceptional set plays no role in practice and these questions reduce to analogous questions concerning TMC's. For periodic points this is not very important in the present instance — points of period n are defined from the congruence $A^n z \equiv z \bmod \mathbb{Z}^2$ and it is easy to study them; but the reduction to symbolic dynamics is very helpful for the remaining questions.

Partitions analogous to $\mathcal{B}$ can also be constructed for other hyperbolic automorphisms g of the two-dimensional torus. They can consist of another collection of parallelograms whose sides are not necessarily perpendicular, but, as before, the boundaries of the parallelograms consist of contracting and expanding arcs (that is, arcs of W^{s} and W^{u}), where the images of the contracting arcs under the action of g lie on the contracting arcs, while the inverse images of the expanding arcs lie on the expanding ones. These partitions are called *Markov partitions.* They were introduced by Adler and Weiss (Adler and Weiss 1967), (Adler and Weiss 1970).

§3. Semilocal Theory

3.1. ϵ-Trajectories. This concept was introduced in (Anosov et al. 1985, Chap. 3, Sect. 2.2). (For flow there are several variants; for the moment any will do.) Let A be a hyperbolic set of the DS $\{g^t\}$. Then for any $\delta > 0$ there exists $\epsilon > 0$ such that each two-sided infinite ϵ-trajectory $\{x(t)\}$ lying entirely in the ϵ-neighbourhood $U_\epsilon(A)$ of A is δ-close to some actual trajectory $\{g^t y\}$ (or, as is also said, it is traced by this trajectory to within a distance δ). In the case of a cascade, the latter means that $\rho(x(i),\ g^i y) < \delta$ for all i, while in the case of a flow, the analogous inequality holds for some reparametrized trajectory $\{g^t y\}$: there exists a surjective homeomorphism h of the real line that preserves its orientation and is such that $\rho(x(t),\ g^{h(t)}y) < \delta$ for all t. (Since a finite or one-sided infinite ϵ-trajectory can be extended to a two-sided infinite one, similar assertions hold for these as well.) The above property (especially its "intrinsic" version for locally maximal A, see below) is called the *ϵ-trajectory tracing* property (pseudo-orbit tracing property, POTP for short, in the English-language literature, where the term "shadowing" is also used). Certain authors (especially the Japanese) call this property *stochastic stability.* The various formulations of this property make sense in the general context of topological dynamics. For a hyperbolic set A it turns out that ϵ and δ have the same order of smallness.

We can always suppose that $\epsilon < \delta$. Clearly $\{g^t y\} \subset U_{\epsilon+\delta}(A) \subset U_{2\delta}(A)$. If A is locally maximal, then it follows that $\{g^t y\} \subset A$ for sufficiently small δ. In other words, if ϵ is sufficiently small, then an ϵ-trajectory lying in $U_\epsilon(A)$ is traced by an actual trajectory lying in A. On the other hand, an ϵ-trajectory $L \subset U_\epsilon(A)$ is clearly traced to within ϵ by some 2ϵ-trajectory $L_1 \subset A$. (In the case of a flow the latter may be discontinuous even if L is continuous.) A trajectory $L_2 \subset A$ tracing L_1 to within δ also traces L to within $\epsilon + \delta$. This means that for a locally maximal A it makes no difference whether we talk about tracing trajectories in $U_\epsilon(A)$ or in A itself. The property of tracing ϵ-trajectories in A by trajectories in A is an "intrinsic" property of the DS $\{g^t|A\}$.

For hyperbolic sets local maximality is equivalent to this "intrinsic" property. The fact is that two trajectories lying in a sufficiently small neighbourhood of a hyperbolic set A and sufficiently close to each other (in the same sense as before) must coincide (see below). Suppose that this A has the above intrinsic property. A trajectory L lying entirely in $U_\epsilon(A)$ is ϵ-close to some 2ϵ-trajectory in A, while the latter is traced to within $\delta = \delta(2\epsilon)$ by some trajectory in A which therefore also traces L to within $\epsilon + \delta$. Hence when $\epsilon + \delta$ is sufficiently small (and this is guaranteed when ϵ is small) it follows that these two trajectories coincide; hence $L \subset A$. But this means local maximality: any trajectory lying entirely in a small neighbourhood of A lies in A itself.

Since we know that local maximality of a hyperbolic set is equivalent to some intrinsic property of it (more precisely, a property of the restriction of our DS to this set), we can conclude that a hyperbolic set A of the DS $\{g^t\}$ that is homeomorphic to a locally maximal hyperbolic set B of the DS $\{h^t\}$ via a homeomorphism establishing an equivalence of the DS's $\{g^t|A\}$ and $\{h^t|B\}$ (for the sake of brevity, we say that A and B are equivalent[15]), is itself locally maximal.[16]

A Bernoulli topological cascade $\{\sigma^i\}$ in Ω_n is realized by means of a Smale horseshoe as a hyperbolic set in a smooth cascade. Therefore it too possesses the above intrinsic property (which in any case is easily verified directly, without having recourse to this realization), and a hyperbolic set A that is equivalent to a Bernoulli cascade is locally maximal. Furthermore a TMC is locally maximal (as an invariant subset Ω' of a Bernoulli cascade in some Ω_n). In fact, let Ω' be selected from Ω_n by the condition that for $\omega = (\omega_i) \in \Omega'$ all pairs (ω_i, ω_{i+1}) belong to some set D of admissible pairs. If the trajectory $\{g^i\omega\}$ is sufficiently close to Ω', then any pair (ω_i, ω_{i+1}) of neighbouring symbols of the sequence $\omega = (\omega_i)$ is the same as the corresponding pair of symbols in some sequence $\omega^{(i)} \in \Omega'$. Hence, all the $(\omega_i, \omega_{i+1}) \in D$ so that $\omega \in \Omega'$. Hence, any hyperbolic set equivalent to a TMC is locally maximal.

A cascade $\{g^i\}$ and a homeomorphism g possess the property of *separating the motions* or the property of *separating the trajectories* (in the present instance they are synonyms) if there exists $\delta > 0$, called the *separation constant* or *separating constant*, such that if

$$\rho(g^i x, g^i y) < \delta \text{ for all } i,$$

then $x = y$. This property is also called *expansiveness*. For flows this definition would literally be vacuous, since if one takes $y = g^s x$ with s sufficiently small, then $g^t x$ and $g^t y$ are close for all t. There are, however, certain "non-vacuous" modifications of this definition. A flow possesses the property of *separation of motions* if for any $\epsilon > 0$ there exists $\delta > 0$ such that if

$$\rho(g^t x, g^t y) < \delta \text{ for all } t, \tag{21}$$

then $y = g^\tau x$ for some $|\tau| < \epsilon$. A flow has the property of *separation of trajectories* if for any $\epsilon > 0$ there exists $\delta > 0$ such that if under some orientation-preserving surjective homeomorphism h of the real line we have $h(0) = 0$ and

[15] And if the homeomorphism conjugates these DS's, then we speak of an isomorphism between A and B. (Of course, isomorphism and equivalence differ only for flows.) This terminology is in accordance with (Anosov et al. 1985, Chap. 1, Sect. 1.5).

[16] It is worth noting that if A is a locally maximal closed invariant subset of the DS $\{g^t\}$ in M and $A_1 \subset A$ is a locally maximal closed invariant subset of the DS $\{g^t|A\}$, then A_1 is locally maximal in M. (This is not related to hyperbolicity.)

$$\rho(g^t x, g^{h(t)} y) < \delta \text{ for all } t, \tag{22}$$

then $y = g^\tau x$ for some $|\tau| < \epsilon$. The second property is stronger than the first (an example showing that they are inequivalent is provided by horocyclic flows well known from geometry). Originally in the formulation of the second property, h could be any continuous function, not necessarily a homeomorphism. In (Oka 1990) it is proved that these two variants are equivalent to each other and to several other versions (where one of them is equivalent to the others in the absence of equilibrium points). By the *separation constant* (of motions or trajectories) it would appear to be natural, in the case of a flow, to mean the number $\delta > 0$ such that when (21) or (22) holds (depending on whether one is dealing with separation of motions or trajectories) x and y of necessity lie on one trajectory.

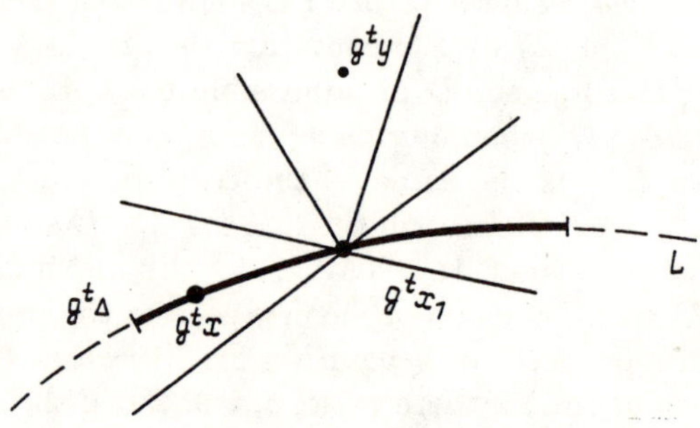

Fig. 19

If A is a hyperbolic set, then the DS $\{g^t|A\}$ has the property of separation of trajectories. Here the idea is that if $x \in A$ and y is close to x, then the position of $g^t y$ with respect to $g^t x$ is sufficiently well described, in classical terminology, by means of a system of variational equations along the trajectory $L = \{g^t x\}$ as long as $g^t y$ is within a certain distance δ from $g^t x$. In the case of a cascade, the non-zero solutions of the system increase exponentially at least on one side with respect to time (it is best to use the Lyapunov metric so that the growth takes place with a "good" estimate over all time) and $\rho(g^t y, g^t x)$ also increases in almost the same way as long as it does not exceed δ. In the case of a flow the discussion is more complicated; as an intermediate link it is advisable to consider the change of the position of $g^t y$ with respect to $g^t \Delta$, where Δ is the arc $\{g^s x;\ |s| \le \epsilon\}$ with small $\epsilon > 0$. If $\rho(x, y) < \delta$ and δ is considerably less than ϵ, then Δ contains the point x_1 closest to y and $g^t y$ moves away from $g^t x_1$, at least in one time direction, while remaining, in terms of local coordinates $\exp^{-1}_{g^t x_1}$ close to $g^t x_1$, inside some cone with axis space $E^s_{g^t x_1} \oplus E^u_{g^t x_1}$ and angle, say, 30° (Fig. 19). This continues until $g^t y$ is at a distance δ from $g^t x_1$ for some $t = T$. In this terminology the arc

$g^T\Delta$ lies in a cone of the same angle with axis space $E^n_{g^Tx_1}$ (this is achieved by making ϵ small). Therefore $\rho(g^Ty, g^T\Delta) > \delta/2$. It can be assumed that in the Lyapunov metric the length of any arc of the trajectory L under the action of the flow changes by not more than twice the amount and differs from its temporal length again by not more than twice the amount. Hence it follows that for $32\delta < \epsilon$ we have $\rho(x, x_1) < 2\delta$, $\rho(g^tx, g^ty) < 4\delta$ and the distance from g^tx_1 to the ends of Δ is at least $\epsilon/8$, which means that as long as g^ty is within a distance δ from g^tx_1, the distance from g^ty to the ends of $g^t\Delta$ is at least 3δ. Suppose now that h is an orientation-preserving surjective homeomorphism of the real line with $h(0) = 0$. If $g^{h(t)}x$ does not leave the segment $g^t\Delta$ as long as g^ty is within a distance δ from g^tx_1, then at this instant we have $\rho(g^Tx, g^{h(T)}y) > \delta/2$. If, on the other hand, $g^{h(t)}x$ leaves the segment before this instant, say when $t = s$, then $\rho(g^{h(s)}x, g^sy) > 3\delta$. Thus, if $\rho(g^{h(t)}x, g^ty) < \delta/2$ for all t, then $y = g^\tau x$ for some $|\tau| \leq \epsilon$. By replacing t by $h^{-1}(t)$, we obtain exactly the original formulation of the property of separation of trajectories, except that instead of δ we now have $\delta/2$.

It follows from the property of separation of trajectories that the trajectory $\{g^ty\}$ tracing an ϵ-trajectory $\{x(t)\} \subset U_\epsilon(A)$ to within δ is unique provided that ϵ and δ are sufficiently small. In fact, when r is sufficiently small, the set $B = \bigcap g^t\overline{U_r(A)}$ is hyperbolic (Sect. 1.2 g)). Let d be the corresponding constant of separation of trajectories. Two trajectories tracing $\{x(t)\}$ to within δ lie inside $U_{\epsilon+\delta}(A)$ and are 2δ-close to each other. If $\epsilon < \delta$ and $2\delta < \min(r, d)$, then we find that they both lie in B and satisfy an inequality of type (22), but with d replacing δ. Hence they coincide.

In view of the uniqueness of the trajectory L' tracing an ϵ-trajectory L, there arises the question whether the dependence of L' on the "initial data" is continuous. In the present instance, "initial data" refers not only to L but to the DS in question. A scarcely untouched formalization of this idea was put forward by me in 1967 and published in (Anosov 1970). See also (Alekseev, Katok, and Kushnirenko 1972), (Katok 1971), (Shub 1978).

By a *family of ϵ-trajectories* of a DS $\{g^t\}$ we mean a triple $(X, \{h^t\}, \phi)$, where X is a topological space, $\{h^t\}$ is a DS in X (with the same time as for $\{g^t\}$, that is, these DS's are either both cascades or both flows), and $\phi : X \to M$ is a continuous map taking each trajectory of $\{h^t\}$ to the ϵ-trajectory of $\{g^t\}$. (Discontinuous ϵ-trajectories of the flows are not considered.) If ϕ takes trajectories to trajectories, then we talk about the *family of trajectories*.

Theorem on the family of ϵ-trajectories. *Let A be a hyperbolic set of the DS $\{f^t\}$. Then*

a) *There exist $\epsilon_0, \delta_0 > 0$ such that for any family of ϵ_0-trajectories $(X, \{h^t\}, \phi)$ of the DS $\{f^t\}$ for which $\phi(X) \subset U_\epsilon(A)$ and for any DS $\{g^t\}$ with $\rho_{C^1}(\{f^t\}, \{g^t\}) < \epsilon_0$, there exists exactly one continuous map $\psi : X \to M$ such that the triple $(X, \{h^t\}, \psi)$ is a family of trajectories of $\{g^t\}$ and $\rho_C(\phi, \psi) < \delta_0$.*

b) *For any* $\delta \in (0, \delta_0)$ *there exists* $\epsilon \in (0, \epsilon_0)$ *such that for any family of* ϵ*-trajectories* $(X, \{h^t\}, \phi)$ *with* $\phi(X) \subset U_\epsilon(A)$ *and any DS* $\{g^t\}$ *with* $\rho_{C^1}(\{g^t\}, \{f^t\}) < \epsilon$ *the map* ψ *in* a) *satisfies the inequality* $\rho_C(\phi, \psi) < \delta$.

We apply this theorem to a description of the motions in a neighbourhood of a transversal homoclinic trajectory. We restrict ourselves to the same situation as in Sect. 2.3 (the DS is a cascade $\{g^i\}$, $gx_1 = x_1$, $L \subset W^{\mathrm{u}}(x_1) \cap W^{\mathrm{s}}(x_1)$). We have no need to bother with the objects of K, U, V type in Sect. 2.3 or the small neighbourhoods of the points $g^{-j}x, g^h x$ that feature there implicitly. We use a metric that is Lyapunov at x_1. We fix some point $y \in L$ and a sufficiently small $r > 0$ such that $r < \rho(y, L \backslash \{y\})$. We take a small $\epsilon \in (0, r)$ and, slightly altering the previous notation, we denote by $\{gx, \ldots, g^{l-1}x\}$ the segment of L lying outside $U_\epsilon(x_1)$. (For sufficiently small ϵ, the points of $L \backslash U_\epsilon(x_1)$ actually form exactly one such segment.) In particular, $x \in U_\epsilon(x_1)$, $g^l x \in U_\epsilon(x_1)$, $y \notin U_\epsilon(x_1)$, $y = g^j x$ for some $j \in \{1, \ldots, l-1\}$.

Let Ω' be the TMC in Ω_{l+2} defined by the same transition graph as in Fig. 17 but with $n+k$ replaced by $l+2$. We consider the map $\phi : \Omega' \to U_\epsilon(\bar{L})$, where for $\omega = (\omega_i)$

$$\phi\omega = \begin{cases} g^{\omega_0 - 1}x & \text{if } \omega_0 \neq 0, \\ x_1 & \text{otherwise.} \end{cases}$$

Clearly $(\Omega', \{\sigma^i | \Omega'\}, \phi)$ is a family of ϵ-trajectories for $\{g^i\}$ that lies in (that is, its image lies in) $U_\epsilon(\bar{L})$ (in fact, even in $\bar{L}$). There exists a family of actual trajectories $(\Omega', \{\sigma^i | \Omega'\}, \psi)$ tracing these ϵ-trajectories to within δ, where $\delta \to 0$ as $\epsilon \to 0$. (The quantity l increases as $\epsilon \to 0$, so that Ω' and ϕ are altered; however, this has no influence on the dependence of δ on ϵ: in the theorem on the family of ϵ-trajectories the same ϵ and δ "service" all the families of ϵ-trajectories.)

We denote $\psi\Omega'$ by D. This is a compact invariant subset of the cascade $\{g^i\}$ (since Ω' is compact). Once $\phi\Omega' \subset L$ it follows that $\psi\Omega' \subset U_\delta(\bar{L})$ and for sufficiently small δ (that is, ϵ) D is hyperbolic (see Sect. 1.2 g)).

We verify that for sufficiently small ϵ the map ψ is injective. Let $\omega = (\omega_i) \in \Omega'$, $\omega' = (\omega'_i) \in \Omega'$, $\omega \neq \omega'$. This means that $\omega_i \neq \omega'_i$ for some i, so that at least one of the numbers ω_i, ω'_i is non-zero. Suppose that it is ω_i. It is clear from the definition of Ω' that if $\omega_i = h > 0$, then for $k = 1, \ldots, l+1$ we have $\omega_{i-h+k} = k$, $\omega'_{i-h+k} \neq k$. For the zero element of the sequence $\sigma^{i-h+j}\omega$ we have

$$(\sigma^{i-h+j}\omega)_0 = \omega_{i-h+j} = j,$$

so that

$$\phi\sigma^{i-h+j}\omega = g^j x = y, \ \phi\sigma^{i-h+j}\omega' \neq y.$$

But since $\phi\sigma^{i-h+j}\omega' \in \bar{L}$, it follows that

$$\rho(\phi\sigma^{i-h+j}\omega, \ \phi\sigma^{i-h+j}\omega') \geq r,$$

$$\rho(g^{i-h+j}\psi\omega, \ g^{i-h+j}\psi\omega') = \rho(\psi\sigma^{i-h+j}\omega, \ \psi\sigma^{i-h+j}\omega') >$$
$$> \rho(\phi\sigma^{i-h+j}\omega, \ \phi\sigma^{i-h+j}\omega') - 2\delta \geq r - 2\delta,$$

which is greater than zero for sufficiently small ϵ. Hence $\psi\omega \neq \psi\omega'$.

In view of the compactness of Ω', ψ is a homeomorphism (between Ω' and D). But since ψ takes the trajectories of $\{\sigma^i\}$ to trajectories of $\{g^i\}$, it follows that ψ is an equivalence of the DS's $\{\sigma^i|\Omega'\}$ and $\{g^i|D\}$. It remains to show that D contains all trajectories lying in some neighbourhood of $\bar{L}$.

First of all, $D \supset L$. In fact, Ω' contains the point $\omega = (\omega_i)$, where $\omega_i = i$ for $1 \le i \le l+1$ and $\omega_i = 0$ for the remaining i. The trajectory of this point is taken under the map ϕ to the ϵ-trajectory

$$\phi\sigma^i\omega = \begin{cases} g^i x & \text{if } 1 \le i \le l+1, \\ \phi\sigma^i\omega = x_1 & \text{otherwise.} \end{cases}$$

The trajectory $\{\psi\sigma^i\omega\}$ must trace this ϵ-trajectory to within δ and lie in $U_\delta(\bar{L})$. But clearly, L has the same properties. Hence it follows that for sufficiently small ϵ and δ we have $L = \{\psi\sigma^i\omega\} \subset D$. Furthermore, D is locally maximal, since it is equivalent to a TMC. But then all trajectories that lie entirely in some neighbourhood of D (which is also a neighbourhood of $\bar{L}$) lie in D.

The theorem on families of ϵ-trajectories enables one to obtain just as easily the following result.

Theorem ((Alekseev 1969) or (Alekseev 1976, Theorem 7.8)). *In the case of a cascade, any neighbourhood of a hyperbolic set A contains a hyperbolic set $B \supset A$ that is a factor (that is, the image under a continuous motion-preserving map) of some TMC.*

Alekseev provides additional information concerning B which can also be obtained in this way. It is similar to the information on local maximal hyperbolic sets given in Sect. 3.2. Subsequently Kurata used Markov partitions to further strengthen this similarity (Kurata 1979b). In this connection we note the following unsolved problem: is it true that every hyperbolic set A is contained in a locally maximal hyperbolic set? Variant: Is it true that any neighbourhood of A contains a locally maximal $B \supset A$? The analogous question for invariant closed subsets of a Bernoulli cascade has an affirmative answer (see Alekseev 1976, Theorem 3.2, taking into account Theorem 3.1).

3.2. Locally Maximal Hyperbolic Sets. The properties of arbitrary hyperbolic sets are unending, as is clear if only from what was said in Sect. 2.2 on closed subsets of a Smale horseshoe. The condition of local maximality, introduced by me at the end of the 60's, ensures a rational restriction of the class of hyperbolic sets: such sets are encountered sufficiently often and at the same time one can obtain sufficiently meaningful results on their structure and on the character of motions in a neighbourhood of them. At that time Smale (Smale 1969) drew attention to another property, namely, the property of the local structure of products which in their formulation relate to hyperbolic sets, and which, as it was soon found out, are equivalent to their local maximality. The equivalence was essentially proved in (Hirsch et al. 1970) (see also (Alekseev, Katok and Kushnirenko 1972) and (Katok 1971)).

A hyperbolic set A has a *local product structure* if there exists $\epsilon > 0$ such that x, $y \in A$ implies that $W^{\mathrm{s}}_{\epsilon}(x) \cap W^{\mathrm{u}}_{\epsilon}(y) \subset A$. Here $W^{\mathrm{s}}_{\epsilon}$, $W^{\mathrm{u}}_{\epsilon}$ can be understood in any of the senses indicated in Sect. 1.3. In any case, the above intersection is either empty or consists of exactly one point; in the case of a cascade it is non-empty when x and y are sufficiently close to each other, while in the case of a flow, when x and y are sufficiently close, precisely one of the intersections $W^{\mathrm{s}}_{\epsilon}(x) \cap W^{\mathrm{u}}_{\epsilon}(g^t y)$ is non-empty, where $|t| < \epsilon$. We denote this unique point of intersection by $[x, y]$; $[\cdot, \cdot]$ is a continuous function taking values in M and defined in some neighbourhood of the diagonal $\{(x, x);\ x \in A\}$ in $A \times A$. The property of the local product structure is that this function takes values in A. It is easily shown that x in A has a neighbourhood U_x that is homeomorphic to the direct product of certain neighbourhoods of x in $W^{\mathrm{s}}(x) \cap A$ and $W^{\mathrm{u}}(L) \cap A$ (the neighbourhoods in the product are taken in the topology corresponding to the topology of immersed submanifolds), where this homeomorphism is given by the map $(x_1, x_2) \mapsto [x_1, x_2]$. This explains the name acquired by this property. In the case of a cascade the second neighbourhood is taken simply on $W^{\mathrm{u}}(x) \cap A$; we say that x_1, x_2 are *canonical coordinates* in U_x (although even without that, the word "canonical" has many meanings). In the case of a flow the second neighbourhood can be considered in natural fashion to be homeomorphic to $(W^{\mathrm{u}}_{\delta}(x) \cap A) \times (-\delta, \delta)$. (In this case we could have considered the intersections $W^{\mathrm{s}}_{\epsilon}(g^t x) \cap W^{\mathrm{u}}_{\epsilon}(x)$ and talked about U_x homeomorphic to a product of neighbourhoods of x in $W^{\mathrm{s}}(L) \cap A$ and $W^{\mathrm{u}}(x) \cap A$. This would have led to other canonical coordinates.)

Theorem. *Let A be a locally maximal hyperbolic set. Then*

1) *The set Ω of nonwandering points of a DS $\{g^t|A\}$ is the same as the closure of the set of its periodic points; hence it is also the same as its Birkhoff centre* (*see* (*Anosov et al.* 1985, *Chap.* 3, *Sect.* 4.1) *concerning nonwandering points and the centre of a DS*).

2) *The zones of attraction and repulsion* (*Anosov et al.* 1985, *Chap.* 3, *Sect.* 2.1) *are given by*

$$\begin{aligned} W^{\mathrm{s}}(A) &= \{x \in M;\ \omega(x) \subset A\} = \bigcup \{W^{\mathrm{s}}(x);\ x \in \Omega\}, \\ W^{\mathrm{u}}(A) &= \{x \in M;\ \alpha(x) \subset A\} = \bigcup \{W^{\mathrm{u}}(x);\ x \in \Omega\}. \end{aligned} \tag{23}$$

3) *Ω is uniquely represented as a finite union of non-intersecting invariant compact sets $\Omega_1, \ldots, \Omega_k$ ("components"), the restriction to each of which of the DS is topologically transitive. This decomposition is called the* spectral decomposition (*by analogy with Smale's terminology in Sect.* 4.1).

4) *We write $\Omega_i > \Omega_j$ if there exists $x \in A$ for which $\alpha(x) \subset \Omega_i$, $\omega(x) \subset \Omega_j$. We then obtain a partial ordering of the collection of sets $\Omega_1, \ldots, \Omega_k$.*

5) *For a cascade $\{g^i\}$ each Ω_i is represented as a finite union of non-intersecting sets $\Omega_{i1}, \ldots, \Omega_{im_i}$ which under the action of g are converted into*

one another in cyclic fashion; here for any $x, y \in \Omega_{ij}$

$$\Omega_{ij} = \overline{W^{\mathrm{s}}(x) \cap W^{\mathrm{u}}(y) \cap \Omega_{ij}},$$

and the restriction $g^{m_i}|\Omega_{ij}$ *has the region mixing property.*

6) *For a flow* $\{g^t\}$, *for each* $i = 1, \ldots, k$ *we have the alternative: either the DS* $\{g^t|\Omega_i\}$ *has the region mixing property or (to within multiplication of time by a constant) it is isomorphic to the Smale suspension (Anosov et al.* 1985, *Chap* 1, *Sect.* 2.4) *over some cascade* $\{f_i^n\}$ *which acts in some compact set* K_i *and has the region mixing property. In the first case, we have for any* $x, y \in \Omega_i$

$$\Omega_i = \overline{W^{\mathrm{s}}(x) \cap W^{\mathrm{u}}(x) \cap \Omega_i},$$

while in the second, for any $x, y \in K_i$

$$K_i = \overline{W^{\mathrm{s}}(x) \cap W^{\mathrm{u}}(y) \cap K_i}.$$

(Here we regard K_i *as being embedded in* M *via the above isomorphism, since* K_i *is naturally embedded in the Smale suspension as the image of the "bottom"* $K_i \times 0$ *of the direct product* $K_i \times [0, 1]$ *followed by an identification of this "bottom" with the "top"* $K_i \times 1$.)

1) − 5) are proved by the author in (Anosov 1970) (where 3) is analogous to Smale's earlier result, see Sect. 4.1); 6) was proved by Bowen in (Bowen 1972).

It follows from 2) that *a fortiori*

$$W^{\mathrm{s}}(A) = \cup\{W^{\mathrm{s}}(x);\ x \in A\},\ W^{\mathrm{u}}(A) = \cup\{W^{\mathrm{u}}(x);\ x \in A\}. \tag{24}$$

Generally speaking, for hyperbolic sets that are not locally maximal (24) may not hold (see (Hirsch et al. 1970), (Nitecki 1971) or the remark on p. 117 of the Russian translation of (Smale 1967b)). Nevertheless, along with local maximality certain other hyperbolic sets A also satisfy (24) and even (23). Something is also said about these in (Anosov 1970).

Locally maximal topologically transitive hyperbolic sets are called *basic.* (This name was originally proposed by Smale in a more special situation, see Sect. 4.1.) As is clear, they play a large role in the hyperbolic theory. Properties 5) and 6) show that of special importance are those basic sets A for which the DS $\{g^t|A\}$ has mixing of regions; the remaining basic sets are, in a way, obtained from them.

The stipulation "in a way" refers only to the case of flows, and then not all of them. For topologically transitive Anosov flows the entire phase manifold $M = A = \Omega$, and in 3) there remains the single basic set $\Omega_1 = M$. In this case the second possibility of the alternative in 6) looks "smoother": to within multiplication of time by a constant $\{g^t\}$ is a Smale suspension over some Anosov cascade (Plante 1972) (in the presence of an invariant measure equivalent to Lebesgue measure, this was established in (Anosov 1967) in the course of the proof of some metric analogue of the above alternative).

In the general case one might also want to say that the K_i in 6) is a hyperbolic set of the cascade $\{f_i^n\}$. However: a) one cannot hope that K_i

will always be situated on some smooth global cross-section of the entire flow $\{g^t\}$ in M; b) although one can assume that $K_i \subset M$, $f_i = f^{T_i}|K_i$ for some constant T_i, the suspension in 6) is taken only over K_i, but K_i is not a hyperbolic set of the diffeomorphism g^{T_i}. We see that, starting from a smooth DS we have arrived at objects lying outside the framework of the hyperbolic theory of smooth DS's. (However, the Bernoulli topological cascade and the TMC already show that it is advisable to go beyond this framework. Although a TMC can always be realized as a hyperbolic set, is it necessary to invoke this realization every time one wishes to note that the TMC is characterized by some hyperbolicity?) Guided by such considerations, we can state certain essential properties of a basic (or, more generally, hyperbolic locally maximal) set and the DS $\{g^t|A\}$ in such a way that this statement is not related to the fact that A lies in M and that the flow $\{g^t\}$ is smooth, but rather that it is related to intrinsic properties of $\{g^t|A\}$ and is meaningful in the general context of topological dynamics. (In fact, the property, considered in Sect. 3.1, of tracing ϵ-trajectories in A has such a character, but it is used in combination with the "extrinsic" condition of hyperbolicity of A.) Starting with Bowen's article (Bowen 1970a), there have appeared several variants of this kind (Bowen 1971), (Alekseev 1976), (Alekseev and Yakobson 1979), (Pollicott 1987), (Ruelle 1978). In all of them the local stable and unstable sets $W^{\mathrm{s}}_\epsilon(x)$ and $W^{\mathrm{u}}_\epsilon(x)$ are used, which are defined by analogy with one of the variants of Sect. 1.3 and where it is postulated that: 1) the approach of $g^t y$ towards $g^t x$, where $y \in W^{\mathrm{s}}_\epsilon(x)$ or $y \in W^{\mathrm{u}}_\epsilon(x)$, has an exponential character as $t \to \infty$ or $t \to -\infty$; 2) the product has a local structure; here in the case of a flow, one also takes into account the role of small time shifts. (The specific formulations in the various articles differ to some extent, this being partly related to the nature of the results that their authors are aiming towards.) Thus, the notion of a hyperbolic set does not go over to topological dynamics, whereas the narrower concept of a locally maximal hyperbolic set does.

Let A be a locally maximal hyperbolic set of a cascade. Applying Alekseev's theorem in Sect. 3.1 to it, we find that A is a factor of a TMC Ω'. This statement can be made more precise. First, if A is zero-dimensional, then one can even construct an isomorphism between some TMC and A. But if $\dim A > 0$, then a surjective homomorphism $\psi : \Omega' \to A$ cannot be injective (in fact, Ω' is zero-dimensional). Hence under the map ψ certain identifications are inevitable: some distinct points are taken to a single point. It may merely be a question that, on the one hand, there are slightly fewer of these "identifications", while on the other hand they might receive some description that would be of assistance in passing from the analysis of the motions in Ω' to the analysis of the motions in A.

This is achieved by means of Markov partitions, an example of which was given in Sect. 2.4. Since the most important of their applications relates to ergodic theory and because these are discussed in (Bunimovich et al. 1965), we shall not even give their definition in the general case. We merely note that the passage from the particular case of Sect. 2.4 to the general case is far from simple. Already for a hyperbolic automorphism of the three-dimensional torus the boundary of an element of a Markov partition cannot consist merely of several regions on W^{u} and W^{s}; to be sure, it is two-dimensional, whereas one of the manifolds W^{u} or W^{s} is one-dimensional. Even in this case, the construction, by contrast to Sect. 2.4, contains several limiting processes, and this is unavoidable since in this case the boundaries of the elements of the Markov partition may not be piecewise-smooth (Bowen 1978). The construction of Markov partitions for n-dimensional Anosov cascades was begun by Sinai (Sinai 1968a), (Sinai 1968b), (Gurevich and Sinai 1969), but the definitive version (now for basic or even locally maximal hyperbolic sets) was put forward by Bowen (Bowen 1970b), (Bowen 1970c). Markov partitions are considered in (Alekseev 1976), (Alekseev and Yakobson 1979), (Bowen 1975b), (Ruelle 1978), (Shub 1978). (For Anosov cascades certain improvements were announced: in (Farrell and Jones 1977) it was pointed out that there are Markov partitions that are cell structures (in the sense as understood in topology). Meanwhile it would seem that it is not clear what additional use can be made of this; see (Farrell and Jones 1993) for a generalization for locally maximal hyperbolic sets of cascades.)

When using Markov partitions the multiplicity N of the map $\psi : \Omega' \to A$ (that is, the maximum number of points in the inverse images of singleton points) is finite. A fuller description of the identifications occurring under the map ψ is obtained in (Bowen 1970c), (Manning 1973), (Yakobson 1976) (see also the accounts in (Alekseev 1976), (Alekseev and Yakobson 1979), (Shub 1978)). These refinements have, in particular, enabled one to prove that N does not exceed the square of the number of elements of the partition, that a minimal subset of a hyperbolic set is zero-dimensional (Kurata 1979b) and the ζ-function of a cascade $\{g^i|A\}$ (see (Anosov et al. 1985, Chap. 2, Sect. 2.4)) with a locally maximal hyperbolic A is rational. However, the last property has been reproved without appealing to Markov partitions, and even with a certain amount of generalization: A is a locally maximal closed invariant set of a smooth cascade $\{g^i\}$, where $\{g^i|A\}$ separates the trajectories, and the periodic points lying in A are non-degenerate (there are no roots of unity among their multiplicators) (see Fried 1987). Various results on periodic points (the asymptotic behaviour of the number of them as the period increases, their distribution) are linked with ergodic theory and are elucidated in (Bunimovich et al. 1985) and (Alekseev and Yakobson 1979); see (Alekseev 1976) for details.

For a TMC $\{\sigma^i|\Omega'\}$ something like our spectral decomposition in 3) has long been known in probability theory. The differences (apart from terminological ones, which we shall not dwell upon) relate to the fact that in probability theory, instead of the phase space of the TMC as the DS it proved possible to operate with the space of its states as a random process, that is, (in our terminology) with the corresponding alphabet or some part of it; this is simpler. All the same, we shall go over from this partition to a partition of the set of nonwandering points of the TMC as a DS into topological transitive components. The connection between this partition and the spectral decomposition for the original hyperbolic set A is not entirely clear (see (Alekseev 1976, Remark 2, Theorem 7.2)). Besides, in practice this does not involve any difficulties since it usually suffices to deal with symbolic models for the basic sets.

In the case of flows it would appear that symbolic models are constructed only for basic sets. Here the model is a *Markov flow*, which is a flow that differs from the Smale suspension over the TMC only by a change of time (in this connection, the latter is topologically transitive, although this plays no role in the definition itself. (We leave aside, of course, the trivial case when the basic set reduces to an equilibrium point.) A one-dimensional basic set of a flow is isomorphic to some Markov flow, and in the general case the basic set is a factor of the Markov flow. Here the character of the identification under a change of time is made more precise (Bowen 1973). If the flow were to have a global cross-section, then everything would reduce to the construction of a Markov partition for the corresponding first-return map. But in general, there is only a system of local cross-sections for which the motion along the trajectories of the flow is defined by some system of partial maps of these local cross-sections from one to the other. Partitions are constructed on these cross-sections which are, so to speak, Markov with respect to these maps. Here the first steps were taken by Ratner who realized these constructions for topologically transitive Anosov flows, first for three-dimensional ones and later for n-dimensional ones (Ratner 1969a), (Ratner 1969b), (Ratner 1973). The construction was done for general basic sets in Bowen's papers mentioned earlier. Pollicott considered the same questions within the framework of topological dynamics (Pollicott 1987). The applications have roughly the same character as for cascades. For example, a minimal subset of a basic set is one-dimensional (Bowen 1973). An unfortunate exception is the ζ-function. Although one can put forward a definition of it for flows which in some cases appears to be quite reasonable, there is an example of a basic set with a non-meromorphic ζ-function (Gallavotti 1976).

So far in this subsection we have been dealing with the construction of locally maximal hyperbolic sets. The following theorem has a different character: a locally maximal hyperbolic set A is locally structurally stable (Robinson 1976). (According to the definition of local structural stability (Sect. 2.2), this means that under a small perturbation in a certain sense, not only is A itself preserved, which would also follow from the theorem on the family of ϵ-trajectories, but also the location of the trajectories in a neighbourhood of A.)

3.3. Remarks

a) Local structural stability of hyperbolic equilibrium points and fixed points are very close to local linearizability, which forms the subject of the Grobmann-Hartman theorem (Arnol'd and Il'yashenko 1985, Chap. 3, Sect. 1.1 and Chap. 6, Sect. 2.1). (Concerning the connection between this theorem and local structural stability see (Anosov 1985).) Kurata has published an analogue of the Grobmann-Hartman theorem for a (not necessarily locally maximal) hyperbolic set A of a cascade $\{g^i\}$ (Kurata 1977). Notation: $D(x)$ is the ball of radius 1 in T_xM with centre at 0_x (the origin of the vector space T_xM), $D \subset TM|A$ is the bundle over A with fibres $D(x)$.

Theorem. *There exists a continuous map $\phi : D \to M$ such that for any $x \in A$ the map $\phi|D(x)$ is a homeomorphism of $D(x)$ onto a neighbourhood of x, $\phi(0_x) = x$ and if $\zeta \in D$, $Tg\zeta \in D$, then $\phi Tg\zeta = g\phi\zeta$.*

Strangely enough this theorem has not achieved much publicity. Possibly the reason is that (by contrast with the Grobmann-Hartman theorem) it does not provide a good "model" for motions around A; in fact, apart from those cases when A has a fairly trivial structure, a point y sufficiently close to A falls in the neighbourhood $\phi(D(x))$ for a continuum of distinct $x \in A$. True, we encountered a similar situation in Sect. 1.3, but there it was a question of stable and unstable manifolds that "existed objectively", so that we had to study the families formed by them, no matter what properties the latter may have possessed. As to linearization, this is to a large degree a question of convenience. But since our inverse images here are not one-to-one, the convenience is not evident.

b) Apart from the statements of a technical nature given above, others can be encountered in the literature. Sometimes they are used instead of the previous ones, and sometimes they are used in the investigation of other questions than those considered here. We state, or at any rate, recall, several such technical statements.

1) *Palis's λ-lemma* (Palis 1969). In fact we have already dealt with the main case of this in Sect. 2.1 when we said that a manifold transversal to $W^s(x_1)$, where $gx_1 = x_1$, under the action of g^n "nestles up to" $W^u(x_1)$. This has already appeared in the textbooks: (Nitecki 1971, Lemma 7.1), (Palis and de Melo 1982, Lemma 6.1), (Newhouse 1980b, Proposition 2.5).

2) *Cloud lemma* (Nitecki 1971). This asserts that the points of a heteroclinic cycle are nonwandering (the name refers to the change of form of a neighbourhood of such a point under the action of the DS). Using the techniques of Sect. 3.1 this was obtained so immediately that it is not even clear that it is worth singling out this lemma.

3) *Specification.* Suppose that we are given positive numbers $\tau, t_1, \ldots, t_k$ (in the case of a cascade, integers) and points $x_1, \ldots, x_k$. A point y *specifies* the set $\{(x_i, t_i)\}$ to within $\epsilon > 0$ (or it ϵ-specifies it) with delay τ if there exist $\tau_i \in (\tau-\epsilon, \tau+\epsilon)$ (in the case of a cascade, $\tau_i = \tau$) such that y is a periodic point with (not necessarily minimum) period $T = \sum(t_i + \tau_i)$ and such that when the interval $[0, T]$ is partitioned into the successive subintervals $[a_i, a_i+t_i+\tau_i]$, we have: $\rho(g^{a_i+t}y, g^t x_i) < \epsilon$ for $0 \le t \le t_i$.

In other words, $\{g^t y\}$ first ϵ-traces $\{g^t x_1\}$ during time t_1, then, after a transitional period of length $\tau_1 \approx \tau$ it ϵ-traces $\{g^{t-a_2} x_2\}$ during time t_2, and so on; after the kth transitional period of length $\tau_k \approx \tau$, the trajectory $\{g^t y\}$ is closed.

Specification Lemma. *If A is a topologically mixing basic set, then for any $\epsilon > 0$ there exists $\tau > 0$ such that the family $\{(x_i, \tau_i);\ x_i \in A;\ i = 1, \ldots, k\}$ (for any k) is ϵ-specified by some point $y \in A$ with delay τ.*

It is easy to construct a periodic δ-trajectory $x(t)$ that coincides with the trajectories $g^{t-a_i} x_i$ on the intervals $[a_i, a_i + t_i]$ and with certain other trajectories on intervals of the form $[a_i + t_i, a_i + t_i + \tau_i]$ with $\tau_i \approx \tau$ (in the case of a flow $x(t)$ is discontinuous). (Here we use the fact that $\overline{W^s(x) \cap A} = A$ for all $x \in A$.) After this, in the case of a cascade, it remains to use the theorem on the family of ϵ-trajectories. In the case of a flow, an additional control, over and above the reparametrization h of the approximating trajectory, is necessary. In this connection, the statements of Sect. 3.1 do not immediately give everything that is needed, even if we strengthen them in the spirit of strong stochastic stability (see the Appendix to Chap. 2). We also need to use the following facts: if in A some reparametrized motion $g^{h(t)} z$ (with $h(0) = 0$) stays close to some motion $g^t x$ for a long time, then $|h(t) - t|$ is small. (Here the hyperbolicity of A is essential; there are flows for which the analogous statement is false.) But our $x(t)$ coincides exactly with the actual motions on intervals of length t_i and τ_i, where the τ_i are large; as regards t_i, they may be small, but then it suffices to invoke strong stochastic stability. (In fact, the lemma is applied in conditions when the t_i are not only large, but even $t_i \gg \tau_i$.)

4) *The surrogate hyperbolic structure close to A* (Hirsch et al. 1970). At a point x of a hyperbolic set A the subspaces E^s_x, E^u_x are defined with appropriate properties. In the semilocal theory they are not defined for $x \notin A$ apart from the following exception: if the positive (or negative) semitrajectory $\{g^t x\}$ is entirely situated sufficiently close to A, then in $T_x M$ there is naturally distinguished the subspace E^s_x (or E^u_x) consisting of those ζ for which $Tg^t \zeta$ decreases exponentially as $t \to \infty$ (or $t \to -\infty$).

It turns out that the fields of the subspaces E^s_x, E^u_x can nevertheless be "artificially" (generally non-uniquely) extended from A to some neighbourhood $U \supset A$ in such a way that along the trajectory $\{g^t x\}$, $x \in U$, inequalities of type a), b) in Sect. 1.1 and the equalities $Tg^t E^s_x = E^s_{g^t x}$, $Tg^t E^u_x = E^u_{g^t x}$ hold as long as it does not leave U. (If the semitrajectory does not leave U, then the corresponding E^s_x or E^u_x coincide with the subspace referred to above.) Furthermore, one can construct continuous fields of small diffeomorphically embedded discs $V^s(x), V^u(x)$ of the same dimension as E^s_x, E^u_x that pass through x, are tangent to E^s_x, E^u_x there and are such that when $t \geq 0$

$$g^t V^s(x) \subset V^s(g^t x), \quad g^{-t} V^u(x) \subset V^u(g^{-t} x),$$

as long as $g^t x$ or $g^{-t} x$ does not leave U. At points of A these manifolds are the same as $W^s_\epsilon(x)$, $W^u_\epsilon(x)$. We do not make precise the meaning of the expression "continuous field" as applied to such objects; this is done in the spirit of Sect. 1.3. We merely note that even though they are small, their sizes are bounded below.

5) *Fundamental regions and neighbourhoods.* We set (but only for item 5))

$$W_\epsilon^s(A) = \bigcup\{W_\epsilon^s(x);\ x \in A\},\ W_\epsilon^u(A) = \bigcup\{W_\epsilon^u(A);\ x \in A\}. \tag{25}$$

A compact subset K of one of these sets is called a *fundamental region* for the latter if

$$W_\epsilon^s(A) \backslash A \subset \bigcup\{g^t K;\ t \geq 0\},\ \text{respectively,}\ W_\epsilon^u(A) \backslash A \subset \bigcup\{g^t K;\ t \leq 0\}.$$

If $K \cap A = \emptyset$, then the fundamental region is said to be *proper*. Such a region exists if A is locally maximal. A *fundamental neighbourhood* V is a compact neighbourhood of a fundamental region (one might therefore need to specify more precisely which of the sets (25) this neighbourhood corresponds to). It turns out that A is contained in the interior of the set

$$\left(\bigcup\{g^t V;\ t \geq 0\}\right) \cup W_\epsilon^u(A),\ \text{respectively,}\ \left(\bigcup\{g^t V;\ t \leq 0\}\right) \cup W_\epsilon^s(A).$$

A fundamental neighbourhood V is *proper* if $V \cap A = \emptyset$.

All this becomes fairly trivial when A reduces to an equilibrium point or a fixed point. But even in this case, the above concepts are of some use. See the geometric proof of the Grobmann-Hartman theorem in (Palis and de Melo 1982). The proof of the local structural stability of locally maximal hyperbolic sets is very reminiscent of this proof (in it one also uses 4)).

c) If the zone of attraction (or repulsion) $W^s(A)$ (or $W^u(A)$) a locally maximal hyperbolic set A of a DS of class C^2 has positive Lebesgue measure, then A is an attractor (or repeller). In particular, A itself has Lebesgue measure zero, except for the case when $A = M$ (Bowen and Ruelle 1975). This is not true for a DS of class C^1; a counterexample of horseshoe type is constructed in (Bowen 1975a).

d) The examples of hyperbolic sets that we have been handling so far were locally constructed as a Cantor set, a Euclidean space or a product of these. But there are basic sets that are constructed by other means. The first such example was pointed out by Guckenheimer (Guckenheimer 1970) (the idea is indicated in the remark on p.155 of the Russian translation of (Smale 1967b). See also (Farrell 1980), (Jones 1983).

e) With the exception of the end of Sect. 3.3, we have so far not said anything about what the situation is with the discovery and localization of hyperbolic sets in specific DS's. Within the framework of pure mathematics there are DS's of a natural geometric or algebraic origin that are Anosov DS's (Sect. 4.2). With regard to DS's described by systems of differential equations arising (perhaps even indirectly) from applications, here one must in the first instance talk about bifurcations at which hyperbolic sets occur fairly frequently (Arnol'd et al. 1986), (Wiggins 1988). (Properly speaking, we also consider the bifurcation of Mel'nikov and Palmer (Sect. 3.3).) But this is a peculiar area which to some extent goes beyond the limits of hyperbolicity. As to the rest, if we leave out the discovery of homoclinic points in numerical experiments, success in the discovery and localization of hyperbolic sets has not been great.

Here the biggest contribution was provided by Alekseev who developed a certain (somewhat unwieldy) technique (if you like, it can be regarded as a far reaching generalization of the horseshoe) and with its help he discovered hyperbolic sets in certain problems of the theory of non-linear vibrations and celestial mechanics (Alekseev 1968), (Alekseev 1981), (Alekseev, Katok and Kushnirenko 1972). There is a simplified account of part of his results (relating to celestial mechanics) in (Moser 1973).

Apart from this one must recall another line of approach. In the 40's and 50's there appeared a number of works by Cartwright and Littlewood on the van der Pol equation and analogues of it (Cartwright and Littlewood 1945), (Cartwright and Littlewood 1947), (Littlewood 1957a), (Littlewood 1957b), (Littlewood 1960). By studying the behaviour of solutions with large friction and under a large periodic perturbation, Cartwright and Littlewood uncovered phenomena which at that time were unexpected, but which now are naturally interpreted as manifestations of hyperbolicity. In attempting to explain the situation, Levinson considered another equation which is simpler to analyse and which, as it turned out, possesses similar properties (Levinson 1949). (In the Russian language a large part of Levinson's studies is set out in (Pliss 1964).) Levinson drew special attention to (and even used) the fact that the phenomena uncovered are related to certain properties of the first-return map and are preserved under small perturbations. In particular, all DS's that are sufficiently close to a Levinson system (with suitable parameters) have infinitely many periodic trajectories. Later when Smale in his first work on the theory of DS's stated his conjecture on the generic character of Morse-Smale DS's (see (Anosov et al. 1985, Chap. 2, Sect. 3) for the definition), Levinson was immediately able to report to him that this was not the case. (Another counterexample, pointed out by Thom, was a hyperbolic automorphism of the two-dimensional torus.) This communication stimulated the discovery of the horseshoe.

Later on, Levinson's equation was analysed from the standpoint of hyperbolic theory and Levinson's results were thereupon supplemented (Osipov 1975), (Osipov 1976a), (Osipov 1976b), (Levi 1981). The more awkward investigations of Cartwright and Littlewood, as far as is known, have so far not been subjected to a reconsideration from the standpoint of the hyperbolic theory.

f) Wild Hyperbolicity. If A, B are two hyperbolic periodic trajectories (possibly, $A = B$), then by means of an arbitrarily small perturbation of the DS in question one can arrange that its stable and unstable manifolds are in general position. The smallness of the perturbation can be understood in the sense of any C^r. The Kupka-Smale theorem asserts even more (Anosov et al. 1985, Chap. 2, Sect. 1.2).

One cannot hope for the analogous assertion, that the stable and unstable manifolds of all the trajectories lying in some hyperbolic sets A, B can be brought into general position all at the same time. Consider a three-dimensional cascade in which A, B are invariant tori, where A is a hyperbolic repeller and B a hyperbolic attractor. Clearly, $\{W^{\mathrm{u}}(x);\ x \in A\}$ and $\{W^{\mathrm{s}}(y);\ y \in B\}$ are families of two-dimensional manifolds and each of these families appears locally as a family of parallel planes in $\mathbb{R}^3$. But in relation to each other they can be so strongly curved that in some places the manifolds of the first family are tangent to those of the second. It is easy to imagine tangencies that are inevitably preserved under small perturbations. For example, this will be the case if locally one of the families appears as the family of planes $x_3 = \text{const}$, while the other family appears as the family of paraboloids of rotation $x_3 = x_1^2 + x_2^2 + \text{const}$ (in the same local coordinates).

In this example each of these families entirely fills out an open set. But this is not essential. Here is another example which, as before, is a three-dimensional cascade. Suppose that in terms of suitable local coordinates a basic set A of index 2 contains a part of the x_1 axis, that the corresponding $W^{\mathrm{s}}_{\mathrm{loc}}(x)$ appear locally as the lines $x_1 = \text{const}$ on the plane $x_3 = 0$ and fill out the region $|x_1|, |x_2| \leq 10$ there, while some $W^{\mathrm{u}}(y)$ intersects it as the surface $x_3 = x_1^2 + (x_2 - 2)^2 - 1$, $|x_3| \leq 1$. Clearly the latter is tangent to certain $W^{\mathrm{s}}_{\mathrm{loc}}$ and this tangency is preserved in an obvious sense under C^1-smooth perturbations. In this example $A = B$ and the points of tangency turn out to be nonwandering points. Thus the set of nonwandering points is not hyperbolic, nor does it become so under C^1-small perturbations. (However in this connection, the main part of our argument is based on the properties of the hyperbolic A and its stable and unstable manifolds.)

In the last example it is essential that some of the $W^{\mathrm{s}}_{\mathrm{loc}}$ fill out some surface that W^{u} intersects along some closed curve; clearly such a picture withstands small perturbations. Suppose now that $A = B$ is a zero-dimensional basic set of index 1 of a two-dimensional cascade (but consisting of infinitely many trajectories). In this case such "continuous fences" do not arise in $W^{\mathrm{s}}(x)$. At first glance it would seem that there is sufficient room where under a suitable perturbation one could shift those arcs of the lines $W^{\mathrm{u}}(y)$ where the directions of their tangents are too close to the directions of the tangents to $W^{\mathrm{s}}(x)$, thus ensuring the transversality of all the $W^{\mathrm{s}}(x)$ and $W^{\mathrm{u}}(y)$, $x, y \in A$. However, Newhouse discovered that this is not the case, at any rate if one considers perturbations that are small in the C^2 sense (Newhouse 1979), (Newhouse 1980b). In this connection, by a *wild hyperbolic set* one means a basic set A of a two-dimensional cascade such that for all C^2-close cascades there are two points x, y in this set with $W^{\mathrm{s}}(x), W^{\mathrm{u}}(y)$ tangent to each other. (As is clear, the formulation relates to what happens under small perturbations and uses the fact that under such perturbations the basic set is preserved in a certain sense, which enables one to refer to A for the perturbed DS.) It turns out that wild basic sets are encountered fairly frequently: if a basic set of a cascade $\{f^n\}$ contains points x, y with $W^s(x), W^u(y)$ tangent to each other, then there are cascades $\{g^n\}$ arbitrarily close (in the C^2 sense) to $\{f^n\}$ with wild hyperbolic sets.

By themselves wild hyperbolic sets and the corresponding $W^s(x)$ and $W^u(x)$ are objects of the usual hyperbolic theory. However, there are phenomena associated with them that go beyond the framework of this theory. If a cascade $\{g^n\}$ has a wild hyperbolic set, then by means of an arbitrarily small perturbation of it one can obtain: 1) a cascade that together with all cascades sufficiently close to it has the property that the stable set at some point of it is not a manifold; 2) a cascade $\{h^n\}$ with a countably infinite set of exponentially stable periodic trajectories, the presence of these being generic (in the usual technical sense of this word, see (Anosov et al. 1985, Chap. 2, Sect. 1.2)) for all cascades sufficiently close to $\{h^n\}$. (It is customary to refer to 2) in particular, as well as the presence close to the above cascade $\{f^n\}$ of the corresponding cascades $\{g^n\}$ and $\{h^n\}$, as the Newhouse phenomenon.) The corresponding assertions for neighbourhoods of the cascade $\{f^n\}$ or $\{g^n\}$ hold not on the whole function space of C^2-smooth cascades, but on any one-parameter family of cascades containing it and satisfying very weak non-degeneracy conditions (here one further requires a higher degree of smoothness).

Several gaps in Newhouse's proofs were noted. However, these have now been made good by Newhouse himself and various other mathematicians (Shil'nikov, Robinson, Davis) at least in the case when the original cascade $\{f^n\}$ has a non-transversal homoclinic point.

g) There are several works in which the semilocal hyperbolic theory is adapted to the situation relating to the application of the ideas and methods of the theory of DS's to physical systems with distributed parameters. (As noted in (Anosov et al. 1985, Chap. 1, Sect. 1.6), here there arise DS's with an infinite-dimensional phase space for which realistic models require certain modifications of the original concepts.) The most recent article of this kind that I know of is (Steinlein and Walther 1990).

§4. Global Theory

4.1. Smale's Axiom A and Structural Stability. In the global hyperbolic theory one studies DS's for which the set of "repetitive" motions is hyperbolic. Here "repetitiveness" of motions can be understood in various ways. There are versions in which the following sets are hyperbolic (their definition is given in (Anosov et al. 1985, Chap. 3)): the set $\mathcal{R}$ of chain-recurrent points (Franke and Selgrade 1977); the set $NW = \Omega$ of nonwandering points; $\overline{L^+}$, where L^+ is the set of ω-limit points of all possible trajectories (it is not necessarily closed, which is why one takes its closure), or $\overline{L^-}$ where L^- is the set of α-limit points of all possible trajectories (Newhouse 1972); $\overline{L}$, where $L = L^+ \cap L^-$; the Birkhoff centre C (Malta 1980); $\overline{\mathrm{Per}}$, where Per is the set of periodic points. These sets are listed in order of decreasing size, although L^+ and L^- are incomparable, in general. There may not be any periodic trajectories, in which case the condition of hyperbolicity of $\overline{\mathrm{Per}}$ is vacuous. (But it is invoked only as an auxiliary notion, see later.) The other sets are indeed always non-empty. In (Shub 1978) there is a uniform discussion of the versions in which hyperbolicity of these sets is postulated.

It is easy to prove that if $\overline{\mathrm{Per}}$ is hyperbolic, then it is locally maximal. And since all its points are nonwandering for the DS $\{g^t|\overline{\mathrm{Per}}\}$, one obtains from Sect. 3.2 the spectral decomposition for $\overline{\mathrm{Per}}$. Furthermore, it turns out that the hyperbolicity of the set $\mathcal{R}$, $\overline{L^+}, \overline{L^-}, \overline{L}$, or C implies that it is the same as $\overline{\mathrm{Per}}$. Hence one obtains the spectral decomposition for this set as well. But the hyperbolicity of Ω does not imply that $\Omega = \overline{\mathrm{Per}}$, although this is the case when $\dim M \leq 2$. (For cascades with $\dim M = 2$ this is a non-trivial result due to Newhouse and Palis (Newhouse and Palis 1973).) For flows with $\dim M = 3$ the question is unsolved, as far as we know. In the remaining cases (that is, for cascades with $\dim M > 2$ and flows with $\dim M > 3$) Dankner (Dankner 1978) constructed examples when Ω is hyperbolic but $\overline{\mathrm{Per}} \neq \Omega$. (Such examples in these dimensions exist on any M and, in the case of discrete time, in any isotopy class.) In this connection we recall the article (Kurata 1979), where there is an example for cascades with $\dim M > 3$.

Presumably the central position in the global theory has so far been occupied by the condition that was first proposed by Smale in his Axiom A. It consists of two parts: A(a) Ω is hyperbolic; A(b) $\Omega = \overline{\mathrm{Per}}$. As is clear from (Dankner 1978), (b) is independent of (a) in general. If a DS satisfies Axiom A, then Sect. 3.2 gives a spectral decomposition of Ω. Thus in this case it was first obtained by Smale (Smale 1967b). With regard to the name, see the explanation in (Smale 1967b) and (Nitecki 1971). It appears a little strained to me, but after all, one has to name this decomposition somehow.

Suppose that one of the sets $\Omega, \ldots, C$ is hyperbolic, where in first case we require that Axiom A(b) hold. Denoting this set by A, we have a spectral decomposition $A = A_1 \cup \ldots \cup A_k$. Departing from the notation of Sect. 3.2, we write $A_i > A_j$ if there is a trajectory in M going "from A_i to A_j", that is, there exists $x \in M$ with $\alpha(x) \subset A_i$, $\omega(x) \subset A_j$. (In Sect. 3.2 it was required that $x \in A$, but now this is not required.) It can happen that the relation $>$ is not a partial ordering, that is, there could exist cycles $A_{i_1} > \ldots > A_{i_k} > A_{i_1}$. If such cycles do not exist, then we say that the DS satisfies the *acyclicity condition*. According to the literal meaning of what has been said, one might have to specify which sets $\Omega, \ldots, C$ one is dealing with (however, see below). Furthermore one could make the same stipulations concerning $\mathcal{R}$, but here the condition of acyclicity is satisfied automatically.

It turns out that the following conditions are equivalent: 1) Hyperbolicity of $\mathcal{R}$. 2) Axiom A and the acyclicity condition for Ω. 3) Hyperbolicity of $\overline{L^+}, \overline{L^-}$ or $\bar{L}$ and the corresponding acyclicity condition. 4) Hyperbolicity of C and the corresponding acyclicity condition. When these conditions hold, all these sets coincide and are the same as $\overline{\text{Per}}$. It follows from 1) – 4) that the DS is Ω-structurally stable (the definition was given in Sect. 2.2). For cascades it has been proved that conversely, Ω-structural stability implies 2) (and hence all the other conditions).

The equivalence of 1) and 2) was proved in (Franke and Selgrade 1977), that of 2) and 3) in (Newhouse 1972), and that of 2) and 4) in (Malta 1980). The conjecture that Ω-structural stability is equivalent to 2) was stated by Smale and Palis (Palis and Smale 1970). For cascades Smale himself proved that 2) suffices for Ω-structural stability (Smale 1970) (some details were completed in (Hirsch et al. 1970); all this is contained in the book (Nitecki 1971)). The analogous assertion for flows was proved by Pugh and Shub (Pugh and Shub 1970). It is also comparatively easy to prove that if a DS satisfies Axiom A, then acyclicity is necessary for its Ω-structural stability. The fact is that in the presence of cycles, for some neighbourhood $U \supset \Omega$ there exists a DS $\{h^t\}$ arbitrarily close to $\{g^t\}$ (in the C^1 sense) for which $NW(\{h^t\}) \not\subset U$. This phenomenon is called an *Ω-explosion* (for the given DS) or, more precisely a C^1-Ω-explosion, since the perturbation is taken to be small in the C^1 sense. By taking it to be small in the C^r sense we obtain the definition of a C^r-Ω-explosion. The absence of a C^0-Ω-explosion is equivalent to the property that $\mathcal{R} = \Omega$, this being the case not only for smooth DS's but also for topological ones, at any rate, for topological cascades; no kind of hyperbolicity is involved here (Shub and Smale 1972), (Nitecki and Shub 1976), (Shub 1978).

A consequence of Pugh's highly non-trivial closure lemma is his density theorem which asserts that in the space of DS's of class C^1 with the appropriate topology, those DS's for which $\Omega = \overline{\text{Per}}$ form a set of the second category (in the restricted sense of the word, that is, its complement is of the first category). References can be found in (Anosov 1985). A new account of the closure lemma in which certain errors of the early accounts are corrected is given in (Pugh and Robinson 1982).

Thus the proof of the necessity of 2) for Ω-structural stability reduces to a proof of A(a). This was only recently achieved for cascades by Palis (Palis 1988) as an improvement of work by Mañé; see below concerning this as well as earlier results of Mañé himself and other authors.

When Axiom A holds one can introduce the *strong transversality condition.* This means that for any trajectories $L_1, L_2 \subset \Omega$ the manifolds $W^{\mathrm{s}}(L_1)$ and $W^{\mathrm{u}}(L_2)$ are in general position, that is, they have only transversal intersections (if at all) (in the sense as it is defined for immersed submanifolds). This is stronger than acyclicity.

Robinson's Theorem. *If a DS satisfies Axiom A and the strong transversality condition, then it is structurally stable* (Robinson 1974), (Robinson 1976).

(Earlier Robbin proved the sufficiency of these conditions for the structural stability of cascades of class C^2 (Robbin 1971).)

Mañé's Theorem. *A structurally stable cascade satisfies Axiom A and the strong transversality condition.*

Thus the conjecture on the equivalence of structural stability and the condition "Axiom A plus strong transversality" (stated by Smale and Palis (Palis and Smale 1970)) remains unproved in one direction only for flows. Judging from certain cursory remarks in the literature, some progress in this direction has been made but, apparently, the authors have striven for definitive results and for this reason have refrained from publishing.

As in the case with Ω-structural stability, the main difficulty lies in the proof of the hyperbolicity of Ω. In (Anosov 1985) information is given on the earlier articles on structurally stable systems and related questions (including various versions of properties of structural stability type; [17] necessary and sufficient conditions are much easier to obtain for some of them. It is worth noting that these conditions turn out to be the same). Here we merely note the following. Of the earlier work, what proved to be essential apart from the investigations of Pugh already referred to, were the papers by Franks (Franks 1971), Pliss (the latter were compiled together in his book (Pliss 1964)) and Mañé (Mañé 1974), (Mañé 1978). In 1980–1982 Liao (Liao 1980), Sanami (Sanami 1983) and Mañé (Mañé 1982) proved the necessity of the hyperbolicity of Ω for the structural stability and even the Ω-structural stability of a cascade in the two-dimensional case. (In Sanami's paper the cascade must be of class C^2. I have not seen Liao's papers and do not know whether he has considered Ω-structural stability. On the other hand, he did study flows, but the accounts of this in the literature are contradictory.) Finally Mañé obtained a proof in the general case (Mañé 1988a), (Mañé 1988b).

In the proof of the necessary conditions for structural stability or Ω-structural stability of cascades a fundamental role is played by the investigation of cascades possessing the following properties: 1) all periodic points of a given cascade are hyperbolic; 2) all cascades sufficiently close to the given one (in the C^1 sense) possess property 1). We denote the set of such cascades by $F(M)$. It has been known for a long time that a structurally stable or Ω-structurally stable cascade belongs to $F(M)$. In the case $\dim M = 2$ it has been proved that cascades in $F(M)$ satisfy Axiom A. For $\dim M > 2$, a weaker assertion has been proved which, however, turns out to be sufficient: cascades satisfying Axiom A and the acyclicity condition form an open dense subset of $F(M)$ (Palis 1988). Flows of Lorenz flow type (Chap. 2, Sect. 2) show that the analogous assertion for flows is false.

In the definition of structural stability (including Ω-structural stability as well) the smallness of the perturbations is understood in the C^1 sense. If one understands it in the C^r sense, $r > 1$, then one obtains another concept, namely, *structural stability in the class C^r*, or *C^r-structural stability.* Necessary conditions for this are known only in small dimensions where completely

[17] See also the Appendix to Chap. 2 with regard to so-called topological stability.

different considerations are applied (Anosov 1985), (Palis and de Melo 1982). There has been no progress on this question in recent years.

The DS's satisfying Axiom A, Axiom A and the acyclicity condition, Axiom A and the strong transversality condition form an open subset of the space of all DS's (in the C^1-topology). But they are not dense in this space (apart from the case of small dimension when Morse-Smale DS's are dense). In general, nobody has come up with a property which, like structural stability in the case of small dimension, holds for a "generic" DS and at the same time defines its properties to a significant degree (Anosov 1985).

The last remark is in connection with structural stability. As is well known, the homeomorphism taking the trajectories of the unperturbed system to trajectories of the perturbed one cannot be a diffeomorphism in general. In the simplest examples of this sort, where the breakdown of smoothness is in relation to equilibrium points and so on, it is nevertheless possible that it is smooth in a large part of the phase space; the singularities are, say, only at certain points or on lines. In the hyperbolic theory there are examples where this homeomorphism must inevitably be "considerably worse": it must take some set of full measure to a set of measure zero (Anosov 1967), (Katok 1970).

4.2. Anosov Systems

a) Introductory Remarks. In the investigation of hyperbolic systems, alongside the usual "dynamical" methods special geometric considerations come into operation; these are concerned with the mutual disposition of the stable W^{s} and unstable W^{u} manifolds. These considerations are most fully developed in the case of Anosov systems, when the manifolds W^{u} and W^{s} pass through each point and form foliations that are in a certain sense "coherent" with each other. We shall consider Anosov DS's from the standpoint of these foliations. In what follows, these DS's are assumed to be smooth of class C^∞ unless otherwise stated. (In fact, less is sufficient, but we do not wish to encumber the exposition by pointing out each time precisely what kind of smoothness is required.)

We recall that by a "foliation" F of dimension q and codimension $m-q$ on a manifold M^m we mean a partitioning of it into "leaves" which are manifolds of dimension q such that in a neighbourhood of each point the induced partition is diffeomorphic (or homeomorphic if the foliation is not smooth) to a decomposition of $\mathbb{R}^m$ into parallel q-planes. The smoothness class of the maps of the neighbourhoods into $\mathbb{R}^m$ is called the smoothness (or smoothness class) of the foliation F. (From the point of view of the theory of differential equations it would be natural to understand the smoothness class of the foliation to mean the smoothness class of the corresponding tangent field associating a point of the tangent space with the leaf passing through it. Such a point of view was adopted in the earlier works (Anosov 1967). The difference here is not very great and we shall not dwell on it.) Concerning foliations, see the corresponding article in "Mathematical Encyclopedia" and the literature cited there.

Theorem. *For an Anosov DS the distributions* E^{u}, $\dim E^{\mathrm{u}} = k$, *and* $E^{\mathrm{s}}, \dim E^{\mathrm{s}} = l$, *are integrable. The invariant foliations* W^{u} *and* W^{s} *(which are tangent to these distributions) are Hölder continuous if the DS is of class* $C^{1+\epsilon}$, *although each of the leaves is a smooth immersed manifold diffeomorphic to* $\mathbb{R}^k$ *or* $\mathbb{R}^l$. *(Here again, "smooth" means "*C^∞*-smooth".)*

For the case of a flow, the distributions $E^{\mathrm{u}} \oplus E^{\mathrm{n}}$ and $E^{\mathrm{s}} \oplus E^{\mathrm{n}}$ are also integrable. (Concerning the corresponding foliations W^{un} and W^{sn} see Sect. 1.3, 1.4.)

b) Holonomy and Transversal Measures. In the theory of foliations great attention is given to "transversal geometry". The foliations associated with an Anosov system also possess such a geometry.

We recall that for each pair of points x and y lying on the leaf L of the foliation F and for the homotopy class γ of paths joining x and y in L we define the germ $h(x, y, \gamma)$ of maps from $(\mathbb{R}^k, 0)$ to $(\mathbb{R}^k, 0)$, where k is the codimension of F.

If the initial and final points coincide ($x = y$), then the germs $h(x, \gamma) = h(x, x, \gamma)$, $\gamma \in \pi_1(L)$ define the *holonomy group* of the leaf.

By a *transversal measure* μ for a foliation F we mean a correspondence that associates with each system of manifolds $\{G_i\}$ transversal to F Borel measures μ_i defined on G_i and invariant with respect to the holonomy map.

Margulis's Theorem (Margulis 1970). *The foliations* W^{u} *and* W^{s} *of an Anosov system have a trivial holonomy and possess a transversal measure if the system is transitive.*

The holonomy group of the leaves of the foliations W^{un} and W^{sn} is trivial for leaves containing no periodic trajectories and isomorphic to $\mathbb{Z}$ for leaves containing periodic trajectories that correspond to generators of this group.

c) Cascades. We begin with a description of examples known at present.

Historically the first and simplest example of an Anosov diffeomorphism is constructed as follows. We choose an integer-valued matrix A with determinant ± 1 and with no eigenvalues equal to unity in modulus. The linear transformation $A : \mathbb{R}^m \to \mathbb{R}^m$ takes the integer lattice $\mathbb{Z}^m \subset \mathbb{R}^m$ into itself. The induced automorphism of the torus $f : T^m = \mathbb{R}^m/\mathbb{Z}^m \to T^m$ is called a *hyperbolic automorphism of the torus.* (An example of this kind was considered in Sect. 2.4.)

We can replace $\mathbb{R}^m$ by a simply connected nilpotent Lie group N, and the subgroup $\mathbb{Z}^m$ by a uniformly discrete subgroup Γ. Then a hyperbolic automorphism N taking Γ onto itself defines an Anosov diffeomorphism of the nil-manifold N/Γ. Finally, there exist Anosov diffeomorphisms on factor manifolds N/Γ by a finite group, that is, infranil-manifolds. This is the broadest class of Anosov diffeomorphisms of algebraic origin (Smale 1967b).

It has been conjectured that, to within topological conjugacy, these examples exhaust all Anosov diffeomorphisms.

This conjecture is false if we replace topological conjugacy by smooth conjugacy, as an example in (Anosov 1967) shows. Furthermore, Farrell and Jones (Farrell and Jones 1978) have constructed an example of an Anosov diffeomorphism on a manifold that is homeomorphic but not diffeomorphic to the torus.

The conjecture has been verified in the case when one of the foliations W^{u} or W^{s} has codimension one. In this case Newhouse and Franks (Franks 1970), (Newhouse 1970) proved the following strengthened version:

Theorem. *An Anosov diffeomorphism of codimension one is topologically conjugate to a hyperbolic automorphism of the torus.*

It turns out that if the manifold M is sufficiently simple, that is, if it is an infranilmanifold, then the conjecture is also true.

Manning's Theorem (Manning 1974). *Each Anosov diffeomorphism of an infranilmanifold is topologically conjugate to an algebraic one.*

We note further that Ruelle and Sullivan (Ruelle and Sullivan 1975) have used transversal measures of the foliations W^{u} and W^{s} to prove that the existence of a transitive cascade of codimension q on a manifold M^m implies the presence of non-trivial cohomology classes of dimensions q and $m - q$.

d) Flows. The simplest example of an Anosov flow is obtained by taking the suspension over an Anosov diffeomorphism.

A second and fundamental example is a geodesic flow on the space of linear elements LM of a manifold M of negative curvature. Both the flow and its foliations are described in detail in (Anosov 1967).

There exist other examples of Anosov flows of algebraic origin that do not reduce to the first two. A complete reference to these is given in (Tomter 1975).

In (Handel and Thurston 1980) Handel and Thurston constructed the first example of a non-algebraic Anosov flow. This is the first of a series of hybrid flows obtained by gluing together two flows on a manifold with boundary.

A fundamental conjecture is that every Anosov flow is topologically conjugate to a flow obtained by gluing together several flows of algebraic origin.

Essentially only the case of flows of codimension one has been studied, that is, those flows for which one of the foliations W^{u} or W^{s} (we shall choose W^{u}) is one-dimensional.

Theorem (Verjovsky 1974); (Solodov 1982); (Anosov 1977). *If for an Anosov flow $\{g^t\}$ defined on a manifold M we have* $\dim W^{\mathrm{u}} = 1$, *then the foliation W^{sn}:* 1) *has codimension one;* 2) *has no one-sided limit cycles;* 3) *has a closed transversal intersecting all its leaves;* 4) *has no null-homotopic closed transversals. Furthermore,* 5) *the embedding of any leaf L of it into M induces a monomorphism of the fundamental groups $\pi_1(L) \to \pi_1(M)$;* 6) *the fundamental group $\pi_1(M)$ contains free subgroups, in particular, it has exponential growth;* 7) *if* $\dim M > 3$, *then the set Ω of nonwandering points of $\{g^t\}$ is the entire manifold.*

Remark. Franks and Williams (Franks and Williams 1980) have constructed an example of an Anosov flow with $\Omega \neq M$ on a three-dimensional manifold M.

Anosov flows of codimension one have been fully studied on manifolds whose fundamental groups are sufficiently simple.

Theorem (Plante 1981). *On a manifold with solvable fundamental group every Anosov flow of codimension one is topologically equivalent to the suspension over a hyperbolic automorphism of the torus.*

Theorem (Ghys 1984). *If the fundamental group of a three-dimensional manifold has a non-trivial centre, then every Anosov flow on it is topologically equivalent to a diagonal flow on* $\widetilde{\mathrm{SL}}(2,\mathbb{R})/\Gamma$, *where* Γ *is the uniform lattice in the universal covering of* $\widetilde{\mathrm{SL}}(2,\mathbb{R})$ *of the special linear group.*

Theorem (Plante 1972). *If for a transitive flow at least one of the foliations* W^{u} *or* W^{s} *is not everywhere dense, then the flow is topologically conjugate to a suspension.*

e) Special Results on the Smoothness of Foliations. Geometrical Rigidity. Anosov uncovered the essential and "unpleasant" fact that no smoothness of an Anosov DS guarantees smoothness of the foliations associated with it. In the most general case, they are of class C^{α} for some $\alpha \in (0,1)$. But if we suppose that the system has codimension one, then a foliation of codimension one would have smoothness of class $C^{1+\alpha}$ for some $\alpha \in (0,1)$. If, in fact, the flow is defined on a three-dimensional manifold and has an invariant measure that is continuous with respect to Lebesgue measure, then the foliations W^{sn} and W^{un} have smoothness $C^{1+\alpha}$ for some α in the interval $(0,1)$. This will be the case when the Anosov flow is a geodesic flow on a smooth two-dimensional manifold; the distribution $E^{\mathrm{u}} \oplus E^{\mathrm{s}}$ is then of class C^{∞} and, consequently, the foliations W^{u} and W^{s} are also of class $C^{1+\alpha}$ for some $\alpha \in (0,1)$.

In the case when the manifold (now of any dimension) has sectional curvatures in the interval $(-4,-1)$, the foliations W^{u} and W^{s} of a geodesic flow are of class C^1.

At the same time, Plante (Plante 1972), in correcting a careless passing observation in (Anosov 1967), showed that there exist arbitrarily small perturbations of a geodesic flow on a two-dimensional manifold of negative curvature for which the foliations W^{u} and W^{s} are not of class C^1.

In recent times the phenomenon known as *geometrical rigidity* has been discovered (geometrical in contrast to measurable or topological rigidity in the case of group actions). This consists in the property that when the foliations W^{u} and W^{s} have an "increased" smoothness there arises an additional geometrical structure on the manifold (that is, one that does not exist *a priori*).

We state the most general of the known theorems of this kind. Let M be a $(2n-1)$-dimensional compact manifold. A one-form λ is said to be canonical for the flow $\{g^t\}$ if

$$\lambda(E^{\mathrm{u}} \oplus E^{\mathrm{s}}) = 0, \ \lambda(\mathbf{v}) = \mathbf{1}.$$

An Anosov flow is called a *contact* flow if $\lambda \wedge (d\lambda)^{n-1}$ does not vanish.

Theorem (Benoist, Foulon, and Labourie 1990). *Suppose that the decomposition $TM = E^{\mathrm{u}} \oplus E^{\mathrm{s}} \oplus E^{\mathrm{n}}$ for an Anosov contact flow $\{g^t\}$ with phase velocity $\mathbf{v}$ is of class C^∞. Then there exists a unique cohomology class $[\alpha] \in H^1(X, \mathbb{R})$ in which there is a closed 1-form $\alpha \in [\alpha]$ such that $1 + \alpha(\mathbf{v}) > 0$ and the flow $\{h^t\}$ with phase velocity vector field $\mathbf{w} = \mathbf{v}/(1 + \alpha(\mathbf{v}))$ has the following property:*

1) *For $n \geq 3$ there exists a finite-sheeted covering of M such that the flow $\{h^t\}$ is lifted to a flow $\{\tilde{h}^t\}$ that is the C^∞ conjugate of a geodesic Riemannian flow of a locally symmetric space of strictly negative curvatures;*

2) *for $n = 2$ there exists a finite-sheeted covering of M such that the flow $\{h^t\}$ is lifted to a flow $\{\tilde{h}^t\}$ that is the conjugate of a finite-sheeted covering of a geodesic flow on the space of linear elements of a surface of genus greater than unity.*

See (Hamenstädt 1990), (Hurder and Katok 1990) concerning other aspects of rigidity and the literature associated with them.

Questions that have been briefly touched upon in this subsection are considered in detail in the survey article (Solodov 1991).

After the Russian edition of this book was published, several new papers on the 3-dimensional case appeared. We can give references only for a few of them (Ghys 1992), (Brunella 1992). Although the theory of 3-dimensional Anosov flows is still far from being complete, it has become substantial.

4.3. Remarks

a) Expanding Maps and Anosov Maps (Coverings). A smooth map g of a closed manifold M into itself is called *expanding* if under its action and the action of its iterates the lengths of all tangent vectors (in the sense of some and therefore any Riemannian metric) increases at an exponential rate, this increase being uniform with respect to all vectors, that is, $|Tg^k\xi| \geq ae^{ck}|\xi|$ for all $\xi \in TM$, $k \geq 0$, where a, c are positive constants that do not depend on ξ, k. An example is given by the map of the circle $\mathbf{S}^1 = \mathbb{R}/\mathbb{Z}$ that is defined in terms of the angular coordinate x mod 1 by the formula $x \mapsto 2x$. As in this example, an expanding map is always non-invertible and from the topological point of view, it is a covering.

These maps were introduced by Shub in (Shub 1969) (of the early papers, see also (Franks 1970)). Their properties are similar to those of Anosov DS's (in particular, they are also structurally stable and are partly even simpler: 1) any expanding map g is conjugate to a map obtained by means of some algebraic construction (with some automorphism of an infranilmanifold) (Gromov 1981); 2) if g is of class C^r, $r \geq 2$, then it has a finite invariant measure defined in terms of local coordinates by a positive density of class C^{r-1} (and in the analytic case, it is analytic) (Krzyzewski 1977), (Krzyzewski 1979), (Krzyzewski 1982). (Property 2) does not follow from 1), since, analogously to what happens for an Anosov DS, the conjugating homeomorphism may be singular in the measure-theoretic sense.)

Franks (Franks 1970) defined *Anosov maps* (*coverings*) as smooth maps $g : M \to M$ for which the tangent bundle splits into a direct sum of two subbundles $TM = E^{\mathrm{s}} \oplus E^{\mathrm{u}}$ that are invariant with respect to Tg and are such that for all $\xi \in E^{\mathrm{s}}$, $\eta \in E^{\mathrm{u}}$, $k \geq 0$, inequalities a) and b) in Sect. 1.1 hold for $|Tg^k\xi|$, $|Tg^k\eta|$. (As previously, the constants a, b, c do not depend on ξ, η, k.) (This is clearly a covering of finite multiplicity.) Apparently one of the aims of Franks was to achieve some unification of Anosov diffeomorphisms and expanding maps by representing them as particular cases of a more general concept. However, the unification attained was fairly relative: these particular cases still remain marked out by their essential properties. Namely, only in these particular cases are the Anosov maps structurally stable and in these cases only under a small perturbation (in the C^1 sense) of an Anosov map does one obtain another Anosov map. However, the definition can be modified so that the Anosov maps (in the new sense) remain as such under small perturbations, but in broad outline the situation remains as before. See (Mañé and Shub 1974), (Manning 1974), (Anosov 1977). Apparently this undermined interest in Anosov maps, although something was done in (Sakai 1987). The following problems appear quite reasonable: to investigate their ergodic properties and their construction in the case of codimension 1 (that is, when the codimension of the leaves of one of the foliations $E^{\mathrm{s}}, E^{\mathrm{u}}$ has this dimension).

b) Smale Systems. The majority of global results of a topological nature relate to opposite types: Anosov DS's (Sect. 4.2)[18] and Smale DS's. The latter satisfy Axiom A, the acyclicity condition and the set Ω of their nonwandering points in the case of a cascade is zero-dimensional, while in the case of a flow it is one-dimensional. The names *Smale cascade, flow, diffeomorphism* have a clear meaning. In a certain sense the Smale DS is the first type with regard to its complexity after Morse-Smale DS's (Anosov et al. 1985, Chap. 2, Sect. 3): we go over from finite Ω to zero-dimensional ones. The general nature of the investigation of systems of these types is partly the same (filtrations of handle-decomposition type which are, in a way, related to DS's and the translation of this geometry in terms of algebraic topology), but when Ω is infinite a new ingredient appears, namely, symbolic dynamics. (For example, one has to deal with conditions for the equivalence of Smale suspensions on TMC's.) An appreciable role is also played by the results of Williams on the classification of TMC's (see Alekseev 1976); it is interesting that for Williams the starting point was another question of hyperbolic theory, namely, the structure of one-dimensional attractors (Chap. 2, Sect. 1).

The name of Smale DS's is connected with the paper (Smale 1970), where it is shown that for any M each diffeomorphism of g is diffeotopic to a struc-

[18] We are referring here mainly to topologically transitive Anosov DS's. The anomalous Anosov DS's in (Franks and Williams 1980) are also Smale DS's.

turally stable one. (Existence of structurally stable flows for any M is proved more simply in (Palis and Smale 1970).) The structurally stable diffeomorphism so constructed was just a Smale diffeomorphism. Shub supplemented Smale's work by showing that a structurally stable Smale diffeomorphism f can be constructed arbitrarily close to g in the C^0 sense (not C^1 !), where f and g are connected by a diffeotopy which is also C^0-small (Shub 1972). The following question is far from settled: when do there exist Smale diffeomorphisms f that are C^0-close to g and are such that the topology entropy $h_{\text{top}}(f) \leq h_{\text{top}}(g)$? It is known that such f sometimes exist and sometimes do not. (However, the latter is not related to the zero-dimensionality of Ω. In each homotopy class of continuous maps of an n-dimensional torus into itself there exists an automorphism a of the torus as a Lie group. It turns out that in some classes the inequality $h_{\text{top}}(g) > h_{\text{top}}(a)$ holds for any diffeomorphism satisfying Axiom A (Fried 1980).)

For a Smale cascade the "basic components" of Ω are isomorphic to some TMC, while for a Smale flow they are equivalent to a Smale suspension over a TMC. In the present case these TMC's are better described by means of the corresponding matrices rather than transition graphs (Anosov et al. 1985, Chap. 1, Sect. 1.2). We note that these matrices are not invariants of the DS but depend on the choice of the decomposition used; in the case of a flow this is also apparent for the reason that the basic set can be represented as a suspension over different TMC's.) The geometry of the construction of the horseshoe type enables one to turn attention to certain other circumstances which the transition matrix does not take into account. In Fig. 7 the contracting and bent strip laid over K goes first upwards, then downwards, whereas in the left hand diagram of Fig. 11, it goes upwards both times. Similarly in the other cases, one can take into consideration whether the orientation is preserved or changed in certain directions under the mapping. To account for the alteration of orientation one can put a "minus" sign in the corresponding positions of the transition matrix. For flows on a three-dimensional sphere there further arise characteristics of the type of linked closed trajectories and invariants of the corresponding knots. Furthermore, for TMC's as for other DS's, zeta-functions are introduced (Anosov et al. 1985, Chap. 2, Sect. 2.4), sometimes modified in some fashion. An important characteristic of a basic set is, of course, its index. The problem is how to connect all these data with the topology of M and (for a cascade) with the action of the map on the homologies, and also to extract from all of this some invariants of the DS. We indicate a number of works of this nature, although we do not claim that this list is complete: (Bowen and Franks 1977), (Franks 1977), (Franks 1981), (Franks 1985), (Maller and Shub 1985), (Shub 1974), (Shub and Sullivan 1975).

c) In symbolic dynamics in the narrow sense of the word (which studies the Bernoulli cascade, its invariant subsets and measures) a new line of approach has recently appeared, this being the investigation of the automorphisms of the corresponding DS's. In (Wagoner 1986), (Zizza 1988) these ideas are invoked for the investigation of the group of homeomorphisms of phase space that commute with the Smale diffeomorphism.

d) There may be very few diffeomorphisms commuting with a given DS. Is it possible that in the group of C^∞-diffeomorphisms the elements commuting only with their own powers form an open dense subset? (This question is ascribed to Smale.) In (Palis and Yoccoz 1989) it is proved that this is the case in that part of this group that consists of diffeomorphisms satisfying Axiom A and the strong transversality condition (that is, structurally stable ones).

e) For cascades $\{g^n\}$ satisfying Axiom A, the acyclicity condition and certain extra conditions, Franks has established certain relations between the homological zeta-function of the map g (Anosov et al. 1985, Chap. 2, Sect. 2.4) or some contiguous function, functions connected with the action of g on the homologies of different dimensions, and the zeta-functions of the map $g|\Omega_i$, where the Ω_i are the components of the spectral decomposition of the set of nonwandering points (Franks 1975), (Franks 1978). For Morse-Smale DS's these relations are equivalent to the Morse-Smale inequalities.

f) Mean Dimension. In some respects cascades on surfaces (also one-dimensional non-invertible maps) are a case of mean dimension which is more complicated than flows on surfaces and DS's on a circle, but simpler than DS's in dimensions ≥ 3. An example is the theorem of Newhouse and Palis noted in Sect. 4.1. In the same article by these authors (Newhouse and Palis 1973) it is shown that a cascade $\{g^n\}$ on a surface with a hyperbolic Ω can be C^1-approximated by an Ω-structurally stable cascade $\{h^n\}$. Here $\{h^n\}$ is constructed having the same Ω. In higher dimensions this may prove impossible (even for finite Ω) (Dankner 1978), (Pugh, Walker, and Wilson 1977).

4.4. Beyond the Limits of Hyperbolicity. Here we list several questions of varying scope which being outside the framework of the hyperbolic theory have some sort of connections with it that are sometimes informal and difficult to express distinctly. (However, their existence is clearly expressed, if only because the majority of the names recalled below have already been encountered in earlier pages.)

a) Entropy Conjecture. This was stated by Shub and asserts that the topological entropy h_{top} of a map from M to itself is not less than the logarithm of the spectral radius of the induced map in the homologies with real coefficients (Shub 1974). The conjecture was proved by Yomdin for maps of class C^∞ that are not necessarily invertible (Yomdin 1987a), (Yomdin 1987b). Here smoothness of class C^∞ is seemingly essential not only for the proof but also for the validity of the assertion itself. For homeomorphisms, counterexamples have been known for some time. Apart from this, in some cases a high degree of smoothness is not needed; for example, if M is a torus then the conjecture is true even for continuous maps. For this reason previous works where infinite smoothness is replaced by the imposition of conditions of quite a different nature and a lower estimate for h_{top} is sometimes given in other terms, still have their value. See the survey (Katok 1977). The last publication in accordance with this tradition that is known to me is (Fried 1987).

Closely related with this are such questions as the existence of an invariant measure for which the metric entropy is the same as h_{top}, the semicontinuous nature of the dependence of h_{top} on the map or of the metric entropy on the invariant measure. Here again it is the C^∞ case that is outlined — the positive results of Newhouse (Newhouse 1988), (Newhouse 1989) relate to this case, whereas for finite smoothness there is the counterexample of Misiurewicz. Yomdin and Newhouse have systematically used a new object, namely, the rate of increase of the volumes of smooth submanifolds of various dimensions under the action of the DS.

Taking account of (Katok 1980), for two-dimensional cascades $\{g^n\}$ of class C^∞ one can obtain even a continuous dependence of the topological entropy on g in the C^∞ topology. In contrast to Sect. 4.3 e), here it is not clear whether this is a specific property of mean dimension.

b) Skew Products. Alongside DS's satisfying Axiom A there are two classes of DS's that form open sets in the space of all DS's with a C^1 topology and for which the qualitative picture is fairly clear. These are the Lorenz flows (Chap. 2) and, of more special interest, a class of skew products of a certain type (Hirsch et al. 1977), (Newhouse 1980a), (Newhouse and Young 1983).

c) Creation of Homoclinic Points under Perturbations. Let x be a hyperbolic periodic point of a cascade $\{g^n\}$ and $y \neq x$, $y \in \overline{W^{\mathrm{s}}(x) \cap W^{\mathrm{u}}(x)}$. Is it true that arbitrarily close to $\{g^n\}$ (in the C^1 sense) there is a cascade $\{f^n\}$ for which y is a homoclinic point whose trajectory is doubly asymptotic to the periodic trajectory obtained from the periodic trajectory $\{g^n x\}$ under this perturbation in the obvious sense? At first glance it appears obvious that the answer is affirmative, but this obviousness is as deceptive as it is in the closure lemma. Indeed, it may be even more deceptive, since the closure lemma has nevertheless been proved in the general case, while the affirmative answer to the present question has been obtained only in certain particular cases, one of which turned out to be essential for the proof of the necessity of Axiom A for structural stability (Mañé 1988a). In Mañé's paper there are references to previous works on the creation of homoclinic points under perturbation in a similar situation. One can add to this the paper (Newhouse 1976).

d) Non-Transversal Homoclinic Points. It has already been said that bifurcation theory provides examples of the occurrence of hyperbolic sets under perturbations. It is no less important that in them one encounters DS's possessing hyperbolic properties which, however, are weaker than in the theory of hyperbolic sets. This has already been spoken about (Sect. 3.3 e)). Here a fundamental role is played by the non-transversal homoclinic points. In all the works mentioned below, two-dimensional cascades (or, equivalently, three-dimensional flows) are considered. Since we are not giving the precise statements, we must again warn the reader about two circumstances that are slurred over in a cursory presentation. 1) In problems of this nature, it is not the C^1-topology that proves to be adequate, but rather the C^r-topology with some $r > 1$, although one can also pose questions relating to the C^1-topology. (Here is a simple analogue: if two curves touch, then by means of a C^1-small perturbation one can make some of their arcs coincide, but for a C^2-small perturbation this cannot be achieved if the tangency is non-degenerate.) 2) It is possible to consider a neighbourhood of a given diffeomorphism g_0 in the function space of diffeomorphisms of a given class C^r, and one can consider a "generic" (in the usual technical sense) one-parameter family $\{g_\lambda\}$ containing the original g_0 for $\lambda = 0$. Sometimes the difference between these two points of view turns out to be fairly appreciable.

Non-transversal homoclinic points were first considered in (Gavrilov and Shil'nikov 1972). For a generic one-parameter family $\{g_\lambda\}$ they showed that if g_0 has such a point, then for any $N > 0$ there exist g_λ with small λ for which there are more than N exponentially stable periodic trajectories. Newhouse strengthened this assertion (see Sect. 3.3 e)). After this, the problem of interpreting the numerical experiments on the appearance of strange attractors was brought to a head. Were not the objects observed in these experiments simply trajectories converging to very long periodic trajectories (Newhouse phenomenon) and not strange attractors (whatever this word might mean)? For example, what takes place with the Hénon map (Hénon 1976)?

This last question has now been partly cleared up, thanks to Benedicks and Carleson. At least in some range of values of the parameters, in the "majority" of them (from the metric standpoint) the trajectories converge to an "actual" attractor[19] which, even if it is not hyperbolic, then at any rate it has sufficiently strong properties of hyperbolicity type (Benedicks and Carleson 1991). Using the ideas and methods of this paper Mora and Viana showed that in certain cases a similar conclusion holds also for a family of diffeomorphisms $\{g_\lambda\}$ containing a diffeomorphism g_0 with a non-transversal homoclinic point (Mora and Viana 1990). (True, they understand an attractor in a certain weaker sense rather than as in Sect. 1.4. Meanwhile it is not clear whether this is related to the essence of the matter.) On the other hand, the "anti-attractor" papers of Palis and Takens (Palis and Takens 1985), (Palis and Takens 1987) appeared even earlier. It turns out that in certain other cases the existence of an attractor is a phenomenon that from the metric point of view is "rare". Both the Mora-Viana conditions and the Palis-Takens conditions are "generic" in certain regions of the function space of families $\{g_\lambda\}$, so that the answer to the question: "Which is more generic, the Newhouse phenomenon or the strange attractor?" sounds as follows: "Sometimes one and sometimes the other, depending on the circumstances". (See also (Tedeschini-

[19] In calling it "actual", I have in mind that it is a more complicated set than a long periodic orbit. It is not yet clear whether it enjoys all the uniformity properties included in the definition of attractor.

Lalli and Yorke 1986), (Tresser 1983), (Wang 1990), (Kan and Yorke 1990).) It would be interesting to clarify which, in fact, are more typical for the DS's considered in (Kolesov et al. 1989).

References*

Adler, R., Weiss, B (1967): Entropy, a complete metric invariant for automorphisms of the torus. Proc. Natl. Acad. Sci. USA *57*, No. 6, 1573–1576. Zbl. 177,80

Adler, R., Weiss, B. (1970): Similarity of Automorphisms of the Torus. Mem. Am. Math. Soc. *98*, 1–43. Zbl. 195,61

Alekseev, V.M. (1968/1969): Quasirandom dynamical systems I. Mat. Sb. Nov. Ser. *76* (1968), 72–134; II Mat. Sb. Nov. Ser. *77* (1968), 545–601; III Mat. Sb. Nov. Ser. *79* (1969), 3–50. [English transl.: I. Math. USSR, Sb. *5*, 73–128 (1969); II. Math. USSR, Sb. *6*, 505–560 (1969); III. Math. USSR, Sb. *7*, 1–43 (1970)] Zbl. 198,569–570

Alekseev, V.M. (1969): Perron sets and topological Markov chains. Usp. Mat. Nauk *24*, No. 5, 227–228. Zbl. 201,566

Alekseev, V.M. (1976): Symbolic Dynamics (Eleventh Mathematical School). Akad. Nauk Ukr. SSR, Kiev.

Alekseev, V.M. (1981): Final motions in the three body problem and symbolic dynamics. Usp. Mat. Nauk *36*, No. 4, 161–176. [English transl.: Russ. Math. Surv. *36*, No. 4, 181–200 (1981)] Zbl. 503.70006

Alekseev, V.M., Katok, A.B., Kushnirenko, A.G. (1972): Smooth dynamical systems. Ninth Summer Mathematical School. Inst. Mat. Akad. Nauk Ukr. SSR, Kiev, 50–348

Alekseev, V.M., Yakobson, M.V. (1979): Symbolic dynamics and hyperbolic dynamical systems. In (Bowen 1979), 196–240

Anosov, D.V. (1962): Roughness of geodesic flows on compact Riemannian manifolds of negative curvature. Dokl. Akad. Nauk SSSR *145*, No. 4, 707–709. [English transl.: Sov. Math., Dokl. *3*, 1068–1070 (1962)] Zbl. 135,404

Anosov, D.V. (1967): Geodesic flows on closed Riemannian manifolds of negative curvature. Tr. Mat. Inst. Steklova *90* (209 pp.). [English transl.: Proc. Steklov Inst. Math. *90* (1969)] Zbl. 163,436

Anosov, D.V. (1970): On a class of invariant sets of smooth dynamical systems. Tr. V Mezhdunar. Konf. po Nelineinym Kolebaniyam. Kachestvenny Metody. Inst. Mat. Akad. Nauk Ukr. SSR *2*, Kiev, 39–45

Anosov, D.V. (ed.) (1977a): Smooth Dynamical Systems. Mir, Moscow (Russian)

Anosov, D.V. (1977b): Introductory article to the collection "Smooth Dynamical Systems", in (Anosov 1977a), 7–31. [English transl.: Transl., II. Ser., Am. Math. Soc. *125*, 1–20 (1985)] Zbl. 569.58025

Anosov, D.V. (1985): Structurally stable systems. Tr. Mat. Inst. Steklova *169*, 59–93. [English transl.: Proc. Steklov Inst. Math. *169*, 61–95(1985)] Zbl. 619.58027

Anosov, D.V., Aranson S. Kh., Bronshtein, I.U., Grines, V.Z. (1985): Dynamical Systems I: Smooth Dynamical Systems. Itogi Nauki Tekh., Ser. Sovrem. Probl. Mat., Fundam. Napravleniya *1*, 151–242. [English transl. in: Encycl. Math. Sci. *1*, 149–230. Springer-Verlag, Berlin Heidelberg New York 1988] Zbl. 605.58001

* For the convenience of the reader, references to reviews in Zentralblatt für Mathematik (Zbl.), compiled using the MATH database, and Jahrbuch über die Fortschritte der Mathematik (FdM.) have, as far as possible, been included in this bibliography.

Anosov, D.V., Sinai, Ya.G. (1967): Some smooth ergodic systems. Usp. Mat. Nauk *22*, No. 5, 107–172. [English transl.: Russ. Math. Surv. *22*, No. 5, 103–167 (1967)] Zbl. 177,420

Arnol'd, V.I., Afraimovich, V.S., Il'yashenko, Yu.S., Shil'nikov, L.P. (1986): Dynamical Systems V: Bifurcation Theory. Itogi Nauki Tekh., Ser. Sovrem. Probl. Mat., Fundam. Napravleniya *5*. [English transl. in: Encycl. Math. Sci. *5*. Springer-Verlag, Berlin Heidelberg New York 1993]

Arnol'd, V.I., Il'yashenko, Yu.S. (1985): Dynamical Systems I: Ordinary Differential Equations. Itogi Nauki Tekh., Ser. Sovrem. Probl. Mat., Fundam. Napravleniya *1*, 5–149. [English transl. in: Encycl. Math. Sci. *1*, 1–148. Springer-Verlag, Berlin Heidelberg New York 1988] Zbl. 602.58020

Arnol'd, V.I., Kozlov, V.V., Nejstadt, A.I. (1985): Dynamical Systems III: Mathematical Aspects of Classical and Celestial Mechanics. Itogi Nauki Tekh., Ser. Sovrem. Probl. Mat., Fundam. Napravleniya *3*, 5–304. [English transl. in: Encycl. Math. Sci. *3*, 1–291. Springer-Verlag, Berlin Heidelberg New York 1987] Zbl. 612.70002

Barge, M. (1987): Homoclinic intersections and indecomposability. Proc. Am. Math. Soc. *101*, 541–544. Zbl. 657.58022

Benedicks, M., Carleson, L. (1991): The dynamics of the Hénon map. Ann. Math., II. Ser. *133*, No. 1, 73–169. Zbl. 724.58042

Benoist, Y., Foulon, P., Labourie, F. (1992): Flots d'Anosov à distributions stables et instables différentiables. J. Am. Math. Soc. *5*, No. 1, 33–74. Zbl. 759.58035

Birkhoff, G.D. (1936/1950): Nouvelles recherches sur les systèmes dynamiques, Mem. Pont. Acad. Sci. Novi Lincaei, III. Sér. 1, 85–216; Collected Math. Papers. Am. Math. Soc., New York, 530–661. Zbl. 16,234

Bohl, P. (1900): On certain differential equations of a general nature applicable to mechanics. Yur'ev 1900; Collected Works. Zinnatne, Riga, 73–198. FdM 31,358

Bohl, P. (1904): Über die Bewegung eines mechanischen Systems in der Nähe einer Gleichgewichtslage. J. Reine Angew. Math. *127*, 179–276. FdM 35,720

Bowen, R. (1970a): Topological entropy and Axiom A. Global Analysis. Proc. Symp. Pure Math. *14*, 23–41. Zbl. 207,544

Bowen, R. (1970b): Markov partitions for Axiom A diffeomorphisms. Am. J. Math. *92*, 725–747. Zbl. 208,259

Bowen, R. (1970c): Markov partitions and minimal sets for Axiom A diffeomorphisms. Am. J. Math. *92*, 907–918. Zbl. 212,291

Bowen, R. (1971): Periodic points and measures for Axiom A diffeomorphisms. Trans. Am. Math. Soc. *154*, 377–397. Zbl. Zbl. 212,291

Bowen, R. (1972): Periodic orbits for hyperbolic flows. Am. J. Math. *94*, 1–30. Zbl. 254.58005

Bowen, R. (1973): Symbolic dynamics for hyperbolic flows. Am. J. Math. *95*, 429–460. Zbl. 282.58009

Bowen, R. (1975a): A horseshoe with positive measure. Invent. Math. *29*, 203–204. Zbl. 306.58013

Bowen, R. (1975b): Equilibrium States and the Ergodic Theory of Anosov Diffeomorphisms. Lect. Notes Math. *470*. Zbl. 308.28010

Bowen, R. (1978): Markov partitions are not smooth. Proc. Am. Math. Soc. *71*, 130–132. Zbl. 417.58011

Bowen, R. (1979): Methods of Symbolic Dynamics. Mir, Moscow (Russian)

Bowen, R., Franks, J. (1977): Homology for zero-dimensional nonwandering sets. Ann. Math., II. Ser.*106*, No.1, 73–92. Zbl. 375.58018

Bowen, R., Ruelle, D. (1975): The ergodic theory of Axiom A flows. Invent. Math. *29*, No. 3, 181–202. Zbl. 311.58010

Bronshtein, I.Yu. (1984): Non-Autonomous Dynamical Systems. Shtiintsa, Kichinev. (Russian). Zbl. 651.58001

Brudno, A.A. (1983): Entropy and complexity of the trajectories of a dynamical system. Tr. Mosk. Mat. O.-va *44*, 124–149. [English transl.: Trans. Mosc. Math. Soc. No. 2, 127–151 (1983)] Zbl. 499.28017

Brunella, M. (1992): On the topological equivalence between Anosov flows on 3-manifolds. Comment. Math. Helv. *67*, 459–470. Zbl. 768.58037

Bunimovich, L.A., Pesin, Ya.B., Sinai, Ya.G., Yakobson, M.V. (1985): Dynamical Systems II: Ergodic Theory of Smooth Dynamical Systems. Itogi Nauki Tekh., Ser. Sovrem. Probl. Mat., Fundam. Napravleniya *2*, 113–232. [English transl. in: Encycl. Math. Sci. *2*, 99–206. Springer-Verlag, Berlin Heidelberg New York 1989]. Zbl. 781.58018

Cartwright, M.L., Littlewood, J.E. (1945): On non-linear differential equations of the second order I: The equation $\ddot{y} - k(1-y^2)\dot{y} + y = b\lambda k\cos(\lambda t + \alpha)$, k large. J. Lond. Math. Soc. *20*, 180–189. Zbl. 61,189

Cartwright, M.L., Littlewood, J.E. (1947): On non-linear differential equations of the second order II: The equation $\ddot{y} + kf(y,\dot{y}) + g(y,k) = p(t) = p_1(t) + kp_2(t)$; $k > 0$, $f(y) \geq 1$. Ann. Math. *48*, 472–494; (1949): Ann. Math. *50*, 504–505. Zbl. 29,126; Zbl. 38,250

Dankner, A. (1978): On Smale's Axiom A dynamical systems. Ann. Math., II. Ser. *107*, 517–533. Zbl. 366.58009

Dankner, A. (1980): Axiom A dynamical systems, cycles, and stability. Topology *19*, 163–177. Zbl. 458.58013

Denker, M., Grillenberger, C., Sigmund, K. (1976): Ergodic Theory on Compact Spaces. Lect. Notes Math. *527*. Zbl. 328.28008

Farrell, F.T., Jones, L.E. (1977): Markov cell structures. Bull. Am. Math. Soc. *83*, 739–740. Zbl. 417.58012

Farrell, F.T., Jones, L.E. (1978): Anosov diffeomorphisms constructed from $\pi_1(\mathrm{Diff}S^n)$. Topology *17*, 273–282. Zbl. 407.58031

Farrell, F.T., Jones, L.E. (1980): New attractors in hyperbolic dynamics. J. Differ. Geom. *15*, No. 1, 107–133. Zbl. 459.58010

Farrell, F.T., Jones, L.E. (1993): Markov cell structures near a hyperbolic set. Mem. Am. Math. Soc. *103*, No. 491. Zbl. 781.58031

Fenichel, N. (1971): Persistence and smoothness of invariant manifolds for flows. Indiana Univ. Math. J. *21*, 193–226. Zbl. 246.58015

Fenichel, N. (1974/1977): Asymptotic stability with rate conditions. Indiana Univ. Math. J. *23*, 1109–1137; II: *26*, 81–93. Zbl. 284.58008; Zbl. 365.58012

Foster, M.J. (1976): Fibre derivatives and stable manifolds: a note. Bull. Lond. Math. Soc. *8*, 286–288. Zbl. 334.55016

Franke, J., Selgrade, J. (1977): Hyperbolicity and chain recurrence. J. Differ. Equations *26*, 27–36. Zbl. 329.58012

Franks, J. (1970): Anosov diffeomorphisms. Global Analysis. Proc. Symp. Pure Math. *14*, 61–94. Zbl. 207,543

Franks, J. (1971): Necessary conditions for stability of diffeomorphisms. Trans. Am. Math. Soc. *158*, 301–308. Zbl. 219.58005

Franks, J. (1975): Morse inequalities for zeta functions. Ann. Math., II. Ser. *102*, 143–157. Zbl. 307.58016

Franks, J. (1977): Constructing structurally stable diffeomorphisms. Ann. Math., II. Ser. *105*, 343–359. Zbl. 366.58007

Franks, J. (1978): A reduced zeta function for diffeomorphisms. Am. J. Math. *100*, 217–243. Zbl. 405.58037

Franks, J. (1981): Knots, links, and symbolic dynamics. Ann. Math., II. Ser., *113*, 529–552. Zbl. 469.58013

Franks, J. (1985): Nonsingular Smale flows on S^3. Topology *24*, 265–282. Zbl. 609.58039

Franks, J., Robinson, C. (1976): A quasi-Anosov diffeomorphism that is not Anosov. Trans. Am. Math. Soc. *223*, 267–278. Zbl. 347.58008

Franks, J., Williams, B. (1980): Anomalous Anosov flows. Global Theory of Dynamical Systems. Lect. Notes Math. *819*, 158–174. Zbl. 463.58021

Franks, J., Williams, R.F. (1985): Entropy and knots. Trans. Am. Math. Soc. *291*, 241–253. Zbl. 587.58038

Fried, D. (1980): Efficiency vs. hyperbolicity on tori. Global Theory of Dynamical Systems. Lect. Notes Math. *819*, 175–189. Zbl. 455.58021

Fried, D. (1983): Transitive Anosov flows and pseudo-Anosov maps. Topology *22*, 299–303. Zbl. 516.58035

Fried, D. (1986): Entropy and twisted cohomology. Topology *25*, 455–470. Zbl. 611.58036

Fried, D. (1987): Rationality for isolated expansive sets. Adv. Math. *65*, No. 1, 35–38. Zbl. 641.58036

Gallavotti, G. (1976): Funzioni zeta ed insiemi basilari. Atti. Accad. Naz. Lincei, VIII. Ser. *61*, No. 5, 309–317. Zbl. 378.58007

Gavrilov, N.K., Shil'nikov, L.P. (1972): On three-dimensional dynamical systems close to systems with a structurally unstable homoclinic curve. I: Mat. Sb., Nov. Ser. *88*, 475–492. [English transl.: Math. USSR, Sb. *17*, 467–485 (1972)] Zbl. 251.58005

Gavrilov, N.K., Shil'nikov, L.P. (1973): On three-dimensional dynamical systems close to systems with a structurally unstable homoclinic curve. II: Mat. Sb., Nov. Ser. *90*, 139–156. [English transl.: Math. USSR, Sb. *19*, 139–156 (1974)] Zbl. 258.58009

Ghys, E. (1984): Flots d'Anosov sur les 3-variétés fibrées en cercles. Ergodic Theory Dyn. Syst. *4*, No. 1, 67–80. Zbl. 527.58030

Ghys, E. (1992): Déformations de flots d'Anosov et de groupes Fuchsiens. Ann. Inst. Fourier *42*, No. 1–2, 209–247. Zbl. 759.58036

Gromov, M. (1981): Groups of polynomial growth and expanding maps. Publ. Math., Inst. Hautes Etud. Sci. *53*, 53–78. Zbl. 474.20018

Guckenheimer, J. (1970): Endomorphisms of the Riemann sphere. Global Analysis. Proc. Symp. Pure Math. *14*, 95–123. Zbl. 205,541

Gurevich, B.M., Sinai, Ya.G. (1969): Algebraic automorphisms of the torus and Markov chains. In Billingsley, P.: Ergodic Theory and Information. I. Mir, Moscow, 205–233. Zbl. 184,433

Hamenstädt, U. (1990): Entropy-rigidity of locally symmetric spaces of negative curvature. Ann. Math., II. Ser. *131*, 35–51. Zbl. 699.53049

Handel, M., Thurston, W. (1980): Anosov flows on new 3-manifolds. Invent. Math. *59*, No. 2, 95–103. Zbl. 435.58019

Hartman, Ph. (1971): The stable manifold of a point of a hyperbolic map of a Banach space. J. Differ. Equations *9*, 360–379. Zbl. 223.58008

Hénon, M. (1976): A two-dimensional mapping with a strange attractor. Commun. Math. Phys. *50*, No. 1, 69–76. Zbl. 576.58018

Hirsch, M., Palis J., Pugh., C.C., Shub, M. (1970): Neighbourhoods of hyperbolic sets. Invent. Math. *9*, No. 2, 121–134. Zbl. 191,217

Hirsch, M., and Pugh, C.C. (1970): Stable manifolds and hyperbolic sets. Global Analysis. Proc. Symp. Pure Math. *14*, 133–163. Zbl. 215,530

Hirsch, M., Pugh., C.C., Shub, M. (1970): Invariant manifolds. Bull. Am. Math. Soc. *76*, 1015–1019. Zbl. 226.58009

Hirsch, M., Pugh., C.C., Shub, M. (1977): Invariant Manifolds. Lect. Notes Math. *583*. Zbl. 355.58009

Hurder, S., Katok, A. (1990): Differentiability, rigidity and Goldbillon-Vey classes for Anosov flows. Publ. Math., Inst. Hautes Etud. Sci. *72*, 5–61. Zbl. 725.58034

Irwin, M.O. (1970): On the stable manifold theorem. Bull. Lond. Math. Soc. *2*, 196–198. Zbl. 198,568

Irwin, M.O. (1972): On the smoothness of the composition map. Q. J. Math., Oxford, II. Ser. *23*, 113–133. Zbl. 235.46067

Jones, L. (1983): Locally strange hyperbolic sets. Trans. Am. Math. Soc. *275*, No. 1, 153–162. Zbl. 508.58033

Kamaev, D.A. (1977): On the topological indecomposability of certain invariant sets of A-diffeomorphisms. Usp. Mat. Nauk *32*, No. 1, 158

Kan, I., Yorke, J.A. (1990): Antimonotonicity: concurrent creation and annihilation of periodic orbits. Bull. Am. Math. Soc., New Ser. *23*, 469–476. Zbl. 713.58026

Katok, A. (1980): Lyapunov exponents, entropy and periodic orbits for diffeomorphisms. Publ. Math., Inst. Hautes Etud. Sci. *51*, 137–173. Zbl. 445.58015

Katok, A.B. (1970): Rotation numbers and U-flows. Usp. Mat. Nauk *25*, No. 5, 243–244. Zbl. 207,227

Katok, A.B. (1971): Local properties of hyperbolic sets. In the Russian translation of (Nitecki 1971), 214–232. Zbl. 311.58001

Katok, A.B. (1977): The entropy conjecture. In (Anosov 1977a), 181–203. [English transl.: Transl., II. Ser., Am. Math. Soc. *133*, 91–108 (1986)] Zbl. 628.58026

Kolesov, A.Yu., Kolesov, Yu.S., Mishchenko, E.F., Rozov, N.Kh. (1989): On the phenomenon of chaos in three-dimensional relaxation systems. Mat. Zametki *46*, 153–155. (Russian) Zbl. 714.34076

Krieger, W. (1972): On unique ergodicity. Proc. 6th Berkeley Symp. Math. Stat. Prob., Berkeley 1970, *2*, 327–345. Zbl. 262.28013

Krzyzewski, K. (1978): Some results on expanding mappings. Astérisque *50* (1977), 205–218. Zbl. 413.58017

Krzyzewski, K. (1979): A remark on expanding mappings. Colloq. Math. *41*, No. 2, 291–295. Zbl. 446.58009

Krzyzewski, K. (1982): On analytic invariant measures for expanding mappings. Colloq. Math. *46*, No. 1, 59–65. Zbl. 508.58028

Kurata, M. (1977): Hartman's theorem for hyperbolic sets. Nagoya Math. J. *67*, 41–52. Zbl. 336.58008

Kurata, M. (1978/1979): Hyperbolic nonwandering sets without dense periodic points. Proc. Japan Acad., Ser. A *54*, No. 7, 206–211. Zbl. 438.58022; (1979) Nagoya Math. J. *74*, 77–86. Zbl. 388.58017

Kurata, M. (1979): Markov partitions of hyperbolic sets. J. Math. Soc. Japan *31*, No. 1, 39–51. Zbl. 399.58008

Kuratowski, K. (1969): Topology. Mir, Moscow, transl. from the English edition, Acad. Press, New York (1966). Zbl. 158,408

Levi, M. (1981): Qualitative Analysis of Periodically Forced Relaxation Oscillations. Mem. Am. Math. Soc. *244*. Zbl. 448.34032

Levinson, N. (1949): A second order differential equation with singular solutions. Ann. Math., II. Ser. *50*, 126–153. Zbl. 41,423

Liao, S.T. (1980): On the stability conjecture. Chin. Ann. Math. *1*, No. 1, 9–30. Zbl. 449.58013

Littlewood, J.E. (1957a): On the non-linear differential equations of the second order: III. The equation $\ddot{y} - k(1-y^2)\dot{y} + y = b\mu k \cos(\mu t + \alpha)$ for large k, and its generalizations. Acta Math. *97*, 267–308. Zbl. 81,84

Littlewood, J.E. (1957b): On the non-linear differential equations of the second order: IV. The general equation $\ddot{y} + kf(y)\dot{y} + g(y) = bkp(\phi)$, $\phi = t + \alpha$. Acta Math. *98*, 1–110. Zbl. 81,84

Littlewood, J.E. (1960): On the number of stable periods of a differential equation of the Van der Pol type. IEE Trans. Circuit Theory *7*, 535–542

Maller, M., Shub, M. (1985): The integral homology of Smale diffeomorphisms. Topology *24*, 153–164. Zbl. 595.58025

Malta, I. (1980): Hyperbolic Birkhoff centers. Trans. Am. Math. Soc. *262*, 181–193. Zbl. 458.58017

Mañé, R. (1975): Expansive diffeomorphisms. Dynamical Systems. Proc. Symp. Univ. Warwick 1974. Lect. Notes Math. *468*, 162–174. Zbl. 309.58016

Mañé, R. (1977): Quasi-Anosov diffeomorphisms and hyperbolic manifolds. Trans. Am. Math. Soc. *229*, 351–370. Zbl. 356.58009

Mañé, R. (1978): Contributions to the stability conjecture. Topology *17*, 383–396. Zbl. 405.58035

Mañé, R. (1982): An ergodic closing lemma. Ann. Math., II. Ser. *116*, 503–540. Zbl. 511.58029

Mañé, R. (1988a): On the creation of homoclinic points. Publ. Math., Inst. Hautes Etud. Sci. *66*, 139–159. Zbl. 699.58024

Mañé, R. (1988b): A proof of the C^1-stability conjecture. Publ. Math., Inst. Hautes Etud. Sci. *66*, 161–210. Zbl. 678.58022

Mañé, R., Pugh, C. (1975): Stability of endomorphisms. Dynamical Systems. Proc. Symp. Univ. Warwick 1974. Lect. Notes Math. *468*, 175–184. Zbl. 315.58018

Manning, A. (1973): Axiom A diffeomorphisms have rational zeta function. Bull. Lond. Math. Soc. *3*, 215–220. Zbl. 219.58007

Manning, A. (1974): There are no new Anosov diffeomorphisms on tori. Am. J. Math. *96*, 422–429. Zbl. 242.58003

Margulis, G.A. (1970): On certain measures associated with U-flows on compact manifolds. Funkts. Anal. Prilozh. *4*, No. 1, 62–76. [English transl.: Funct. Anal. Appl. *4*, 55–67 (1970)] Zbl. 245.58003

Martin, N.F.G., England, J.W. (1981): Mathematical Theory of Entropy. Addison-Wesley, Reading, Mass. Zbl. 456.28008

Mather, J. (1968): Characterization of Anosov diffeomorphisms. Indagationes Math. *30*, 479–483. Zbl. 165,570

Mel'nikov, V.K. (1963): On the stability of the centre under time-periodic perturbations. Tr. Mosk. Mat. O.-va *12*, 3–52. [English transl.: Trans. Mosc. Math. Soc. 1963, 1–56 (1965)] Zbl. 135,310

Meyer, K.R., Sell, G.R. (1989): Melnikov transforms, Bernoulli bundles and almost periodic perturbations. Trans. Am. Math. Soc. *314*, 63–105. Zbl. 707.34041

Mora, L., Viana, M. (1990): Abundance of Strange Attractors. Preprint IMPA, Rio de Janeiro

Moser, J. (1973): Stable and random motions in dynamical systems. With special emphasis on celestial mechanics. Ann. Math. Stud. *77*. Zbl. 271.70009

Newhouse, S.E. (1970): On codimension one Anosov diffeomorphisms. Am. J. Math. *92*, 761–770. Zbl. 204,569

Newhouse, S.E. (1972): Hyperbolic limit sets. Trans. Am. Math. Soc. *167*, 125–150. Zbl. 239.58009

Newhouse, S.E. (1976): On homoclinic points. Proc. Am. Math. Soc. *60*, 221–224. Zbl. 344.58015

Newhouse, S.E. (1979): The abundance of wild hyperbolic sets and non-smooth stable sets for diffeomorphisms. Publ. Math., Inst. Hautes Etud. Sci. *50*, 101–152. Zbl. 445.58022

Newhouse, S.E. (1980a): Dynamical properties of certain non-commutative skew products. Global Theory of Dynamical Systems. Lecture Notes Math. *819*, 353–363. Zbl. 477.58019

Newhouse, S.E. (1980b): Lectures on dynamical systems. Dynamical Systems, C.I.M.E. Lect., Bressanone 1978. Prog. Math. *8*, 1–114. Zbl. 444.58001

Newhouse, S.E. (1988): Entropy and volume. Ergodic Theory Dyn. Syst. (Charles Conley memorial volume) *8*, 283–299. Zbl. 638.58016

Newhouse, S.E. (1989): Continuity properties of entropy. Ann. Math., II. Ser. *129*, 215–235. Zbl. 676.58039

Newhouse, S.E., Palis, J. (1973): Hyperbolic nonwandering sets on two-dimensional manifolds. Dynamical Systems, Proc. Symp. Univ. Bahia, Salvador 1971, 293–301. Zbl. 279.58010

Newhouse, S.E., Young, L.S. (1983): Dynamics of certain skew products. Geometric Dynamics. Lect. Notes Math. *1007*, 611–629. Zbl. 535.58036

Nitecki, Z. (1971): Differentiable Dynamics. An Introduction to the Orbit Structure of Diffeomorphisms. MIT Press, Cambridge (Mass.) London. [Russian transl.: Mir, Moscow 1975] Zbl. 246.58012

Nitecki, Z., Shub, M. (1976): Filtrations, decompositions and explosions. Am. J. Math. *97*, 1029–1047. Zbl. 324.58015

Oka, M. (1990): Expansiveness of real flows. Tsukuba J. Math. *14*, No. 1, 1–8. Zbl. 733.54030

Osipov, A.V. (1975): The hyperbolicity of a basic family for Levinson's equation. Differ. Uravn. *11*, 1795–1800. [English transl.: Differ. Equations *11*, 1340–1344 (1976)] Zbl. 329.34036

Osipov, A.V. (1976a): On the behaviour of solutions of the Levinson equation. Differ. Uravn. *12*, 2000–2008. [English transl.: Differ. Equations *12*, 1401–1407 (1977)] Zbl. 349.34024

Osipov, A.V. (1976b): Ω-stability of the Levinson equation. Vestn. Leningr. Univ. 1976, No. 7, 156–157. Zbl. 357.34054

Palis, J. (1969): On Morse-Smale dynamical systems. Topology *8*, 385–405. Zbl. 189,239

Palis, J. (1970): A note on Ω-stability. Global Analysis. Proc. Symp. Pure Math. *14*, 221–222. Zbl. 214,508

Palis, J. (1988): On the C Ω-stability conjecture. Publ. Math., Inst. Hautes Etud. Sci. *66*, 211–215. Zbl. 648.58019

Palis, J., de Melo, W. (1982): Geometric Theory of Dynamical Systems. An Introduction. Springer-Verlag, New York Berlin Heidelberg. Zbl. 491.58001

Palis, J., Smale, S. (1970): Structural stability theorems. Global Analysis. Proc. Symp. Pure Math. *14*, 223–231. Zbl. 214,507

Palis, J., Takens, F. (1985): Cycles and measure of bifurcation sets for two-dimensional diffeomorphisms. Invent. Math. *82*, 397–422. Zbl. 579.58005

Palis, J., Takens, F. (1987): Hyperbolicity and the creation of homoclinic orbits. Ann. Math., II. Ser. *125*, 337–374. Zbl. 641.58029

Palis, J., Yoccoz, J.C. (1989): Rigidity of centralizers of diffeomorphisms. Ann. Sci. Ec. Norm. Super., IV. Ser. *22*, No. 1, 81–98. Zbl. 709.58022

Palmer, K.J. (1984): Exponential dichotomies and transversal homoclinic points. J. Differ. Equations *55*, 225–256. Zbl. 508.58035

Patterson, S., Robinson, C. (1988): Basins for general non-linear Hénon attracting sets. Proc. Am. Math. Soc. *103*, 615–623. Zbl. 651.58028

Plante, J.F. (1972): Anosov flows. Am. J. Math. *94*, 729–754. Zbl. 257.58007

Plante, J.F. (1981): Anosov flows, transversely affine foliations, and a conjecture of Verjovsky. J. Lond. Math. Soc., II. Ser. *23*, No. 2, 359–362. Zbl. 465.58020

Pliss, V.A. (1964): Non-Local Problems of the Theory of Vibrations. Nauka, Moscow-Leningrad. [English transl.: Acad. Press, New York 1966] Zbl. 123,58

Pliss, V.A. (1977): Integral Sets of Periodic Systems of Differential Equations. Nauka, Moscow. (Russian) Zbl. 463.34002

Pollicott, M. (1987): Symbolic dynamics for Smale flows. Am. J. Math. *109*, No. 1, 183–200. Zbl. 628.58042

Przytycki, F. (1976): Anosov endomorphisms. Stud. Math. *58*, No. 3, 249–285. Zbl. 357.58010

Pugh, C.C., Robinson, R.C. (1982): The C^1 closing lemma, including Hamiltonians. Preprint, Berkeley, Univ. of California. Appeared in: Ergodic Theory Dyn. Systems *3*, 261–313 (1983). Zbl. 548.58012

Pugh, C.C., Shub, M. (1970): The Ω-stability theorem for flows. Invent. Math. *11*, No. 2, 150–158. Zbl. 212,291

Pugh, C.C., Walker, R.B., Wilson, F.W., Jr. (1977): On Morse-Smale approximation — a counterexample. J. Differ. Equations *23*, 173–182. Zbl. 346.58006

Ratner, M.E. (1973): Markov partitions for Anosov flows on n-dimensional manifolds. Isr. J. Math. *15*, No. 1, 92–114. Zbl. 269.58010

Ratner, M.E. (1969a): Invariant measure with respect to an Anosov flow on a three-dimensional manifold. Dokl. Akad. Nauk SSSR *186*, 261–263. [English transl.: Sov. Math., Dokl. *10*, 586–588 (1969)] Zbl. 188,266

Ratner, M.E. (1969b): Markov splitting for U-flows in three-dimensional manifolds. Mat. Zametki *6*, 693–704. [English transl.: Math. Notes *6*, 880–886 (1969)] Zbl. 198,568

Robbin, J.W. (1971): A structural stability theorem. Ann. Math., II. Ser. *94*, 447–493. Zbl. 224.58005

Robinson, R.C. (1975): Structural stability of C^1 flows. Dynamical Systems. Proc. Symp. Univ. Warwick, 1974. Lect. Notes Math. *468*, 262–277. Zbl. 307.58012

Robinson, R.C. (1976): Structural stability of C^1 diffeomorphisms. J. Differ. Equations *22*, 28–73. Zbl. 343.58009

Robinson, R.C. (1983): Bifurcation to infinitely many sinks. Commun. Math. Phys. *90*, 433–459. Zbl. 531.58035

Robinson, R.C. (1988): Horseshoes for autonomous Hamiltonian systems using the Melnikov integral. Ergodic Theory Dyn. Syst. (Charles Conley Memorial Volume) *8*, 395–409. Zbl. 666.58039

Ruelle, D. (1978): Thermodynamic Formalism. Addison-Wesley, Reading, Mass. Zbl. 401.28016

Ruelle, D., Sullivan, D. (1975): Currents, flows and diffeomorphisms. Topology *14*, 319–327. Zbl. 321.58019

Sakai, K. (1987): Anosov maps on closed topological manifolds. J. Math. Soc. Japan *39*, 505–519. Zbl. 647.58039

Sanami, A. (1981/1983): The stability theorem for discrete dynamical systems on two-dimensional manifolds. Proc. Japan. Acad., Ser. A *57*, 403–407. Zbl. 509.58028; Nagoya Math. J. *90*, 1–55. Zbl. 515.58022

Shil'nikov, L.P. (1967): On a problem of Poincaré and Birkhoff. Mat. Sb. *74*, 378–397. [English transl.: Math. USSR, Sb. *3*, 353–371 (1969)] Zbl. 244.34033

Shub, M. (1969): Endomorphisms of compact differentiable manifolds. Am. J. Math. *91*, 175–199. Zbl. 201,563

Shub, M. (1972): Structurally stable diffeomorphisms are dense. Bull. Am. Math. Soc. *78*, 817–818. Zbl. 265.58003

Shub, M. (1974): Dynamical systems, filtrations and entropy. Bull. Am. Math. Soc. *80*, 27–41. Zbl. 305.58014

Shub, M. (1978): Stabilité globale des systèmes dynamiques. Astérisque *56*. Zbl. 396.58014

Shub, M., Smale, S. (1972): Beyond hyperbolicity. Ann. Math., II. Ser. *96*, 587–591. Zbl. 247.58008

Shub, M., Sullivan, D. (1975): Homology theory and dynamical systems. Topology *14*, 109–132. Zbl. 408.58023

Sinai, Ya.G. (1968a): Markov partitions and C-diffeomorphisms. Funkts. Anal. Prilozh. *2*, No. 1, 64–89. [English transl.: Funct. Anal. Appl. *2*, 61–82 (1968)] Zbl. 182,550

Sinai, Ya.G. (1968b): Construction of Markov partitions. Funkts. Anal. Prilozh. *2*, No. 3, 70–80. [English transl.: Funct. Anal. Appl. *2*, 245–253 (1968)] Zbl. 194,226

Smale, S. (1963): A structurally stable differentiable homeomorphism with an infinite number of periodic points. Tr. Mezhdunar., Teor. Prikl. Mekh., Kiev 1961, 2, 365–366. Zbl. 125,116

Smale, S. (1965): Diffeomorphisms with many periodic points. Differential and Combinatorial Topology. Symp. honour M. Morse, Princeton Univ. Press. *27*, 63–80. Zbl. 142,411

Smale, S. (1967a): Dynamical systems on n-dimensional manifolds. Differential Equations and Dynamical Systems, Proc. Int. Puerto Rico 1965, 483–486. Zbl. 183,98

Smale, S. (1967b): Differentiable dynamical systems. Bull. Am. Math. Soc. *73*, 747–817. Zbl. 202,552

Smale, S. (1970): The Ω-stability theorem. Global Analysis. Proc. Symp. Pure Math. *14*, 289–297. Zbl. 205,541

Smale, S. (1973): Stability and isotopy in discrete dynamical systems. In: Dynamical Systems, Proc. Symp. Univ. Bahia, Salvador 1971, 527–530. Zbl. 284.58013

Smale, S. (1980): The Mathematics of Time. Essays on Dynamical Systems, Economic Processes, and Related Topics. Springer-Verlag, New York Berlin Heidelberg. Zbl. 451.58001

Solodov, V.V. (1982): Homeomorphisms of the line and foliations. Izv. Akad. Nauk SSSR Ser. Mat. *46*, 1047–1061. [English transl.: Math. USSR, Izv. *21*, 341–354 (1982)] Zbl. 521.57022

Solodov, V.V. (1991): Topological questions in the theory of dynamical systems. Usp. Mat. Nauk *46*, No. 4, 93–114. [English transl.: Russ. Math. Surv. *46*, No. 4, 107–134 (1991)] Zbl. 739.58049

Steinlein, H., Walther, H.-O. (1990): Hyperbolic sets, transversal homoclinic trajectories, and symbolic dynamics for C^1-maps in Banach spaces. J. Dyn. Differ. Equations *2*, 325–365. Zbl. 702.58051

Tedeschini-Lalli, L., Yorke, J.A. (1986): How often do simple dynamical processes have infinitely many coexisting sinks? Commun. Math. Phys. *106*, 635–657. Zbl. 602.58036

Tomter, P. (1975): On the classification of Anosov flows. Topology *14*, 179–189. Zbl. 365.58013

Tresser, Ch. (1983): Un théorème de Sil'nikov en $C^{1,1}$. C. R. Acad. Sci., Paris, Ser. 1 *296*, No. 13, 545–548. Zbl. 522.58034

Verjovsky, A. (1974): Codimension one Anosov flows. Bol. Soc. Mat. Mex., II. Ser. *19*, No. 2, 49–77. Zbl. 323.58014

Wagoner, J. (1986): Realizing symmetries of a subshift of finite type by homeomorphisms of spheres. Bull. Am. Math. Soc. *14*, 301–303. Zbl. 603.58039

Wang, X.-J. (1990): The Newhouse set has a positive Hausdorff dimension. Commun. Math. Phys. *131*, 317–332. Zbl. 708.58017

Wiggins, S. (1988): Global Bifurcations and Chaos. Analytical Methods. Appl. Mat. Sci. *73*, Springer-Verlag, New York Berlin Heidelberg. Zbl. 661.58001

Yakobson, M.V. (1976): On some proofs of Markov partitions. Dokl. Akad. Nauk SSSR *226*, 1021–1024. [English transl.: Sov. Math., Dokl. *17*, 247–251 (1976)] Zbl. 344.58016

Yomdin, Y. (1987a): Volume growth and entropy. Isr. J. Math. *57*, 285–301. Zbl. 641.58036

Yomdin, Y. (1987b): C^k-resolution of semialgebraic mappings. Addendum to "Volume growth, and entropy". Isr. J. Math. *57*, 301–317. Zbl. 641.54037

Zeeman, E.C. (1975): Morse inequalities for diffeomorphisms with shoes and flows with solenoids. Dynamical Systems, Proc. Symp. Univ. Warwick 1974, Lect. Notes Math. *468*, 44–47. Zbl. 308.58012

Zizza, F. (1988): Automorphisms of hyperbolic dynamical systems and K_2. Trans. Am. Math. Soc. *307*, 773–797. Zbl. 672.34065

Chapter 2
Strange Attractors

R.V. Plykin, E.A. Sataev, S.V. Shlyachkov

Introduction

The term "strange attractor" was introduced in (Ruelle and Takens 1971) to denote a limit set of a dynamical system that is not a manifold and is consequently not a fixed point, limit cycle, invariant torus and so on. It immediately gained wide prevalence among physicists. Nowadays the notion of "strange attractor" has more the nature of a paradigm than a rigorously defined mathematical object. A dynamical system is considered to have a strange attractor if the phase space of the system has a limit set consisting of trajectories with chaotic behaviour. With regard to chaos, there are a number of "physical" criteria for chaotic behaviour. Among these we can single out the presence of homoclinics, the continuity of the spectrum, the presence of a positive Lyapunov exponent, the fractionality of some dimension, the presence of an infinite series of bifurcations, and period doubling. There are an enormous number of works in which for a given physical system one of the above criteria or some other criterion is used to deduce the presence of a strange attractor in the system. An extensive bibliography of works of this kind can be found in the book (Neimark and Landa 1987).

Alongside the existence of "physical" criteria for chaotic motion, rigorous mathematical results have been obtained only for attractors that to some extent possess properties of uniform hyperbolicity, such as hyperbolic attractors and attractors of Lorenz type.

In this book we restrict ourselves to a discussion of attractors of maps. Certain results (for example, for the Lorenz attractor) go over to the case of flows. However flows have their own special properties, which we shall not go into. We shall be considering a map of some open subset Q of a manifold M into the manifold M. As a particular case, Q can be the whole of M. The manifold M can be the space $\mathbb{R}^n$.

The usual definition of an attractor of a DS (Chap. 1, Sect. 1.3) easily carries over to the "semilocal" case in which one studies iterations of the map $T : Q \to M$. As was noted in Chap. 1, Sect. 1.3, this definition has its drawbacks. We therefore call attractors in this sense *quasi-attractors.* More suitable for us is the definition of a probabilistic limit set in the sense of Milnor (Milnor 1985).

Definition. A *probabilistic attractor* of a dynamical system is the smallest closed set containing ω-limit sets for almost all points of the phase space.

An attractor in this sense cannot be a quasi-attractor. An example is given in Fig. 1. Another example is a minimal set in the example of the Poincaré-Denjoy cascade on S^1 or a flow on the torus.

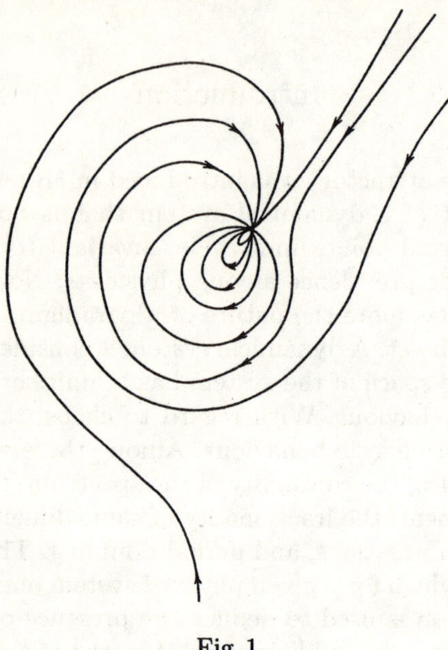

Fig. 1

For the special case of the Smale horseshoe $\{F^n\}$ (Chap. 1, Sect. 2.2) a probabilistic attractor is a fixed attracting point b, while both b and $\overline{W^u(C)}$ are quasi-attractors.

§1. Hyperbolic Attractors

1.1. Quasi-attractors Related to the Horseshoe. In Chap. 1, Sect. 2.2 it was shown how to extend the map f defining the Smale horseshoe (Chap. 1, Fig. 7) to a homeomorphism $F : S^2 \to S^2$ with certain properties (the main step of the extension is shown in Chap. 1, Fig. 8). Modifications of the horseshoe were also indicated (Chap. 1, Figs. 11, 12). For some of these an equally simple extension with similar properties is possible (Figs. 2, 3). We shall explain why, in the case of Fig. 2, F is constructed with two attractive fixed points. The number of connected components of the set $K \cap fK$ can be any positive integer n.

We denote by S_n the Smale quasi-attractor corresponding to the map of the horseshoe for which the number of connected components of $K \cap fK$ is

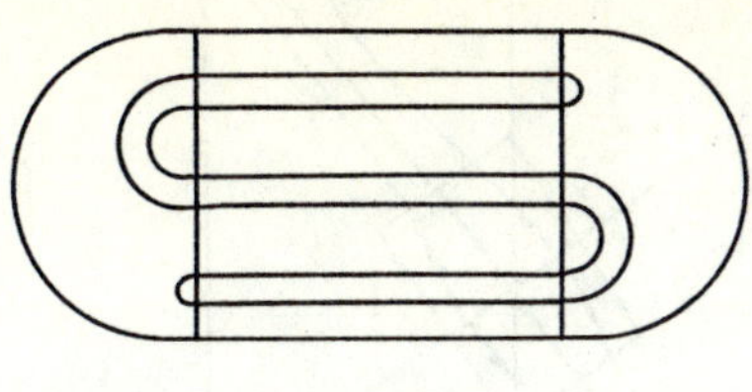
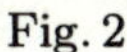

Fig. 2

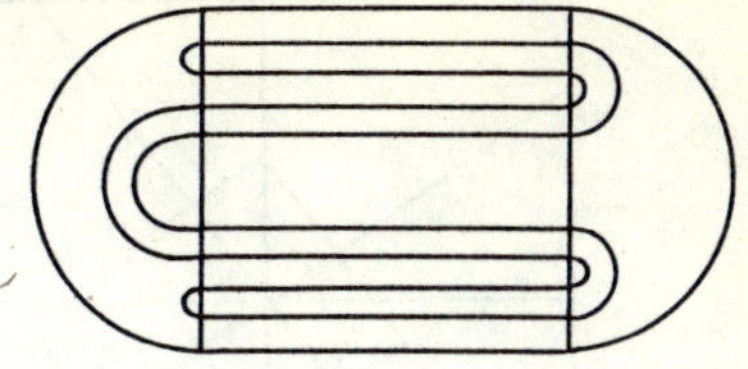

Fig. 3

equal to n. The following theorem due to Barge solves the problem of the homeomorphism of the continua S_n.

Theorem (Barge 1986). *The quasi-attractors S_n and S_m are homeomorphic if and only if the numbers n and m have the same set of prime factors.*

As was pointed out in Chap. 1, Sect. 2.2, S_2 is an indecomposable Brouwer-Janiszewski continuum. In (Kamaev 1977) a more general assertion concerning $\overline{W^s(A)}$ and $\overline{W^u(A)}$ is proved for a number of zero-dimensional hyperbolic sets A.

1.2. Attractors of Codimension 1. The following hyperbolic attractor was constructed by Smale (Smale 1970); it is the so-called *DA-diffeomorphism* ("derived from Anosov").

We outline the idea of Smale's construction, which consists in a "surgical" modification of a hyperbolic automorphism of the two-dimensional torus in a neighbourhood of the fixed point (0, 0). One can alter a hyperbolic automorphism of the torus in a neighbourhood of (0, 0) in such a way that the stable foliation for the new diffeomorphism is the same as the stable foliation for the original automorphism (with the exception of the leaf passing through (0, 0); this leaf splits into two leaves corresponding to the two new leaves that arise as a result of the bifurcation of the fixed points of hyperbolic type located on the old stable manifold of the point (0, 0) on different sides of it; the point (0, 0) itself is converted to a fixed repeller point). The closure of the unstable manifold of any of the two new fixed hyperbolic points is a hyperbolic attractor and is the same as the set of nonwandering points from which the point (0, 0) has been removed (see Fig. 4).

The above construction has a broad generalization enabling one to obtain a DA-diffeomorphism from a hyperbolic automorphism of the torus of dimension n, for which the stable manifold is one-dimensional and the unstable one $(n-1)$-dimensional.

An example of a hyperbolic attractor of a diffeomorphism of the 2-sphere was constructed by Plykin in (Plykin 1974).

The basis of the construction of the corresponding map is a diffeomorphism of a region D homeomorphic to a disc with three holes into itself. The region D can be represented as a union of three regions D_1, D_2, D_3, each of which has

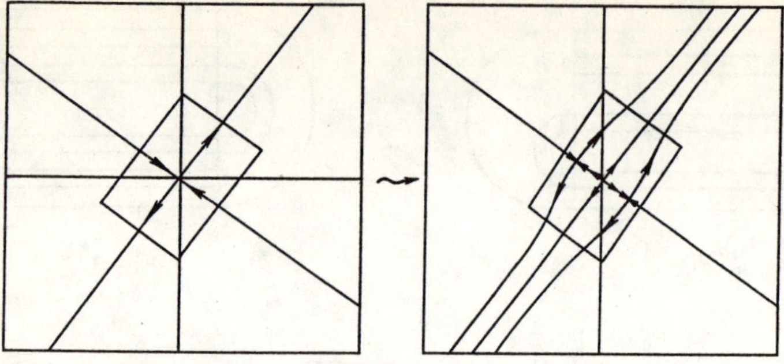

Fig. 4

the form of a half-disc from which a concentric half-disc has been removed. The half-discs D_2 and D_3 are attached along their diameters to the diameter of D_1. (The sum of the diameters of D_2 and D_3 must be less than the diameter of D_1.) The regions D_i are fibred into sectors and the map $f: D \to D$ takes each radius of the above partition, which is a foliation everywhere except for the leaf along which all three regions D_1, D_2, D_3 are attached, to a radius of the same partition and subjects it to a compression with a suitably chosen coefficient m.

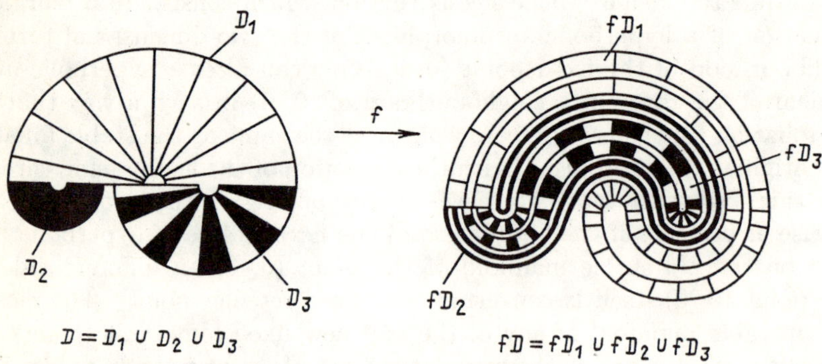

Fig. 5

The diffeomorphism f acts on the fundamental group $\pi_1(D, X_0)$ which is isomorphic to a free group with generators $\sigma_1, \sigma_2, \sigma_3$:

$$f_*\sigma_1 = \sigma_1^{-1}\sigma_3^{-1}\sigma_1\sigma_3\sigma_1,$$
$$f_*\sigma_2 = \sigma_2\sigma_1^{-1}\sigma_3\sigma_1\sigma_2\sigma_1^{-1}\sigma_3^{-1}\sigma_1\sigma_2^{-1},$$
$$f_*\sigma_3 = \sigma_2\sigma_1^{-1}\sigma_3\sigma_1\sigma_2^{-1}.$$

It turns out that hyperbolic attractors of diffeomorphisms of the 2-sphere are indecomposable Wada continua which partition the sphere into $k \geq 4$ regions (Plykin 1974).

The partitioning of D into subregions D_1, D_2, D_3 in the example described above is a particular case of a general construction for partitioning regions containing hyperbolic attractors diffeomorphic to a 2-sphere into irreducible regions (Plykin 1984).

In the papers (Zhirov 1989) a classification is given of hyperbolic attractors of diffeomorphisms of the 2-sphere for which the rank of the one-dimensional Aleksandrov-Čech homologies are equal to 3 in terms of the matrix of the geometric intersections. These classifying results make essential use of the arithmetic properties of integer-valued unimodular matrices.

The examples described above admit a generalization to the case of so-called expanding attractors of codimension 1. In (Plykin 1984) there is a description of these attractors to within topological conjugacy.

Definition. A hyperbolic attractor (or repeller) Λ of a diffeomorphism f is said to be *expanding* (or *contracting*) if

$$\dim \Lambda = \dim W^{\mathrm{u}}(x) = k \text{ (or } \dim \Lambda = \dim \mathrm{W^s(x)}).$$

Let Λ be an expanding attractor of codimension 1 (that is, $\dim \Lambda = \dim M - 1$). We denote by $W^{\mathrm{u}}_{\delta}(x)$ (or $W^{\mathrm{s}}_{\delta}(x)$) the connected component of the intersection of the manifold $W^{\mathrm{u}}(x)$ (or $W^{\mathrm{s}}(x)$) with the δ-neighbourhood of the point x containing x.

Suppose that f is topologically mixing on Λ, and let $x_1, x_2 \in \Lambda$. We choose $\epsilon > 0$, $\delta > 0$ such that the neighbourhood $U(x_1) = \bigcup_{x \in W^{\mathrm{u}}_{\epsilon}(x_1)} W^{\mathrm{s}}_{\delta}(x)$ is homeomorphic to $\mathbb{R}^n$. Let $r > 0$ be sufficiently large so that

$$W^{\mathrm{s}}_r(x_2) \cap W^{\mathrm{u}}_{\epsilon}(x_1) \neq \emptyset.$$

We choose a direction on $W^{\mathrm{s}}_r(x_2)$ and in accordance with this direction we enumerate the points $z_1, z_2, \ldots, z_m$ of the successive intersections of $W^{\mathrm{s}}_r(x_2)$ with $W^{\mathrm{u}}_{\epsilon}(x_1)$. We choose the orientation in $U(x_1)$ induced by the tangent vector to $W^{\mathrm{s}}_r(x_2)$ at z_1 (which defines the direction of the curve $W^{\mathrm{s}}_r(x_2)$) and some $(n-1)$-frame of the tangent vectors to $W^{\mathrm{u}}_{\epsilon}(x_1)$ at the point z_1.

At each of the points z_i, $i > 1$, we have an n-frame induced by the tangent vector $W^{\mathrm{s}}_r(x_2)$ at this point, which is defined by the direction of $W^{\mathrm{s}}_r(x_2)$ and the $(n-1)$-frame of the tangent vectors to the manifold $W^{\mathrm{u}}_{\epsilon}(x_1)$ at this same point and has the same orientation as the $(n-1)$-frame at z_1. If all the frames occurring in this way define the same orientation of $U(x_1)$ and the validity of this assertion does not depend on the arbitrariness of the choice of the points $x_1, x_2 \in \Lambda$ and the numbers $\epsilon > 0, \delta > 0, r > 0$, then the attractor Λ is said to be *orientable*. An attractor that is not orientable is called *non-orientable*.

Orientability (or *non-orientability*) can also be defined for a contracting repeller of codimension 1.

Let Λ be an expanding attractor of a diffeomorphism $f : M \to M$ of codimension 1 and let $D(M\backslash\Lambda)$ be an accessible boundary of the region $M\backslash\Lambda$ (that is, the set of all points x of the boundary of $M\backslash\Lambda$ for which there exists an open arc belonging to $M\backslash\Lambda$ based on the point x). The set of connected components $C_1, \ldots, C_m$ of $D(M\backslash\Lambda)$ is called a *bunch* of degree m (or an *m-bunch*) if C_1 and C_m, and also C_i and C_{i+1}, $i \in \{1, \ldots, m-1\}$, can be joined by open arcs of the one-dimensional stable manifolds contained in $M\backslash\Lambda$.

We note that each connected component of $D(M\backslash\Lambda)$ is homeomorphic to $\mathbb{R}^{n-1}$.

In the case when $M = n \geq 3$, the degrees of the bunches entering in $D(M\backslash\Lambda)$ do not exceed two, while in the case of an oriented attractor they are equal to two exactly.

We consider a set Γ (or $\tilde{\Gamma}$) of pairs (A, P) (or triples (A, P, Θ)), where A is a hyperbolic automorphism of the n-dimensional torus T^n, $n \geq 2$, in the spectrum of which the unique eigenvalue is less than unity in modulus, P is a finite A-invariant subset of the torus, Θ is a linear involution of the same torus that commutes with A and has a finite set of fixed points $\mathrm{Fix}\Theta$, which for the case of a triple occurs in P. We introduce in Γ (or $\tilde{\Gamma}$) an equivalence relation by setting $(A_1, P_1) \sim (A_2, P_2)$ (or $A_1, P_1, \Theta_1) \sim (A_2, P_2, \Theta_2)$) if there exists a linear map $\Psi : T^n \to T^n$ such that

$$A_2 = \Psi A_1 \Psi^{-1},\ P_2 = \Psi P_1 \text{ (or } \mathrm{A}_2 = \Psi \mathrm{A}_1 \Psi^{-1},\ \mathrm{P}_2 = \Psi \mathrm{P}_1,\ \Theta_2 = \Psi \Theta_1 \Psi^{-1}).$$

The equivalence class of the pair (A, P) (or the triple (A, P, Θ)) is called the *class of the orientable* (or *non-orientable*) attractor.

Classification Theorem (Plykin 1984). *With every orientable (or non-orientable) expanding attractor Λ of codimension 1 of a diffeomorphism $f : M \to M$, $\dim M \geq 2$, such that the accessible boundary of $D(M\backslash\Lambda)$ consists of bunches of degrees not exceeding 2, there is associated a class of an orientable (or non-orientable) attractor. The restrictions to the domains of attraction of the diffeomorphisms f_1, f_2 acting on manifolds of the same dimension $n \geq 2$ and having expanding attractors of codimension 1 with accessible boundaries with the properties mentioned above are topologically conjugate if and only if the classes of attractors are the same. For each class there is a manifold M and a diffeomorphism $f : M \to M$ possessing an attractor of this class.*

The question of the classification of expanding attractors of codimension 1 to within homeomorphism has not been solved in any dimension. It is not even known whether there exist two non-homeomorphic expanding attractors with the same orientability and the same Čech homologies.

It is known (Plykin 1984) that the existence on a manifold of a cascade with an expanding attractor of codimension 1 imposes restrictions on the manifold. In particular, its fundamental group must be infinite.

If the nonwandering set of a dynamical system induced by a diffeomorphism $f : M \to M$, $\dim M \geq 3$, consists only of orientable attractors and repellers of codimension 1, then the manifold M is a connected sum of a certain number of tori (Mamaev and Plykin, see (Plykin 1990)).

1.3. One-dimensional Attractors. Smale noted that the so-called Vietoris-van-Danzig solenoids can be endowed with a hyperbolic structure and converted into strange attractors of smooth dynamical systems for which the restriction of the dynamical system to the attractor is topologically similar to a shift map of the solenoid (Smale 1967).

We describe an example of an expanding map of given type as a limit of an inverse spectrum of circles

$$S^1 \leftarrow S^1 \leftarrow S^1 \leftarrow \ldots \leftarrow S^1 \leftarrow \ldots$$

with projections $g : S^1 \to S^1$ of the form $g(z) = z^k,\ k \geq 2$, where

$$S^1 = \{z \in \mathbb{C} : |z| = 1\}.$$

If we represent the points of such a solenoid Ξ by sequences $(z_1, z_2, \ldots, z_n, z_{n+1}, \ldots)$, $z_n = z_{n+1}^k$, then the restriction to Ξ of the diffeomorphism constructed by Smale is topologically similar to the shift map $G : \Xi \to \Xi$, where

$$G(z_1, z_2, \ldots, z_n, \ldots) = (gz_1, z_1, z_2, \ldots, z_n, \ldots).$$

In a series of papers (Williams 1967), (Williams 1970a), (Williams 1970b), (Williams 1974), it was proved by Williams that expanding attractors of smooth dynamical systems are *generalized solenoids*, that is, limits of an inverse spectrum of maps of complexes endowed with a smooth structure. Here the restrictions of the dynamical systems to the attractors are topologically similar to shift maps of generalized solenoids.

In the paper (Williams 1967) Williams gave a classification of one-dimensional expanding attractors represented in the form of generalized solenoids X_∞ consisting of sequences

$$(x_0, x_1, \ldots, x_n, x_{n+1}, \ldots),\ x_n = f(x_{n+1}),$$

where the element x_n belongs to some complex X endowed with a smooth structure, the so-called branch manifold, on which a map $f : X \to X$ is defined such that some iterate of it takes a sufficiently small neighbourhood of each branch point into an interval.

In the same paper Williams proves that the shift maps $f_\infty : X_\infty \to X_\infty$, $h_\infty : Y_\infty \to Y_\infty$ of two generalized solenoids are topologically similar if and only if there exist continuous maps $p : X \to Y$, $q : Y \to X$ and a positive integer m such that the diagrams

$$\begin{array}{ccc} X & \xrightarrow{f} & X \\ {\scriptstyle p}\downarrow & & \downarrow{\scriptstyle p} \\ Y & \xrightarrow{h} & Y \end{array} \qquad \begin{array}{ccc} X & \xrightarrow{f} & X \\ {\scriptstyle q}\uparrow & & \uparrow{\scriptstyle q} \\ Y & \xrightarrow{h} & X \end{array} \qquad \begin{array}{ccc} X & \xrightarrow{f^m} & X \\ {\scriptstyle p}\downarrow & \overset{q}{\nearrow} & \downarrow{\scriptstyle p} \\ Y & \xrightarrow{h^m} & Y \end{array}$$

are commutative. In this case the maps $f: X \to X$ and $h: Y \to Y$ are said to be s-equivalent ($f \underset{s}{\sim} h$).

Let $x_0 \in X$, $y_0 \in Y$ be fixed points of the maps $f: X \to X$, $h: Y \to Y$ and $p(x_0) = y_0$. If $f \underset{s}{\sim} h$, then the induced homomorphisms of the fundamental groups $f_*: \pi_1(X, x_0) \to \pi_1(X, x_0)$, $h_*: \pi_1(Y, y_0) \to \pi_1(Y, y_0)$ are also s-equivalent. If X and Y are bouquets of circles, then according to Williams, $f_* \underset{s}{\sim} h_*$ implies $f \underset{s}{\sim} h$. In (Ustinov 1987) Ustinov constructed algebraically invariant s-equivalences and applied this to show the absence of topological similarity of shift maps of two generalized solenoids induced by the two maps f and h, described below, of a bouquet of two circles; and this is how the problem posed by Williams in (Williams 1978a) was solved.

Each of the maps f and h of the bouquet of circles $K = S_1 \vee S_2$ is similar to a map that is a uniform expansion on each of the circles of the bouquet S_i, $i = 1, 2$. In addition, the fixed common point O of the bouquet is preserved and the map is uniquely defined by its action on the fundamental group $\pi_1(K, O)$, which is a free group with two generators. The actions of Williams's maps on the fundamental group with generators a and b are defined as follows: $f_*(a) = a^2b^2a$, $f_*(b) = a$, $h_*(a) = ababa$, $h_*(b) = a$.

We give a brief description of a version of the theory of the solenoidal representation of expanding attractors (Plykin 1984) which differs from Williams's original theory by doing away entirely with the smoothness assumptions on the branch manifolds. These assumptions are extremely burdensome in the verification and implementation of the constructions of the attractors.

Definition. We say that a connected compact set K is endowed with a *disc structure* if the following conditions hold:

1) there exist maps $h_i: D \to K_i \subset K$ that are homeomorphisms of the open n ball D onto its image K_i with $K = \bigcup_{i=1}^{e} K_i$;

2) each point $\xi \in K$ has a basis of neighbourhoods, each of which is a finite (at most l) union of disc-like images of open n-balls lying in D, under the action of the maps h_i, $i \in \{1, 2, \dots, l\}$.

Definition. A point ξ of a compact set endowed with a disc structure is called *regular* if there exists a basis of neighbourhoods of it, each element of which is a disc. Otherwise the point ξ is called a *branch point.*

The set of regular points is open and the set of branch points is closed.

Definition. By a *branch manifold* we mean a triple (K, Δ, g), where K is a connected compact set endowed with a disc structure, Δ is a metric compatible with the topology of K, and $g: K \to K$ is a continuous map; moreover there exist constants $\epsilon > 0$, $\lambda > 1$ such that the following conditions hold:

1°. Each disc of diameter not exceeding $\epsilon > 0$ occurring in the basis of neighbourhoods of some point of K is the g-image of a disc of some basis neighbourhood.

2°. The restriction of g to a disc of diameter not exceeding ϵ is an expansion with expansion coefficient not exceeding λ and the image of such a disc is situated in some disc.

3°. There exists an integer $m > 0$ such that the g^m-image of a basis neighbourhood of a point composed of discs of diameter not exceeding ϵ is situated in a disc of some basis neighbourhood.

4°. For each open set $U \subset K$ there is an integer $k \geq 0$ such that $g^k U = K$.

It follows from the definition of the map $g : K \to K$ that it is of finite multiplicity and that the periodic points are dense in K.

By the *generalized solenoid* Ξ induced by the triple (K, Δ, g) we mean the sequences $(x_0, x_1, \ldots, x_n, x_{n+1}, \ldots)$, where $x_n = gx_{n+1} \in K$; as before, the shift map $G : \Xi \to \Xi$ is defined by the formula

$$G(x_0, x_1, \ldots, x_n, \ldots) = (gx_0, x_0, x_1, \ldots, x_n, \ldots).$$

We have the following results.

Theorem on the Local Structure of a Direct Product. *There exists a basis of neighbourhoods for each point of Ξ that are homeomorphic to the product of a Cantor discontinuum and a disc. The periodic points of the map G are dense in Ξ.*

Theorem on the Representability of Expanding Attractors as Generalized Solenoids. *Each expanding attractor is homeomorphic to a generalized solenoid. Furthermore, the restriction to the attractor of the diffeomorphism f inducing the attractor is topologically similar to a shift map of the generalized solenoid.*

We give two theorems which provide methods for constructing one-dimensional generalized solenoids.

Theorem A. *Let $K = \vee_{i=1}^n S_i$ be a bouquet of n oriented circles of the same circumference endowed with the standard metric and inducing a metric Δ on K. Let $\pi_1(K, O)$ be the fundamental group of K with system of generators $s_1, s_2, \ldots, s_n$ and distinguished point O, the common point of all the circles of the bouquet. The map $g : K \to K$ is a uniform expansion on each circle S_i ($i \in \{1, 2, \ldots, n\}$), preserves the fixed point O and is uniquely defined by its action on the fundamental group of the form*

$$g^* s_i = s_1 h_i i s_2, \; i \in \{1, 2, \ldots, n\},$$

where the words h_i are formed from letters of the alphabet $\{s_1, s_2, \ldots, s_n\}$ and for any s_i, s_j there exists an integer $m > 0$ such that the word $g_*^m s_i$ contains s_j. The triple $\{K, \Delta, g\}$ defines a one-dimensional branch manifold.

Theorem B. *Let $K = \vee_{i=1}^n S_i$ be a bouquet of n circles endowed with the same metric as in Theorem* A. *We partition the set $\mathfrak{S}$ of letters s_1, s_1^{-1}, s_2, s_2^{-1}, …, s_n, s_n^{-1} formed from the generators of $\pi_1(K, O)$ into two disjoint*

subsets $\mathfrak{S}^1, \mathfrak{S}^2$ *such that the inclusions* $s_i \in \mathfrak{S}^1$(*or* $s_i \in \mathfrak{S}^2$) *entail the inclusions* $s_i^{-1} \in \mathfrak{S}^1$ (*or* $s_i^{-1} \in \mathfrak{S}^2$) *and, in addition, one of the following two possibilities holds:* 1) $s_1 \in \mathfrak{S}^1$, $s_2 \in \mathfrak{S}^2$; 2) $s_1 \in \mathfrak{S}^2$, $s_2 \in \mathfrak{S}^1$.

The map $g : K \to K$ fixed at the point O, which is a uniform expansion on each circle S_i, $i \in \{1, 2, \ldots, n\}$, is uniquely defined by indicating its action on $\pi_1(K, O)$ as follows:

$$g_* s_i = \begin{cases} \alpha h_i \alpha^{-1}, & s_i \in \mathfrak{S}^1, \\ \beta h_i \beta^{-1}, & s_i \in \mathfrak{S}^2, \end{cases}$$

where one of the following possibilities holds for the pair α, β:

$$\alpha = s_{l_1}^{k_1}, \quad \beta = s_{l_2}^{k_2}, \quad l_1 \neq l_2, \quad l_1, l_2 \in \{1, 2\}, \quad k_1, k_2 \in \{-1, 1\}.$$

Adjacent letters of each word $g_* s_i$ ($i \in \{1, \ldots, n\}$) belong to different sets $\mathfrak{S}^1, \mathfrak{S}^2$ and for any s_i, s_j there exists an integer $m_i \geq 0$ such that the word $g_*^{m_i} s_i$ contains either s_j or s_j^{-1}.

The triple (K, Δ, g) defines a branch manifold.

Additional information on the solenoidal representation of expanding attractors of diffeomorphisms of the 2-sphere is contained in the following two theorems.

Fedotov's Theorem (Fedotov 1980). *Let* K *be a bouquet of* n *circles and* $g : K \to K$ *a map leaving the common point of the bouquet fixed. For the realization of the limit of the inverse spectrum*

$$K \overset{g}{\leftarrow} K \overset{g}{\leftarrow} K \leftarrow \ldots \leftarrow K \leftarrow \ldots$$

as a hyperbolic attractor of a diffeomorphism of the sphere, it suffices that the following conditions hold:

1. $g_* : \pi_1(K, O) \to \pi_1(K, O)$ *is an endomorphism.*
2. *For some* m, $1 \leq m < n$, *we have*

$$g_* s_i = \begin{cases} (s_1 \alpha_i) s_{\pi(i)} (s_1 \alpha_i)^{-1}, & 1 \leq i \leq m; \\ (s_n^{-1} \alpha_i) s_{\pi(i)} (s_n^{-1} \alpha_i)^{-1}, & m < i \leq n; \end{cases}$$

here, s_i, $i \in \{1, 2, \ldots, n\}$, *are generators of* $\pi_1(K, O)$, $\alpha_i \in \pi_1(K, O)$ *and* $(\pi(1), \pi(2), \ldots, \pi(n))$ *is a permutation.*

3. *The conditions of Theorem* B *hold for* g.
4. $g_*(s_1 s_2 \ldots s_n) = s_1 s_2 \ldots s_n$.

Theorem on the Solenoidal Representation of Hyperbolic Attractors of a Diffeomorphism of the 2-Sphere (Plykin 1984). *Every one-dimensional hyperbolic attractor* Λ *of a diffeomorphism* f *of the 2-sphere* S^2 *for which the rank of its one-dimensional Aleksandrov-Čech cohomology group is equal to* n *is a generalized solenoid induced by a map of a bouquet of* n *circles satisfying the conditions of Theorem* B. *Here the shift of the solenoid is topologically similar to some iterate of* $f|\Lambda$.

We note that for one-dimensional hyperbolic attractors of diffeomorphisms of the 2-sphere the number n is greater than 3.

§2. The Lorenz Attractor

In (Lorenz 1963) Lorenz carried out a numerical investigation of the system

$$\begin{aligned}\dot{x} &= -\sigma x + \sigma y,\\ \dot{y} &= rx - y - xz,\\ \dot{z} &= -bz + xy,\end{aligned}$$

which he borrowed from Saltzman (Saltzman 1962). Lorenz discovered that for $\sigma = 10$, $b = 8/3$, $r = 28$, the trajectories of the system have clearly reflected features of stochastic behaviour.

For several years this paper did not attract the attention of people dealing with numerical experimentation or theoretical mathematicians. The turning point occurred in the middle of the 70s; at that time the ground had already been prepared for a theoretical interpretation of Lorenz's results.

As Williams writes in (Williams 1977), "several years ago Yorke carried out some computations with Lorenz's equations which attracted the interest of other mathematicians. He gave several lectures on this .... Ruelle, Lanford, Guckenheimer, who took an interest in this, attempted to study these equations." To this it can be added that the papers (McLaughlin and Martin 1974), (McLaughlin and Martin 1975) aroused further interest in (Lorenz 1963). In these two papers results of (Lorenz 1963) and more recent data of numerical experiments were broadly compared with theoretical ideas in the theory of smooth DSs.

We give some geometric and ergodic properties of the Lorenz attractor (that is, an attractor having a Lorenz system) and attractors similar to it.

Our description of the properties is not complete. We do not give a complete description of bifurcations, we do not describe symbolic dynamics, and we do not give properties relating to dimensional characteristics.

2.1. Physical Models Leading to a Lorenz System. First we describe the original model, which leads to an open Lorenz system. We consider the convective motion of a fluid that occurs between two parallel horizontal planes under the assumption that the temperature difference between the upper and lower planes is kept constant.

The motion of the fluid in this layer is described by a system of two partial differential equations (Rayleigh-Bénard problem) (see Saltzman 1962). If we assume that the solution does not depend on the y coordinate, then by representing it as Fourier series in the variables x and z and cutting off these series so that only three terms remain, then it can be shown that these remaining coefficients of the series satisfy a Lorenz system, which is therefore a Galerkin approximation of the problem.

There exist a whole series of other models of hydrodynamical origin (Brindley and Moroz 1980), (Gibbon and McGuines 1980), (Pedlosky and Frenzen 1980) that reduce to a Lorenz system by means of a similar procedure, that is, by passing to a Galerkin approximation. Of course, we still have the problem on the extent to which the properties of the Galerkin approximation are related to the properties of the original system. However, there also exist models that reduce to a Lorenz system via another route. For example, convection in a circular tube placed in a vertical position and heated from below is described by a system of integro-differential equations (Welander 1967). The infinite-dimensional dynamical system induced by these equations has a three-dimensional invariant attractive subspace. It can be proved that on this subspace the dynamical system is conjugate to the flow induced by the system of equations (1).

We will go into one of the models in a little more detail. It is of interest in that a Lorenz system is obtained from it by a direct route, and also because there is a physical device described by this model enabling one to visualize chaotic oscillations in a Lorenz system (Lorenz 1979).

Suppose that we have a wheel that can rotate about a horizontal axis. The rim of the wheel is made of a porous material which absorbs water. The distribution of the mass along the rim is given by the function $m(\theta, t)$, where t is time and θ the angular coordinate of a point on the rim. We suppose that the rate of inflow of water at a given point is proportional to the height of the point above some fixed level (for example, the water enters only at the upper point of the rim at a constant rate). Suppose that the loss of water proceeds at a rate proportional to m. Taking friction into account, the motion of the wheel is described by the system of equations

$$\frac{d(a^2\bar{m}\Omega)}{dt} = -ga\overline{m\cos\theta} - ka^2\bar{m}\Omega,$$

$$\frac{\partial m}{\partial t} + \Omega\frac{\partial m}{\partial \theta} = (A + 2B\sin\theta) - hm,$$

where Ω is the angular velocity, g, a, k, A, B, h are constants and $\bar{m}$ and $\overline{m\cos\theta}$ are the means with respect to θ.

Since the total rate of influx of the water is constant and the rate of loss is proportional to the total mass, it follows that the total mass asymptotically approaches the value $2\pi\frac{A}{h}$. If we assume that the mass attains this value, then the first equation takes on the form

$$\frac{d\Omega}{dt} = -k\Omega - \frac{gh}{2\pi aA}\overline{m\cos\theta}.$$

But

$$\frac{d\overline{m\cos\theta}}{dt} = \int_0^{2\pi} \frac{d}{dt}(m\cos\theta)d\theta = \int_0^{2\pi}\left[\left(\frac{\partial m}{\partial t} + \Omega\frac{\partial m}{\partial \theta}\right)\cos\theta - \Omega m\sin\theta\right]d\theta$$
$$= \int_0^{2\pi}(A\cos\theta + 2B\sin\theta\cos\theta - hm\cos\theta - \Omega m\sin\theta)d\theta$$
$$= -h\overline{m\cos\theta} - \Omega\overline{m\sin\theta}.$$

The quantity $d(\overline{m\sin\theta})/dt$ is calculated in similar fashion. The system

$$\frac{d\Omega}{dt} = -k\Omega - \frac{gh}{2\pi aA}\overline{m\cos\theta},$$
$$\frac{d\overline{m\cos\theta}}{dt} = -\Omega\overline{m\sin\theta} - h\overline{m\cos\theta},$$
$$\frac{d\overline{m\sin\theta}}{dt} = \Omega\overline{m\cos\theta} - h\overline{m\sin\theta} + 2\pi B$$

can be reduced by a linear change of variables to a Lorenz system with $b = 1$.

Finally one must not forget that certain models of lasers reduce to a Lorenz system (Haken 1975), and also disc dynamos (Knobloch 1981), (Robbins 1977). Thus, as is clear, it is not only the strange attractor in general that has a bearing on a fairly wide circle of physical problems, but also the specific case of the Lorenz attractor.

2.2. Bifurcation in a Lorenz System. We describe a sequence of bifurcations in a Lorenz system under a change of the parameter r which leads to the appearance of a strange attractor. The values of the constants σ and b are fixed: $\sigma = 10$, $b = 8/3$.

For $0 < r < 1$ the Lorenz system has a unique equilibrium point, namely the origin, which is globally stable, that is, all the trajectories are attracted to the point O $(0,0,0)$. The eigenvalues of the system linearized near the point O are real and negative. As r increases, one of the eigenvalues passes through 0 at the instant $r = 1$. Thus the origin loses its stability and two other stable equilibrium points O_1 and O_2 branch from it; these points move away from the starting point O as r increases. The unstable manifolds of O, which are separatrices Γ_1 and Γ_2, wind round O_1 and O_2 respectively provided that r does not exceed the value $r_1 \approx 13.926$. When $r = r_1$, both separatrices become doubly asymptotic to a saddle, that is, two homoclinic trajectories occur. When r passes through the value r_1, each separatrix gives rise to periodic trajectories L_1 and L_2, respectively. The separatrix Γ_1 now converges to the equilibrium point O_2, and the separatrix Γ_2 to the equilibrium point O_1. Furthermore, with the birth of the cycles L_1 and L_2 there appears an invariant limit set Ω which is unstable and whose trajectories are in one-to-one correspondence with a set of doubly infinite sequences of two symbols, the periodic trajectories of saddle type corresponding to periodic sequences. In this interval of values of the parameter r, so-called "metastable chaos" is observed (Yorke and Yorke 1979); these are trajectories that start in a sufficiently small neighbourhood of points of Ω, behave chaotically for a while by performing irregular oscillations, but then these oscillations quickly die down. We note that there also exist trajectories performing chaotic oscillations over an infinite interval of time; however, the measure of the set of such trajectories is equal to zero. This "preturbulent" regime is maintained during the interval $r \in (r_1, r_2)$, where $r_2 \approx 24.06$. The value of r_2 is characterized by the property that if earlier Γ_1 and Γ_2 were asymptotic to the stable foci O_2 and O_1, then when $r = r_2$ they now converge to the saddle trajectories L_1 and L_2. At the same time, the "Lorenz attractor" proper arises, that is, a stable limit set Σ to which all the trajectories of some neighbourhood of it are attracted. The properties of Σ are described below.

However, for $r_2 < r < r_3$, where $r_3 \approx 24.74$, there are two other attractors apart from Σ in the phase space of the system, namely, the fixed points O_1 and O_2. As r passes through the bifurcation value r_3 the points O_1 and O_2 lose their stability, and when $r \in (r_3, 28]$ the Lorenz attractor is the only attracting set in the phase space of the system (1).

It needs to be pointed out that the bifurcations described above and the Lorenz attractor arising in connection with them are not the only ones worthy of attention. For $\sigma = 10$, $b = 8/3$ the system exhibits very varied and interesting properties when r is large. Thus, as the parameter r is decreased from 100.795 to 99.524 one observes an infinite series of bifurcations of period doubling of the stable periodic orbit, analogous to the series of bifurcations of period doubling in the Feigenbaum family of one-dimensional maps $f_\mu = 1 - \mu x^2$. There are also other intervals of variation of the parameter r in which one can observe successive bifurcations of period doubling (Sparrow 1982).

One further interesting phenomenon is observed when $r \approx 166.3$, namely, "intermittent chaos", that is, intervals of regular behaviour of the trajectories alternating with periods of chaotic motion.

2.3. First-Return Map. To study the flow induced by a Lorenz system it is convenient to go over to the first-return map T which arises in some region D on the plane Π defined by the equation $z = r - 1$. The points O_1, O_2 lie in the plane Π. The trajectories of the flow are tangent to the plane Π at the curve Q defined by the equation $xy = b(r-1)$. The curve Q has two branches: one branch is situated in the region $(x > 0,\ y > 0)$ and the other in the region $(x < 0,\ y < 0)$. When $r \in [r_3, 28]$ we can choose for D the region bounded by the branches of the curve Q and the lines $x - y = c$, $x - y = -c$ for a suitable choice of the constant $c > 0$. If, on the other hand, $r \in (r_2, r_3)$, then we can choose for D the region of the plane Π bounded by the lines $x - y = c$, $x - y = -c$ and the stable manifolds of the cycles L_1, L_2.

It is not difficult to show that if $p \in D$, then Tp is a second point of intersection of the half-trajectory $\{S_t p,\ t > 0\}$ with Π (the first point of intersection lies outside D).

Computer calculations show that it is possible to choose new coordinates in the region D (which we still denote by x, y; no confusion will arise since we shall no longer be using the old coordinates) such that the following conditions hold:

(L.1) In the new coordinates the region D is the square

$$D = \{(x,y) : |x| < 1,\ |y| < 1\}.$$

(L.2) The map T is defined at all points of D except on the line S defined by the equation $x = 0$; the line S is the line of intersection of the stable manifold of O and the region D.

The line S divides D into the two subregions

$$D_1 = \{(x,y) \in D : x > 0\},$$
$$D_2 = \{(x,y) \in D : x < 0\}.$$

(L.3) The map T is smooth in each of the subregions D_1, D_2 (it is even analytic; we shall merely use the fact that it is smooth of class C^2).

(L.4) The map T is defined by functions $f(x,y),\ g(x,y)$ such that

$$T(x,y) = (f(x,y), g(x,y)).$$

The following conditions hold for the functions $f(x,y),\ g(x,y)$:

a) $\|f_x^{-1}\| < 1$;

b) $\|g_y\| < 1$;

c) $1 - \|f_x^{-1}\|\,\|g_y\| > 2\sqrt{\|f_x^{-1}\|\,\|f_y\|\,\|g_x f_x^{-1}\|}$;

d) $\|g_x f_x^{-1}\| \cdot \|f_y\| < (1 - \|f_x^{-1}\|)(1 - \|g_y\|)$

(here $f_x, g_y, \ldots$ denote the partial derivatives

$$\frac{\partial f}{\partial x}, \frac{\partial g}{\partial y}, \ldots; \quad \|\cdot\| = \sup_D |\cdot|\Big).$$

In what follows we shall denote the restrictions of the map T and the functions $f(x,y), g(x,y)$ to the region D_i, $i \in \{1,2\}$, by $T_i, f_i(x,y), g_i(x,y)$.

(L.5) In a neighbourhood of S the functions $f_i(x,y), g_i(x,y)$ have the form

$$f_i(x,y) = F_i(y|x|^\beta,\ |x|^\alpha),$$
$$g_i(x,y) = G_i(y|x|^\beta,\ |x|^\alpha),$$

where $F_i(u,v), G_i(u,v)$ are smooth functions in a neighbourhood of the origin; the quantities $0 < \alpha < 1$, $\beta > 0$ are given by

$$\alpha = |\lambda_2/\lambda_1|; \quad \beta = |\lambda_3/\lambda_1|$$

($\lambda_1, \lambda_2, \lambda_3$ are the eigenvalues of the linearized system in a neighbourhood of O and ordered so that $\lambda_3 < \lambda_2 < 0 < \lambda_1$).

It follows from (L.5) that there exist one-sided limits

$$p_1 = \lim_{x \to +0} T_1(x, y); \quad p_2 = \lim_{x \to -0} T_2(x, y)$$

that are independent of y. In other words, each of the maps T_1, T_2 can be extended by continuity to the line S, where

$$T_1(S) = p_1; \quad T_2(S) = p_2. \tag{1}$$

The behaviour of the phase trajectories of the Lorenz system is depicted in Fig. 6.

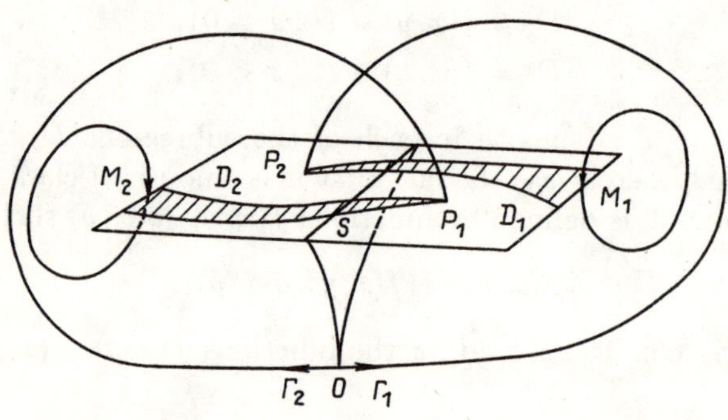

Fig. 6

As is clear, T is not quite the usual first-return map: we can either consider it on the non-compact set $D_1 \cup D_2$, or we can extend it to S, but there it is discontinuous, non-injective and non-single-valued (see (1)).

Definition. A map $T : D \backslash S \to D$ satisfying conditions (L.1) – (L.5) is called a *map of Lorenz type*.

The conditions (L.1) – (L.5) are the same as those featuring in (Afraimovich, Bykov, and Shil'nikov 1977), (Afraimovich, Bykov, and Shil'nikov 1982) (apart from a slight rephrasing). In certain other papers more restrictive conditions are imposed on T. There is no doubt concerning the existence of maps satisfying these more restrictive conditions and the existence of flows for which the first-return maps satisfy these conditions. Hence whenever we talk about the possibility in principle of certain phenomena being sorted out, the corresponding investigations will be completely meaningful from this point of view. However, we are not always sure that such strengthened conditions for the map T associated with the Lorenz system are satisfied.

We distinguish three types of Lorenz map: the orientable type, the semi-orientable type and the non-orientable type. In Fig. 7 we illustrate the images TD_i, $i = 1, 2$, for each of these three cases.

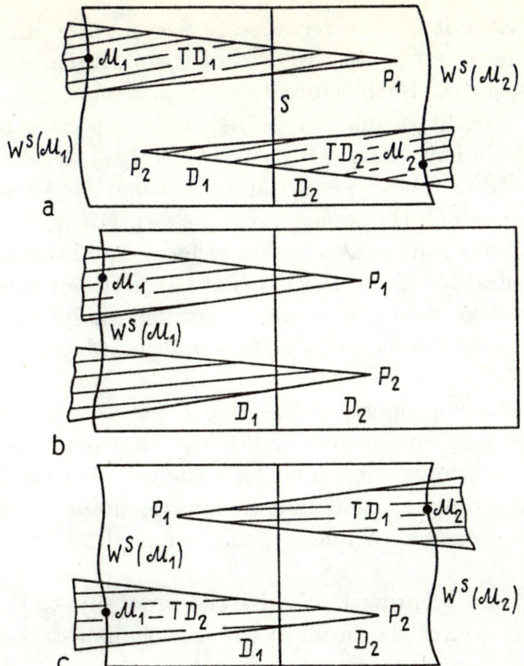

Fig. 7. First-return maps for systems of Lorenz type: a) orientable case, b) semi-orientable case, c) non-orientable case

For a system of Lorenz type with parameter values $\sigma = b = 8/3$, $r \in (r_2, 28)$, an orientable type is realized. A non-orientable type is realized for the Marioka-Shimizu model:

$$\dot{x} = y; \quad \dot{y} = -\lambda y + x - xz; \quad \dot{z} = -\alpha z + x^2$$

for $\alpha \cong 0.45$, $\lambda \cong 0.59$ (Shil'nikov 1986). Examples of systems of "physical" origin in which a semi-orientable type is realized are not known to the authors of this chapter. It is not difficult to construct artificially such a system.

Properties (L.1), (L.2), (L.3) follow from the form of the phase portrait of the flow depicted in Fig. 6. This form was established by means of numerical calculations on a computer.

The validity of properties (L.1) – (L.5) for a Lorenz system can be regarded as fairly reliably established (Sinai and Vul 1981). For all that, there is not complete confidence in them.

In the paper (Robinson 1989) Robinson used a slightly modified system of Rychlik equations (Rychlik 1990) to show that for the system

$$\dot{x} = y; \quad \dot{y} = x - 2x^3 + \alpha y + \beta x^2 y - \nu yz; \quad \dot{z} = -\gamma z + \delta x^2$$

there exists an open set of parameter values for which there exists an attractor of Lorenz type. The advantage of this system is that conditions (L.1) – (L.5) can be verified without having recourse to a computer. In the main this is connected with the fact that in the present case bifurcations of codimension 2 are being investigated. Such bifurcations in a two-parameter family in general position (say, with parameters μ_1, μ_2) had been considered even earlier by Shil'nikov (Shil'nikov 1981). One parameter (μ_1) in this bifurcation corresponds to the value $(|\lambda_2/\lambda_1| - 1)$, and the other ($\mu_2$) corresponds to the side of $W^s_{\mathrm{loc}}(O)$ on which the separatrix Γ passes. For $\mu_2 = 0$ both separatrices are doubly asymptotic to O. Certain other inequalities must be satisfied; these require integration round the loops of the figure-of-eight curve. For the Robinson system these inequalities can be verified analytically. Here both the orientable and non-orientable cases can be realized, depending on the sign of ν.

We note one further property of the map T for a Lorenz system. It follows from the symmetry of the system under the change of variables $(x, y) \mapsto (-x, -y)$ that T is symmetric under this change of variables. We shall be considering both symmetric and non-symmetric maps satisfying properties (L.1) – (L.5). If T possesses symmetry, then we shall say that the symmetric case holds.

Property (L.5) can be proved. We give the proof due to Il'yashenko, which is an improved version of the proof of the analogous assertion in (Shil'nikov 1963).

According to (Arnol'd et al. 1986), in a neighbourhood of the point O the Lorenz system of equations reduces under a smooth transformation to the following normal form:

$$\dot{x}_0 = \lambda_1 x_0; \quad \dot{y}_0 = \lambda_3 y_0; \quad \dot{z}_0 = \lambda_2 z_0,$$

where x_0, y_0, z_0 are the new coordinates after the reduction has been carried out. Consider the first-return map of the area element $(z_0 = \epsilon,\ 0 < y_0 < \epsilon)$ to the area element $(x_0 = \epsilon, 0 < z_0 < \epsilon)$. A straightforward calculation shows that it is defined by the formulae

$$\bar{y}_0 = y_0 x_0^{\beta}; \quad \bar{z}_0 = x_0^{\alpha}.$$

We take new coordinates on the area element D in the neighbourhood of the line S by choosing coordinates of the form $x = Ax_0,\ y = By_0$, where A and B are suitably chosen constants.

The map associating the coordinates (x, y) on the plane D with the point $(\bar{y}_0, \bar{z}_0)$ is at the very least C^2-smooth, by virtue of the standard theorems on the smooth dependence of the solution of a differential equation on the initial conditions.

The following assertions are a consequence of (L.4):

(H) There exist constants $L > 0, \sigma > 0, q > 1, \lambda \in (0, 1)$ such that the following properties hold:

(H.0) $L\sigma < 1$.

(H.1) If I, the curve in D that is the graph of $y = H(x)$, does not intersect S, where $|H'(x)| < L$, then TI is a smooth curve that is the graph of the function $y = \tilde{H}(x)$ with $|\tilde{H}'(x)| < L$.

(H.2) If J is the curve in D that is the graph of the function $x = h(y)$, where $|h'(y)| < \sigma$, then $T^{-1}(J \cap TD_i)$ is also a smooth curve in D_i which is the graph of the function $x = \tilde{h}(y)$ with $|\tilde{h}'(y)| < \sigma$ $(i = 1, 2)$.

(H.3) (a) If the points p, p' lie on the same curve I satisfying the conditions in (H.1) and not intersecting S, then

$$\rho(Tp, Tp') > q\rho(p, p').$$

(b) If the points p, p' lie on the same curve J satisfying the conditions in (H.2) and p, p' lie in the same region TD_i, then

$$\rho(T^{-1}p, T^{-1}p') > \lambda^{-1}\rho(p, p').$$

The constant q in condition (H.3) is called the coefficient of expansion; the constant λ is called the coefficient of contraction (see Figs. 8, 9).

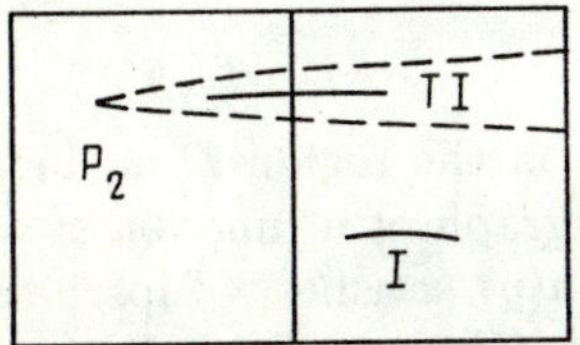

Fig. 8

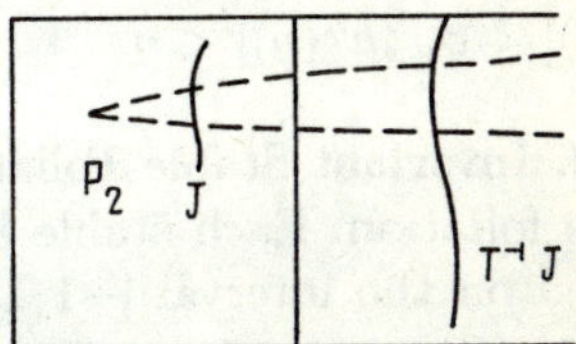

Fig. 9

Definition. By the limit set, or quasi-attractor, we mean the set

$$\Sigma = \text{clos} \bigcap_{n=1}^{\infty} T^n D,$$

where clos denotes the closure.

For some $p \in \Sigma$ the quantity $T(p)$ or $T^{-1}(p)$ may not be defined. But if it is defined, then it lies in Σ.

We consider as the tangent spaces $T_pD = \mathbb{R}^2$, $p \in D$, the cones $K_p^{\rm s}$ and $K_p^{\rm u}$ with vertical and horizontal axis spaces and angles $\tan^{-1} L$, $\tan^{-1} \sigma$. It follows from conditions (H) that they have properties analogous to those indicated in Chap. 1, Sect. 1.2, except that the maps T, T^{-1} and their "differentials" (in the terminology of Chap. 1 they are g, g^{-1}, G and G^{-1}) are not everywhere defined (so that the corresponding inclusions and inequalities must be understood as follows: if the objects featuring in them are defined, then these inclusions and inequalities are valid). As we shall see below, there is a T-invariant set $M \subset \Sigma$ for the points of which there exists a decomposition of the tangent space $\mathbb{R}^2$ into a sum of stable and unstable subspaces $E_p^{\rm s}, E_p^{\rm u}$ $(p \in M)$. However,

M is not closed and it has not been proved that condition (H) ensures the continuous dependence of the subspaces $E^{\mathrm{s}}_p, E^{\mathrm{u}}_p$ on the point p, although it is true that no counterexample has been constructed. We venture to make the following conjecture:

Conjecture. *There exists a map T of Lorenz type of class C^∞ for which there do not exist subspaces $E^{\mathrm{s}}_p, E^{\mathrm{u}}_p$ defined at each point $p \in \Sigma$, continuously dependent on p in Σ and invariant with respect to T.*

The validity of this conjecture for a map T is equivalent to that of the analogous conjecture for a flow.

In connection with this conjecture we note that even if a flow defined by a Lorenz system has a decomposition $\mathbb{R}^3 = E^{\mathrm{s}}_p \oplus E^{\mathrm{u}}_p \oplus E^0_p$ continuously dependent on p, it does not satisfy Axiom A since the fixed point O is not isolated in the set of nonwandering points.

One can guarantee the presence of an invariant stable subspace E^{s}_p that is continuously dependent on p in the set Σ under the following condition:

(H.4) There exists a constant η such that if the curve J is the graph of the function $x = h(y)$ with $|h'(y)| < \sigma$, $|h''(y)| < \eta$, then the curve $T^{-4}(J \cap TD_i)$, $i = 1, 2$, is the graph of the function $x = \tilde{h}(y)$ with $|\tilde{h}''(y)| < \sigma$, $|\tilde{h}''(y)| < \eta$.

2.4. Invariant Stable Foliation. There exists in the region D an invariant stable foliation. Each stable leaf $W^{\mathrm{s}}(p)$ is the graph of a function $x = h(y)$ defined on the interval $[-1, 1]$.[1] The function $h(y)$ satisfies a Lipschitz condition with constant $\sigma > 0$ defined in condition (H.2). If $Tp' \in W^{\mathrm{s}}(p)$, then $T(W^{\mathrm{s}}(p')) \subset W^{\mathrm{s}}(p)$.

For $p \in D$ the leaf $W^{\mathrm{s}}(p)$ can be constructed in the following way.

(a) If $T^{\tilde{n}}p \in S$ for some $\tilde{n} \in \mathbb{Z}^+$, then $W^{\mathrm{s}}(p)$ is the same as the connected component of $T^{-\tilde{n}}S$ containing the point p.

(b) Suppose that $T^n p \notin S$ for any $n \in \mathbb{Z}^+$. The set $T^{-n}S$ divides D into a certain number of subregions; we denote the one containing the point p by $D_n(p)$. Clearly $D_1(p) \supset D_2(p) \supset \ldots$.

Since T expands in the horizontal direction, it follows that the set $W^{\mathrm{s}}(p) = \bigcap_{n=1}^{\infty} D_n(p)$ is a curve that is the graph of a function $x = h(y)$ defined on $[-1, 1]$ (see Fig. 10).

The Lipschitz property of the stable leaves follows from the following trivial assertions:

Assertion 1. *Each connected component of the set $T^{-n}S$ is a smooth curve which is the graph of a function $x = h(y)$, where $|h'(y)| < \sigma$.*

Assertion 2. *Each leaf $W^{\mathrm{s}}(p)$ is either a connected component of $T^{-n}S$ or the limit in the C^0-topology of curves of such a form.*

[1] We have no way of extending the stable leaf to the boundary of the region $|y| \leq 1$, so that $W^{\mathrm{s}}(p)$ is less than the stable set in the sense of Chap. 1, Sect. 1.2. (For this reason, the notation is slightly different.)

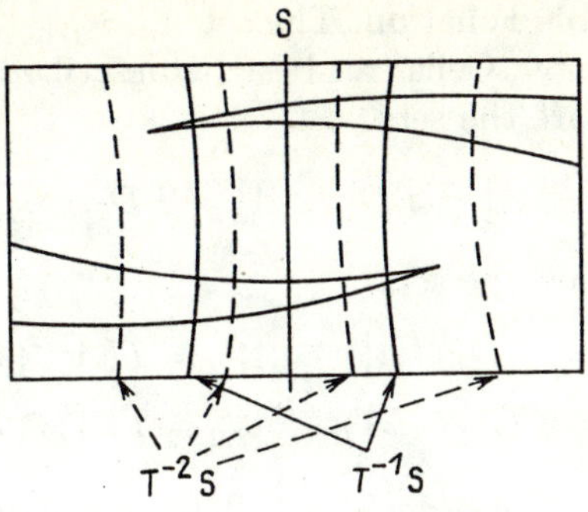

Fig. 10

If the points p, p' lie on the same stable leaf, then

$$\rho(Tp, Tp') < \lambda\rho(p, p').$$

If condition (H.4) holds, then each leaf $W^s(p)$ is smooth.

By virtue of the presence of an invariant stable foliation, one can factor the region D with respect to this foliation and go over to the map of an interval. More precisely, we identify all the points of D lying in the same stable leaf W^s with the point p that is the point of intersection of W^s with the interval $I = \{(x, y) : y = 0, \ |x| \leq 1\}$. The one-dimensional map $\hat{T} : I \to I$ maps each point $p \in I$ to the point $p' = W^s(Tp) \cap I$. The map $\hat{T}$ has a single point of discontinuity $x = 0$ and is continuous and monotone increasing on the half-intervals $[-1, 0)$ and $(0, 1]$. Furthermore the diagram

$$\begin{array}{ccc} D \backslash S & \xrightarrow{T} & D \\ \pi\downarrow & & \downarrow\pi \\ I\backslash\{0\} & \xrightarrow{\hat{T}} & I \end{array}$$

is commutative, where $\pi : D \to I$ is the projection along the stable manifold. Although the map T expands in the horizontal direction, the factored map $\hat{T}$ is not in general expanding in the standard Euclidean metric. Nevertheless, the map $\hat{T}$ is expanding in a certain specially chosen metric related to the map T. This enables one to carry over many of the dynamical properties of expanding maps of an interval to two-dimensional maps of Lorenz type. At the same time, results requiring smoothness on the intervals of monotonicity $[-1, 0)$ and $(0, 1]$ cannot be immediately carried over to such two-dimensional maps since the hypotheses that have been made merely guarantee the continuity of the stable foliation and hence of the maps $\hat{T}_1 : [-1, 0) \to I$, $\hat{T}_2 : (0, 1] \to I$. This is the main reason why the well developed ergodic theory of one-dimensional maps turns out to be practically useless when considering two-dimensional maps of Lorenz type (without, of course, taking into account its heuristic role).

2.5. Invariant Unstable Foliation. The set $\Lambda_n = \bigcap_{k=1}^{n} T^k D$ consists of subsets, called cells of order n. Cells can best be described by induction.

The cells of order 1 are the sets

$$\Lambda_1^1 = TD_1; \quad \Lambda_2^1 = TD_2.$$

The cells of order 2 are the sets

$$\Lambda_{1,1}^2 = T(D_1 \cap \Lambda_1^1); \quad \Lambda_{1,2}^2 = T(\Lambda_1^1 \cap D_2);$$
$$\Lambda_{2,1}^2 = T(\Lambda_2^1 \cap D_1); \quad \Lambda_{2,2}^2 = T(\Lambda_2^1 \cap D_2).$$

Let $\Lambda_{i_1,\ldots,i_{n-1}}^{n-1}$ be a cell of order $n-1$. Then a cell of order n is the set

$$\Lambda_{i_1,i_2,\ldots,i_n}^n = T(\Lambda_{i_1,i_2,\ldots,i_{n-1}}^{n-1} \cap D_{i_n}).$$

Some sets may be empty.

We denote the cell of order n containing the point p by $\Lambda^n(p)$.

Clearly cells of order n lie in the interior of cells of order $n-1$. Each cell of order $n-1$ contains at most two cells of order n.

All the cells, apart from two exceptional ones, have the form of spindles stretched along the x axis. The boundary of each cell, apart from the two exceptional ones, consists of two curves I_1, I_2; the ends of the curves are the points $T^{j_1}p_1, T^{j_2}p_2$ for some j_1, j_2; the maximum of the numbers j_1, j_2 is called the rank of the cell. The curves I_1, I_2 are tangent to each other at their end points. We will add the points $T^{j_1}p_1, T^{j_2}p_2$ to the cells.

In the orientable case the set of cells of order n comprises two "bunches"; one of them lies in the set TD_1 and the other in TD_2. Hence it follows that Σ consists of two connected components. The same is true of the non-orientable case. In the semi-orientable case the number of connected components of Σ is finite (see Afraimovich, Bykov and Shil'nikov 1982).

By the unstable leaf of a point $p \in \Sigma$ we mean the set

$$W^{\mathrm{u}}(p) = \bigcap_{n=1}^{\infty} \Lambda^n(p).$$

In view of the contraction in the vertical direction (condition (H.3)) the set $W^{\mathrm{u}}(p)$ is either a curve which is the graph of a function $y = H(x)$ satisfying a Lipschitz condition with constant L, or a point. If the lengths of the cells $\Lambda^n(p)$ are bounded below by a positive constant not depending on n or p, then each unstable leaf is a curve whose length is bounded below by the same constant. If, on the other hand, the lengths of the cells $\Lambda^n(p)$ can be arbitrarily small, then there has to be a leaf that degenerates to a point.

The geometric structure of an attractor can be described in the following way.

We define a graph whose vertices are the points $T^n p_1, T^m p_2$, $n, m = 0, 1, \ldots$. Two vertices $T^n p_i, T^m p_j$ are joined by an edge if there exists a leaf $W^{\mathrm{u}}(p)$ whose ends are the points $T^n p_i, T^m p_j$.

If the pair of vertices $T^n p_i, T^m p_j$ is joined by an edge of the graph, then there exist unstable leaves whose ends are the points $T^n p_i, T^m p_j$. We have the following alternative: either there exist only finitely many unstable leaves with ends $T^n p_i, T^m p_j$, or the pair of points $T^n p_i, T^m p_j$ is joined by a "Cantor set" of leaves. The latter means that the union of all unstable leaves whose ends are the points $T^n p_i, T^m p_j$ is homeomorphic to the direct product of an interval and a Cantor set in which all the right hand ends of the intervals are glued to one point, and the left hand ends to another. It is not known whether the first of the alternatives is realizable.

It is comparatively easy to visualize the structure of the set of unstable leaves in the rational case, that is, when there exist numbers m, n such that $T^m p_1 \in S$, $T^n p_2 \in S$. In this case the number of vertices of the graph is finite and each pair of vertices is joined by a "Cantor set" of unstable leaves (provided there exists an edge of the graph joining these two vertices). A neighbourhood of points $T^j p_i$ in an attractor is homeomorphic to a cone over a Cantor set (Fig. 11). Accordingly, a neighbourhood of each point on the separatrix Γ_1 (or Γ_2) has the form of a "Cantor book", that is, it is homeomorphic to $F \times I'$, where F is a Cantor set and I' is an interval (see Williams 1977, Williams 1979).

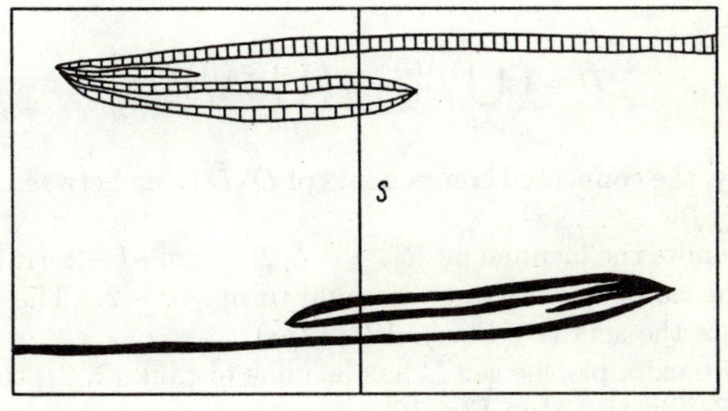

Fig. 11

2.6. Lacunae. We consider an example of a map of Lorenz type whose action on the regions D_1, D_2 is schematically illustrated in Fig. 12. This example is a slight modification of an example in (Afraimovich, Bykov, and Shil'nikov 1982).

In this example we have a special case of the following situation.

There is the set $\tilde{D} = \tilde{D}_1^{(0)} \cup \tilde{D}_2^{(0)} \cup S$; the boundary of the set $\tilde{D}_1^{(0)}$ is the line S and the manifold $W^s(T^k p_2)$ for some natural number k. The boundary of the region $\tilde{D}_2^{(0)}$ is the line S and the curve $W^s(T^m p_1)$ for some natural number m.

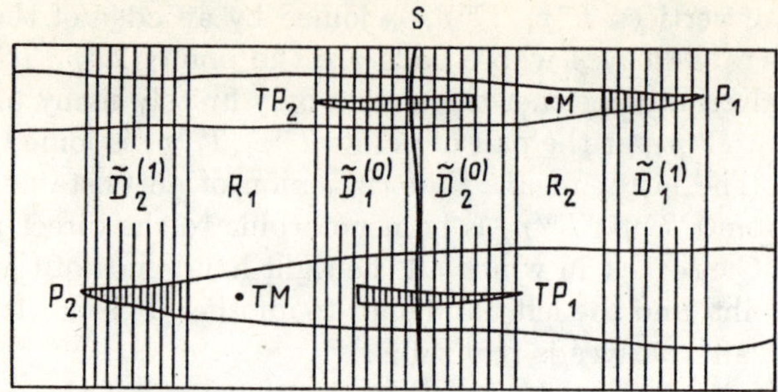

Fig. 12

We construct the regions $\tilde{D}_1^{(i)}$, $i = 1, \dots, m-1$, bounded by the curves $W^s(T^{i-1}p_1)$, $W^s(T^{k+i}p_2)$. Similarly, the regions $\tilde{D}_2^{(i)}$, $i = 1, \dots, k-1$, are bounded by the curves $W^s(T^{i-1}p_2)$, $W^s(T^{m+i}p_1)$. If it turns out that the regions $\tilde{D}_1^{(i)}, \tilde{D}_2^{(j)}$, $i = 0, \dots, m-1$; $j = 0, \dots, k-1$, are pairwise disjoint and

$$T^m \tilde{D}_1^{(0)} \subset \tilde{D}; \quad T^k \tilde{D}_2^{(0)} \subset \tilde{D},$$

then we say that the set Σ has lacunae of rank (m, k). The lacunae are "holes" in the set

$$\hat{D} = \Big(\bigcup_{i=0}^{m-1} \tilde{D}_1^{(i)}\Big) \cup \Big(\bigcup_{j=0}^{k-1} \tilde{D}_2^{(j)}\Big) \cup S,$$

more precisely, the connected components of $D \setminus \hat{D}$ lying between the manifolds $W^s(p_1)$, $W^s(p_2)$.

We shall denote the lacunae by R_j, $j = 1, 2, \dots, m+k-2$. (It is not difficult to see that the number of lacunae is equal to $m + k - 2$.) The boundaries of the lacunae are the sets $W^s(T^i p_1)$, $W^s(T^j p_2)$.

In the above example the set Σ has lacunae of rank (2, 2); the lacunae are the sets R_1, R_2 illustrated in Fig. 12.

In the example given in Fig. 13 there are four lacunae.

In the first example the lacunae contain the invariant set Σ^- consisting of the two points M, TM ($T^2M = M$). These two points are the only nonwandering points contained in the lacunae. In the second example the lacunae contain the composite invariant set Σ^- consisting of all the nonwandering points contained in the lacunae. The map T on the set Σ^- is hyperbolic, while the set Σ^- itself is locally maximal for the map T restricted to a neighbourhood of Σ^-. The points of Σ^- are in one-to-one correspondence with the sequences of the topological Markov chain described by the transition graph (Fig. 14). Fig. 15 illustrates an example of a map with three lacunae in the asymmetric case.

We give sufficient conditions under which there are no lacunae (in the theorems given below, q is the coefficient of expansion).

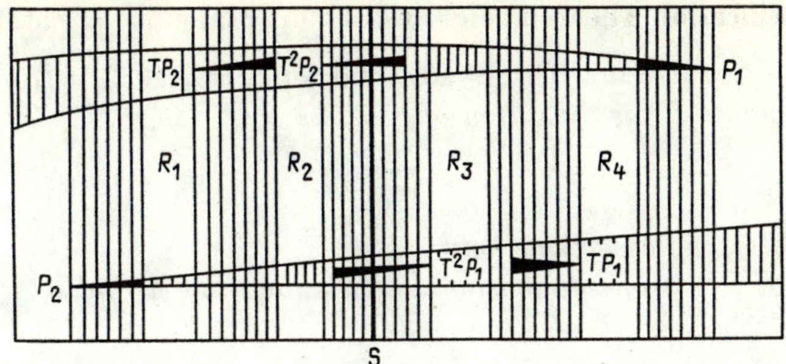

Fig. 13

Fig. 14

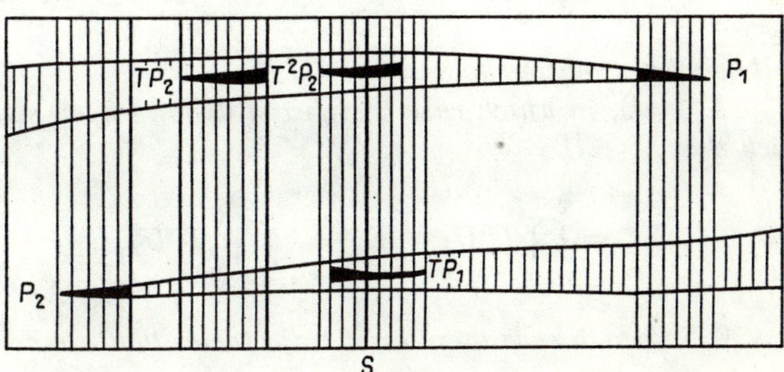

Fig. 15

Theorem 1. *If $q > \sqrt{2}$, then there are no lacunae.* (Afraimovich, Bykov, and Shil'nikov 1982).

Theorem 2. *Let N_1, N_2 be positive integers such that the points $p_1, T_{p_1}, \ldots, T^{N_1-1}p_1$ lie in D_2 and the points $p_2, \ldots, T^{N_2-1}p_2$ lie in D_1. If*

$$q^{N_1+N_2} > q^{N_1} + q^{N_2},$$

then there are no lacunae. (Afraimovich, Bykov, and Shil'nikov 1982).

In connection with Theorems 1, 2 we note that numerical calculations show that the coefficient of expansion for a map corresponding to a Lorenz system is close to unity (Sinai and Vul 1981).

2.7. Classification Theorem. Here we shall consider the orientable case.

Theorem (See Afraimovich, Bykov, and Shil'nikov 1982). *The set of non-wandering points Σ^0 is the union of two sets*

$$\Sigma^0 = \Sigma^+ \cup \Sigma^-,$$

where the following assertions hold:

(a) *the set Σ^+ is one-dimensional and is the same as the set*

$$\Sigma^+ = \operatorname{clos} \bigcap_{n=1}^{\infty} T^n \hat{D};$$

the periodic trajectories are everywhere dense in Σ^+; the restriction of T to Σ^+ is topologically transitive;

(b) *the set Σ^- is zero-dimensional and lies in the set $R = D \backslash \hat{D}$; the periodic trajectories are everywhere dense in Σ^-; the restriction of T to Σ^- satisfies Axiom* A; *the set Σ^- is the same as the set*

$$\Sigma^- = \bigcap_{n=-\infty}^{\infty} T^n R.$$

(c) *the following alternative holds:*

either $\Sigma^- \cap \Sigma^+ = \emptyset$, in which case there exist disjoint open sets $U_1 \supset \Sigma^+$, $U_2 \supset \Sigma^-$ such that

$$\Sigma^+ = \bigcap_{n=1}^{\infty} T^n U_1; \quad \Sigma^- = \bigcap_{n=-\infty}^{\infty} T^n U_2,$$

or $\Sigma^- \cap \Sigma^+ \neq \emptyset$, in which case there exist trajectories that are doubly asymptotic to Σ^+;

(d) *the induced map $\tilde{T}$ of the map T on the set $\tilde{D}^{(0)}$ has the attractor $\tilde{\Sigma} = \Sigma^+ \cap \tilde{D}^{(0)}$; the map $\tilde{T}$ is topologically mixing on $\tilde{\Sigma}$.*

In connection with (d) we explain that the derived map $\tilde{T}$ on the set A associates with the point $a \in A$ the point $T^i a$ that hits A for the first time $i > 0$. Topologically mixing means that for any subsets $A, B \subset \tilde{\Sigma}$ that are open in $\tilde{\Sigma}$ we have $T^n A \cap B \neq \emptyset$ for all sufficiently large n.

According to (Sataev 1992), Σ^+ is a probabilistic attractor. It is therefore natural to call the set Σ^+ an attractor of a map of Lorenz type or a planar Lorenz attractor. Similarly, it is natural to regard the set of orbits of a point $p \in \Sigma^+$ as an attractor of a flow.

If from the very beginning we take the region $\tilde{D}^{(0)}$ as the original area element, then the map T has no lacunae. Hence it follows that an attractor of a flow is always connected.

Example. Consider the map of the interval $[-1, 1]$ to itself defined by the function $f_a(x) = \text{sign}x + ax$, where $a \in (1, 2)$ is a constant (f_a is not defined at the point 0). We denote by I_0 the interval $[-a+1, a-1]$.

The validity of the following assertions can be verified by straightforward calculations:

(a) $f_a(I_0)$ consists of the two intervals

$$I_1 = (-1, a^2 - a - 1]; \quad I_2 = [1 + a - a^2, 1);$$

(b) for $a \in (1, \sqrt{2})$ the intervals I_0, I_1, I_2 are disjoint and $f_a(I_1) \subset I_0$, $f_a(I_2) \subset I_0$.

Thus the map f_a has the two lacunae:

$$R_1 = (\tilde{a}^2 - a - 1, 1 - a), \quad R_2 = (a - 1, 1 + a - a^2).$$

(c) $f_a(R_1) \supset R_2, \quad f_a(R_2) \supset R_1$.

Hence it follows that there exists a periodic trajectory $\{x_1, x_2\}$, where $x_1 \in R_1$, $x_2 \in R_2$, $f_a(x_1) = x_2$, $f_a(x_2) = x_1$. It follows from the inequality $a > 1$ that for every point $x \in R_1 \cup R_2$ there exists a natural number n such that $f^n(x) \in I_0 \cup I_1 \cup I_2$.

It is not difficult to modify the map in the above example so that it becomes a map of a square of Lorenz type. For this we must consider a function $g(x)$ sufficiently close to $f_a(x)$ in the C^0-topology, where $g(x)$ behaves like $ax + \text{sign}x(1 + |x|^\alpha)$, $\alpha \in (0, 1)$, in a neighbourhood of 0. We then consider the map T

$$T : (x, y) \mapsto (g(x),\ h(x, y)),$$

where $h(x, y) = \frac{1}{2}\text{sign}x + \frac{1}{6}x^\beta y$, $\beta > \alpha$.

2.8. Structurally Stable and Structurally Unstable Properties. We define the topology in the set of maps of Lorenz type. It is natural to consider a topology such that two maps of Lorenz type are close if they are obtained as first-return maps for flows defined by systems of equations with right hand sides that are close in C^1. The following definition satisfies these requirements.

Definition. Let $T : D \backslash S \to D$ be a map of Lorenz type. A δ-neighbourhood of T consists of maps T such that

(1) The following inequalities hold for the quantities α, β (or $\tilde{\alpha}, \tilde{\beta}$) defined in condition (L.5) for T (or for $\tilde{T}$):

$$|\alpha - \tilde{\alpha}| < \delta; \quad |\beta - \tilde{\beta}| < \delta,$$

(2) The following inequalities hold for the functions F_i, G_i (or $\tilde{F}_i, \tilde{G}_i$) defined in condition (L.5) for the map T (or for $\tilde{T}$):

$$\|F_i - \tilde{F}_i\|_{C^1} < \delta; \quad \|G_i - \tilde{G}_i\|_{C^1} < \delta.$$

The map $T : D \backslash S \to D$ is called *structurally stable* if for each map $\tilde{T}$ sufficiently close to T in the sense of the above-defined topology, there exists a homeomorphism $h : D \to D$, $h(S) = S$ conjugating the maps T and $\tilde{T}$, that is, such that the diagram

$$\begin{array}{ccc} D\backslash S & \xrightarrow{T} & D \\ h\downarrow & & \downarrow h \\ D\backslash S & \xrightarrow{\tilde{T}} & D \end{array}$$

is commutative.

A property of the map $T : D\backslash S \to D$ is said to be structurally stable if this property holds for any map $\tilde{T}$ sufficiently close to T.

In spite of their fairly strong hyperbolicity properties, it is the discontinuity of maps of Lorenz type that causes these maps not to be structurally stable (Afraimovich, Bykov, and Shil'nikov 1982), (Williams 1977). Furthermore, in any neighbourhood of T there exists an uncountable set of pairwise non-conjugate maps (Williams 1979). This fact is very closely related to the fact that for maps of Lorenz type the theorem on ϵ-trajectories does not hold in its classical formulation (Shlyachkov 1985), (Komuro 1985). We recall that all the structural stability properties of hyperbolic diffeomorphisms are in one way or another consequences of the theorem on ϵ-trajectories.

However, a certain modification of the theorem on ϵ-trajectories is valid for maps of Lorenz type. In (Shlyachkov 1985), (Shlyachkov 1988), Shlyachkov proved that for any $\delta > 0$ there exists $\epsilon > 0$ and a map T_δ lying in a δ-neighbourhood of T such that for every ϵ-trajectory $\{p_n\}_{-\infty}^{\infty}$ of T there exists an actual trajectory of the map $T_\delta\{\tilde{p}_n\}_{-\infty}^{\infty}$ such that $\rho(p_n, \tilde{p}_n) < \delta$ for all $n \in \mathbb{Z}$. In other words, every ϵ-trajectory of a Lorenz map T is traced by an actual trajectory of a map close to T.

All the same, maps of Lorenz type possess some kind of structural stability. In (Guckenheimer and Williams 1979) it is proved that there exists a structurally stable two-parameter family of maps of Lorenz type, that is, every map sufficiently close to some map of the family is conjugate to another map of this same family. In the proof essential use is made of the fact that the coefficient of expansion of maps of the family is greater than $\sqrt{2}$ which, as has already been noted, does not hold for a real Lorenz system. The following theorem, which refines the results of (Guckenheimer and Williams 1979), is proved in (Shlyachkov 1985), (Shlyachkov 1988).

Theorem. *Every map of Lorenz type can be included in a two-parameter structurally stable family of pairwise non-conjugate maps of Lorenz type that are continuously dependent on the parameter.*

One can say that maps of Lorenz type have structural stability of codimension 2.

Existence of a Quasi-attractor. This is a structurally stable property. From the physical point of view, this means that a Lorenz attractor is observable.

Presence of Lacunae. If T contains lacunae $R_1, \dots, R_s$ and $\Sigma^+ \cap \Sigma^- = \emptyset$, then there exists $\delta > 0$ such that the maps in a δ-neighbourhood of T have lacunae $\tilde{R}_1, \dots, \tilde{R}_s$. Here the lacunae $\tilde{R}_j$ are close to the lacunae R_j in the sense that the $\tilde{R}_j$ lie in a δ-neighbourhood of the sets R_j and the sets R_j lie in a δ-neighbourhood of the sets $\tilde{R}_j$. If $\Sigma^+ \cap \Sigma^- \neq \emptyset$, then in any neighbourhood of T there exists $\tilde{T}$ having no lacunae (Afraimovich, Bykov, and Shil'nikov 1982).

Irrational and Rational Cases. A map T is of rational type if for some natural numbers n, m

$$T^n p_1 \in S; \quad T^m p_2 \in S.$$

A map T is of irrational type if for any natural numbers $n, m \in \mathbb{Z}^+$

$$T^n p_1 \notin S, \quad T^m p_2 \notin S.$$

It turns out that in any neighbourhood of any maps T there exist $\tilde{T}_1, \tilde{T}_2, \tilde{T}_3$ such that $\tilde{T}_1$ is of rational type, $\tilde{T}_2$ is of irrational type and $\tilde{T}_3$ is of neither type. Hence it follows, in particular, that neither rationality nor irrationality is a structurally stable property.

Long and Short Cells. A map T has long cells if the lengths of the cells are bounded below by a positive constant. A map T has short cells if the lengths of the cells can be arbitrarily small.

A map T has long cells if there exist $n, m \in \mathbb{Z}^+$ such that $T^n p_1 \in S$, $T^m p_2 \in S$. Another sufficient condition for the presence of long cells is that for any natural numbers $n, m, T^n p_1, T^m p_2$ lie on the stable manifolds of saddle-type periodic trajectories.

It turns out that for any map T and any $\delta > 0$ there exist maps $\tilde{T}_1, \tilde{T}_2$ δ-close to T such that $\tilde{T}_1$ has short cells, $\tilde{T}_2$ has long cells and none of the points of $\tilde{T}_2^n p_1, \tilde{T}_2^m p_2$ lies on the line S or on the stable manifold of a periodic point (Klinshpont 1989).

Continuity of Topological Entropy. The property of continuity of topological entropy as a function of a map may serve as a distinctive characterization of structural stability of a dynamical system. Loosely speaking, the continuity of entropy shows that the complexity of a dynamical system cannot change abruptly under a small perturbation of the map. It turns out that maps of Lorenz type possess the property of continuity of topological entropy with respect to small C^0-perturbations. This was proved in (Malkin 1982) in terms of one-dimensional maps for the orientable case, while the corresponding result for the semi-orientable and non-orientable cases follows from (Shlyachkov 1988).

2.9. A Continuum of Non-homeomorphic Attractors. Williams showed that for various flows of type such as that described by a Lorenz system, there exists a continuum of attractors that cannot be converted into one another by a homeomorphism close to the identity. His starting point was the discussion of attractors of such flows set forth in the next subsection.

There remained an open question on the homeomorphism of limit sets of maps of Lorenz type. The answer was provided by Klinshpont. We give a brief account of his paper.

We denote by A_T the set of limit points of the trajectories $\{T^i p_1\}, \{T^i p_2\}$ from which the trajectories themselves are discarded (if they occur among them). For each point $z \in A_T$ all its inverse images $T^{-i}z$ are uniquely defined. We set $\alpha(z) = (\alpha_1, \alpha_2, \ldots)$, where $\alpha_i = 0$ if $T^{-i}z \in D_1$ and $\alpha_i = 1$ if $T^{-i}z \in D_2$. Let Ω_T be the set of sequences $\alpha(z)$. For the sequence $\beta \in \Omega_T$ we set

$$N_T(\beta) = \{z \in A_T;\ \alpha(z) = \beta\},$$
$$|N_T(\beta)| = \text{cardinality of } N_T(\beta),$$
$$\chi_T = \{|N_T(\beta)|;\ \beta \in \Omega_T\}.$$

Theorem (Klinshpont). *If Σ_{T_1} is homeomorphic to Σ_{T_2}, then $\chi_{T_1} = \chi_{T_2}$.*

The set χ_T has the following properties:

1. If χ_T contains a natural number, then $1 \in \chi_T$.

2. If the positive integer $n \in \chi_T$, $n > 1$, then χ_T contains two numbers n_1, n_2 such that either $n = n_1 + n_2$ or $n = n_1 + n_2 + 1$.

Furthermore, let $n_1, n_2, \ldots$ be any increasing sequence of natural numbers satisfying the conditions:

a) $n_1 = 1$.

b) For any $j > 1$ there exist $i, k < j$ such that either $n_j = n_i + n_k$ or $n_j = n_i + n_k + 1$.

It then turns out that there exists a map T for which

$$\chi_T = \{n_1, n_2, \ldots, \aleph_0, \mathfrak{c}\}$$

($\mathfrak{c}$ is the power of the continuum). Hence it turns out that there are continually many pairwise non-homeomorphic quasi-attractors of maps of Lorenz type.

2.10. An Alternative Treatment of the Attractor in a Lorenz System. Much of the information on attractors and flows described by a Lorenz system and systems related to it is obtained by studying maps of Lorenz type. This approach was applied in the early work in this area. However, in some of the early papers an alternative interpretation of the same numerical experiments was suggested and used (Williams 1977), (Williams 1979).

Suppose that in Fig. 6, especially in its lower part, there is a strong contraction in the direction of the y axis (across the diagram). In spite of this the limiting motions are not concentrated in one plane, so that below the point D_1 they move off to the right of $W^s(O)$, and below D_2, to the left. All these points then move to parts of the phase space that are a long way from each other, and therefore this contraction does not prevent them from arriving at D displaced from each other in the direction of the y axis. Eventually there is an expansion in the direction of the x axis. Thus there are three factors: a strong contraction in one direction, a separation of the points into those going to the right and those going to the left, and the occurrence of an expansion in another direction.

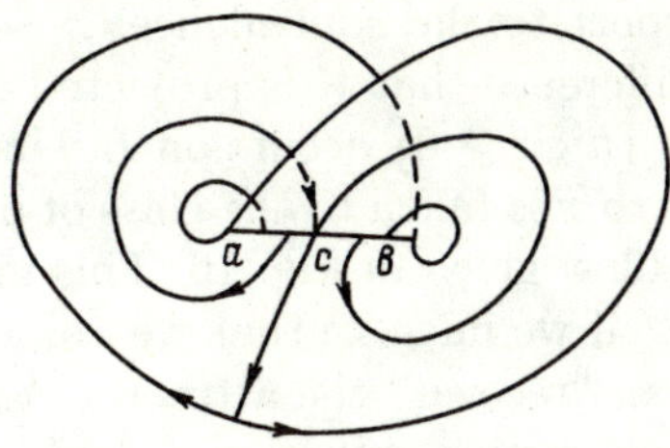

Fig. 16

Fig. 16 illustrates this situation in exaggerated form. Underneath, the contraction along the y axis is infinitely strong, so that all the points turn out to be in the (x, z) plane. As a result, the motion occurs not in three-dimensional space but only in the two-dimensional branch manifold L as illustrated in the diagram. The branching occurs along the segment ab. The motion proceeds as shown by the arrows. Once two different points arrive in the same place on ab in the course of the motion, they are for ever "stuck together" at the same point, so that it is not a flow that acts on L, but a semiflow $\{h^t,\ t \geq 0\}$.

This is something of an imaginary situation. Can Fig. 16 correspond to something that happens in a real system, where the contraction in the y direction, although strong, is not in fact infinitely strong, and different points are never "stuck together"?

Imagine that Fig. 16 is superimposed on Fig. 6 so that L is embedded in $\mathbb{R}^3$. (Near O the boundary of L goes along Γ_1 and Γ_2 and then deviates from them along the direction of the y axis.) We can expect that the trajectories enter a small neighbourhood U of this L. We can now make our assumption concerning the strong contraction in the direction of the y axis more precise: we suppose that it occurs in U (and therefore not just in the lower part of the diagram but partly in the upper part as well). In U the trajectories tend to some attractor A containing the equilibrium point O. But we do not expect that O will be an isolated point of A, so that A will not be a hyperbolic attractor in the precise sense of the word; but all the same we assume that there are one-dimensional local "very stable" manifolds $W^{ss}_{loc}(x)$ passing through the point $x \in A$ in the direction of the strong contraction. It seems plausible that this family of arcs can be extended to a one-dimensional foliation of the entire region U and even in a somewhat larger region that is invariant with respect to the flow $\{S^t\}$, in the sense that

$$S^t W^{ss}_{loc}(x) \subset W^{ss}_{loc}(S^t x) \text{ for } t \geq 0. \tag{2}$$

The leaves are short arcs whose ends go beyond the limits of U. The leaves are transversal to L and each of them intersects L just once. The union of these leaves fills out a large part V of U; there still remain narrow "semitubes" along the boundary of L.

We factor V with respect to the equivalence: $x \sim y$ if x and y lie in the same leaf. It makes no difference that V is projected onto L along the leaves. In view of (2) a semiflow $\{h^t;\ t \geq 0\}$ occurs on L. Under the action of h^t the class of points equivalent to x is taken to the class of points equivalent to $S^t x$. This is precisely the semiflow given in Fig. 16. This means that in Fig. 16 we are looking sideways at U if we imagine that we are looking along the leaves.

It turns out that we can "recover" the attractor A and the motion $\{S^t/A\}$ in it (to within topological conjugacy) from L and $\{h^t\}$. Namely, the maps $h^t : L \to L$ form an inverse spectrum of topological spaces. The attractor A is the inverse limit of this spectrum and S^t/A is defined as follows. A point $u \in A$ is a class $(x_\tau;\ \tau \geq 0)$ of equivalent points $x_\tau \in L$ (with respect to this spectrum); equivalence means that $h^\theta x_\tau = x_{\tau-\theta}$ for $\tau \geq \theta \geq 0$ (if convenient, $\tau \mapsto x_\tau$ is a continuous branch of the negative half-trajectory of x_0). Then for $\tau \geq 0$ we have

$$S^\tau u = v = (y_\theta;\ \theta \geq 0),$$

where $\tilde{y}_\theta = x_{\theta-\tau}$ for $\theta \geq \tau$, $y_\theta = h^{\tau-\theta} x_0$ for $\theta \leq \tau$, $S^{-\tau} u = v$, $v = (y_\theta)$, $y_\theta = x_{\tau+\theta}$.

All this could be rigorously justified if the trajectories $S^\tau x$ really did occur in some region U of suitable type, and if inside this region in a direction approximately parallel to the y axis there really were a contraction, its rate being greater than the rate of approach of the points in the other directions. If the first rate were considerably in excess of the second, then the smoothness of the foliation W^{ss} would be ensured, which would allow us to use this approach in questions of a topological and even metric (ergodic) character.

There is no doubt about the existence of smooth flows with such properties. Hence there exist smooth flows with attractors described by the above construction. This is of prime importance since these attractors differ from previously known ones and their properties are of interest. In particular, these attractors are preserved under small perturbations (in the sense that in the perturbed system there is an attractor close to the attractor of the unperturbed system in the Hausdorff metric) but their intrinsic structure is altered in general. They are sometimes called *geometric Lorenz attractors*.

It is not known, however, to what extent all this is applicable to a Lorenz system and systems related to it. There are no systematic numerical data that would confirm that for a Lorenz system the contraction in one direction is significantly in excess of the possible contractions in the other directions. On the contrary, it would appear that there is no such excess. This makes it all the more doubtful that the existence or the smoothness of the foliation W^{ss} can be guaranteed. Finally, if this were in fact the case for some system, then the volume of calculations involved would in all probability be far in excess of that required for establishing appropriate properties of the first-return map with sufficient confidence.

§3. Metric Properties of One-dimensional Attractors of Hyperbolic Maps with Singularities

3.1. Objects of Investigation. The results of this section relate not only to maps of Lorenz type, but also to a wider class of maps. Without giving its precise definition (see Sataev 1992), we merely note that it includes two further examples.

Belykh Map (Belykh 1982). It is defined in the square

$$D = \{(x, y) \in \mathbb{R}^2 : |x| \le 1,\ |y| \le 1\}.$$

It is defined on the interval $[-1, 1]$ by a differentiable function $s_1(y)$ such that $|s_1(y)| < 1$. On the set

$$D_1 = \{(x, y) \in D : x < s_1(y)\}$$

the map T is defined as follows:

$$(x, y) \mapsto \Big(-1 + q_1(x+1),\ -1 + \frac{1}{\lambda_1}(y+1)\Big).$$

On the set

$$D_2 = \{(x, y) \in D : x > s_1(y)\}$$

the map T is defined as follows:

$$(x, y) \mapsto \Big(1 + q_2(x-1),\ 1 + \frac{1}{\lambda_2}(y-1)\Big).$$

Here $q_1, q_2, \lambda_1, \lambda_2$ are constants greater than 1. The constants and the function $s_1(y)$ must be chosen so that the inclusions $TD_1 \subset D$, $TD_2 \subset D$ hold.

We denote by q the quantity $\min(q_1, q_2)$ (it is required in what follows) (see Fig. 17).

Lozi Map (Lozi 1978). It is defined in the strip

$$D = \{(x, y) \in \mathbb{R}^2 : |y| \le 2b,\ |x| \le 1 + 2b\}$$

and takes the point (x, y) to the point $(1+y-a|x|, bx)$, where a, b are constants, $a \in (1, 2)$ and b is sufficiently small (see Fig. 18).

In what follows the reader may assume that we are considering either a map of Lorenz type or one of the maps given above. As before, a quasi-attractor is denoted by Σ.

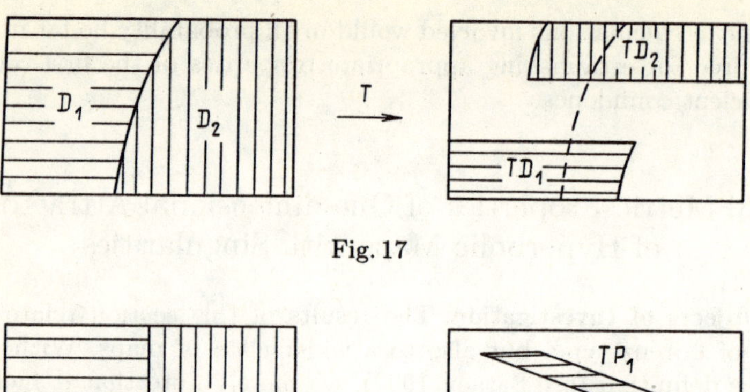

Fig. 17

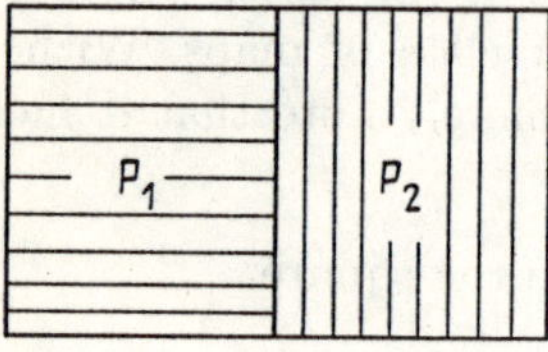

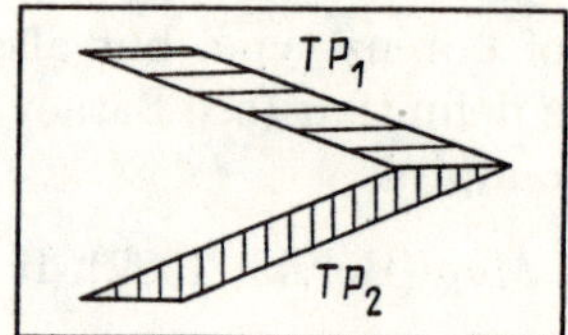

Fig. 18

For the rest of our investigations, an important role is played by local stable and unstable manifolds. Referring the reader to (Sataev 1992), (Katok and Strelcyn 1986) and the literature cited therein for details, we merely note certain peculiarities which distinguish the present situation from that of Chap. 1 and Sect. 1 of the present chapter. In general, these manifolds are not constructed for all points of the quasi-attractor.

We fix a number $\kappa < 1$ sufficiently close to 1 (in the Lorenz case we need to assume that $\kappa > 1/q$). For each $r > 0$ we define the sets

$$\begin{aligned} M^-(r) &= \{p \in \Sigma : \rho(T^{-n}p, S) \geq r\kappa^n, \ n = 0, 1, \dots\}, \\ M^+(r) &= \{p \in \Sigma : \rho(T^{n}p, S) \geq r\kappa^n, \ n = 0, 1, \dots\}, \\ M^0(r) &= M^+(r) \cap M^-(r). \end{aligned}$$

(As before, S is a line of discontinuity or a system of such lines.) These sets are closed. Furthermore, if r is sufficiently small, then they are empty. Finally, we set

$$M^+ = \bigcup_{r>0} M^+(r); \quad M^- = \bigcup_{r>0} M^-(r); \quad M^0 = M^+ \cap M^-.$$

$W^{\mathrm{u}}_{\mathrm{loc}}(p)$ is constructed for $p \in M^-$, and $W^{\mathrm{s}}_{\mathrm{loc}}(p)$ for $p \in M^+$. Here their size on $M^{\pm}(r)$ depend on r.

We then extend the local leaves to global ones, as was done in Chap. 1, Sect. 1.3. In our case, the latter may be discontinuous or non-smooth. The semilocal unstable manifold $W^{\mathrm{u}}_{\mathrm{sl}}(p)$ consists of those points $T^{-n}p'$ for which $T^{-n}p'$ lies in the same D_j as $T^{-n}p$ for all $n \geq 0$. (This definition would also be suitable for certain points $p \notin M^-$, but we do not use it.) The semilocal unstable manifold turns out to be partially a global unstable manifold at the point p. It can be proved that for $p \in M^-$ the manifold $W^{\mathrm{u}}_{\mathrm{sl}}(p)$ is a smooth curve lying in Σ. (For a Lorenz map $W^{\mathrm{u}}_{\mathrm{sl}}(p)$ is the same as the manifold $W^{\mathrm{u}}(p)$ in Sect. 2.5.)

3.2. Invariant *u*-Gibbs Measures. To describe the ergodic properties of an attractor (or quasi-attractor) it is necessary to have an invariant measure that in a certain sense is unique. For Anosov DSs and hyperbolic attractors of maps without singularities the natural invariant measures are the u-Gibbs measures. The scheme for constructing an invariant measure for a Lorenz attractor is outlined in (Bunimovich and Sinai 1980). In this paper, extra conditions requiring the absence of lacunae are laid down. For attractors of Lozi type, invariant measures are constructed in (Collet and Levy 1984), (Young 1985). The scheme for constructing the corresponding invariant measures largely follows that for constructing u-Gibbs measures for Anosov systems and hyperbolic attractors without singularities (see Sinai 1972).

We describe the scheme for constructing a u-Gibbs measure.

Let ν be an arbitrary measure that is continuous with respect to a smooth measure in some neighbourhood of Σ. Consider the sequence of measures

$$\mu_n = \frac{1}{n}(\nu + T_*\nu + \ldots + T_*^{n-1}\nu),$$

where T_* is the action on the measure induced by the map T. It turns out that for these maps the sequence of measures μ_n converges in the weak * topology to a measure μ that is invariant with respect to T. The measure μ must be considered as a natural one. Generally speaking, μ is not unique; it depends on the initial measure ν. We must therefore consider the class of measures obtained by the method just described.

An invariant measure can also be constructed in another way. Let $p \in M^-$. We denote by ν_W the measure concentrated on $W = W^{\mathrm{u}}_{\mathrm{loc}}(p)$ and absolutely continuous with respect to length (that is, the measure l_W such that for any set $A \subset D$ $l_W(A) = l(A \cap W)/l(W)$, where $l(W)$, $l(A \cap W)$ are the lengths of the curve W and its subset $A \cap W$ respectively).

If T is a hyperbolic map with singularities having a one-dimensional quasi-attractor Σ, then the sequence of measures

$$\tilde{\mu}_n = \frac{1}{n}(\nu_W + T_*\nu_W + \ldots + T_*^{n-1}\nu_W)$$

converges in the weak * topology to a measure μ_W that is invariant with respect to T.

The measures μ, μ_W relate to the class of so-called u-Gibbs measures.

Definition. A T-invariant measure μ is called a u-Gibbs measure if the following conditions hold:

1. There exist constants $C > 0$, $\beta > 0$ such that for any $r > 0$ we have the estimate

$$\mu(M^0(r)) \geq 1 - Cr^\beta.$$

2. The partition Σ into semilocal unstable manifolds is measurable and the conditional measure is absolutely continuous with respect to the measure $l_{W^{\mathrm{u}}_{\mathrm{sl}}(p)}$ on almost every manifold $W^{\mathrm{u}}_{\mathrm{sl}}(p)$.

Pesin-Sataev Theorem (Sataev 1992). *Let T be a hyperbolic map with singularities that has a one-dimensional quasi-attractor Σ. Then there exist finitely many sets $\Sigma_1, \ldots, \Sigma_k \subset \Sigma$ and u-Gibbs measures μ_j, $j = 1, \ldots, k$, concentrated on Σ_j such that:*

1. *For any $r > 0$, $j \in \{1, \ldots, k\}$ the sets $\Sigma_j \cap M^0(r)$ are closed.*

2. *The restriction of T to Σ_j is topologically transitive and ergodic with respect to the measure μ_j.*

3. *Each set Σ_j, $j = 1, 2, \ldots, k$, can be represented in the form*

$$\Sigma_j = \Sigma_j^1 \cup \Sigma_j^2 \cup \ldots \cup \Sigma_j^{s(j)},$$

where $\Sigma_j^1, \ldots, \Sigma_j^{s(j)}$ are disjoint sets; the intersection of each set Σ_j^i with any set $M^0(r)$ is closed; finally,

$$T\Sigma_j^i = \Sigma_j^{i+1}, \ i = 1, \ldots, s(j) - 1; \quad T\Sigma_j^{s(j)} = \Sigma_j^1.$$

4. *Each u-Gibbs measure μ decomposes into a sum*

$$\mu = p_1\mu_1 + \ldots + p_k\mu_k,$$

where p_j, $j = 1, \ldots, k$, are non-negative numbers, $p_1 + \ldots + p_k = 1$.

5. *The restriction of $T^{s(j)}$ to a set Σ_j^i with invariant measure μ_j^i (which is the same as the normalized restriction of μ_j to Σ_j^i) is topologically mixing and isomorphic to a Bernoulli shift.*

6. *The periodic points are everywhere dense in $\Sigma_1 \cup \ldots \cup \Sigma_k$.*

7. *For each u-Gibbs measure μ the formula for the entropy*

$$h_\mu(T) = \int_\Sigma \log J^u(p) d\mu(p)$$

holds, where $J^u(p)$ is the Jacobian of the map T restricted to the subspace E^{u}_p.

8. *For any function $\phi(p)$ that is defined and continuous in D and for almost every (with respect to Lebesgue measure on D) point $p \in D$ there exists the limit*

$$\lim_{n\to\infty} \frac{1}{n}[\phi(p) + \ldots + \phi(T^{n-1}p)],$$

which is equal to one of the quantities

$$\int \phi(p) d\mu_j(p) \quad (j \in \{1, \ldots, k\}).$$

Theorem (Bunimovich 1983). *Let T be a map of Lorenz type having no lacunae, μ an invariant u-Gibbs measure (which, under our assumptions, is unique), and $\phi(p)$ a function on the domain D satisfying a Hölder condition. Then there exist constants $a, \sigma > 0$ such that for any number s*

$$\lim \mu\left\{p : \frac{\Sigma_{k=0}^{n-1}\phi(T^k p) - na}{\sigma\sqrt{n}} < s\right\} = \frac{1}{\sqrt{2\pi}}\int_{-\infty}^{s} e^{-u^2/2}du.$$

(The above formula is called the "central limit theorem".)

Application. ϵ-Trajectories and Stability Properties of Dynamical Systems[2]

As is clear from Chap. 1, Sect. 3, ϵ-trajectories are an important instrument for the investigation of DSs with hyperbolic behaviour. The corresponding results are also of interest in connection with the interpretation of numerical experiments. A trajectory obtained by means of some digital process is an ϵ-trajectory. Because of the accumulation of unavoidable computational errors (which for a DS with hyperbolic behaviour proceeds at an exponential rate) the calculated length of the interval of a trajectory can strongly deviate from the real trajectory with the same initial point. Nevertheless, provided suitable conditions are satisfied, it still represents some actual trajectory (with a small error). Finally, it can still be said that the intrinsic stochastic behaviour of a DS occasioned by hyperbolicity is stronger than that brought about by small external perturbations.

The notions of ϵ-trajectory, ϵ-tracing and separation of the motions makes sense for a DS in a metric space. In this general situation there is, as before, a connection between stochastic stability (tracing of an ϵ-trajectory) with a certain stability of the DS with respect to small perturbations. Here we must talk about the smallness of a perturbation only in the C^0 sense.

A homeomorphism $f : M \to M$ is said to be *topologically stable* if for any $\epsilon > 0$ there exists $\delta > 0$ such that for every homeomorphism $g : M \to M$ with $\rho_C(f, g) < \delta$ there exists a continuous map $h : M \to M$ such that $h \circ f = g \circ h$ and $\rho_C(h, 1_M) < \epsilon$.

It turns out that every stochastically stable trajectory-separating homeomorphism of a metric space in which every bounded set is relatively compact is topologically stable (Morimoto 1981), (Walters 1978). More precisely, the following assertion is valid: if f is a stochastically stable trajectory-separating homeomorphism with separation constant $e(f)$, then for any $\epsilon \in (0, e/3)$ there exists $\delta > 0$ such that if $g : M \to M$ is a homeomorphism and $\rho_C(f, g) < \delta$, then there exists a unique continuous map $h : M \to M$ for which $h \circ g = f \circ h$

[2] This application was written by S.V. Shlyachkov.

and $\rho_C(h, 1_M) < \epsilon$. Furthermore, if the homeomorphism g is trajectory-separating with separation constant $e(g) \geq 2\epsilon$, then the map $h : M \to M$ is a homeomorphism. This last property enables one to prove the structural stability of certain homeomorphisms with respect to a certain type of perturbation. In particular, it follows that Anosov diffeomorphisms are structurally stable. The relation also holds in the opposite direction: if $f : M \to M$ is a topologically stable homeomorphism of a compact manifold of dimension $\dim M \geq 2$, then f is stochastically stable (Walters 1978). (See also Morimoto 1981).

With regard to the connection between topological stability and the other notions in Chap. 1, for cascades it has been proved that it follows from Axiom A and strong transversality (that is, according to Chap. 1, Sect. 4.1, from structural stability). Similar publications concerning flows are unknown. There is the related notion of topological Ω-structural stability; as regards the latter, it has been proved that it follows from Axiom A and acyclicity (that is, from Ω-structural stability) both for cascades and flows (Nitecki 1971), (Kato and Morimoto 1979).[3] The qualitative picture of the behaviour of trajectories for a topologically stable DS is similar to that for structurally stable DSs (Hurley 1984), (Hurley 1986), although it is not known whether a topologically stable DS must be topologically equivalent to a structurally stable one.

The connection between the properties of stability and the property of tracing ϵ-trajectories justifies the efforts directed towards sorting out whether some homeomorphism is stochastically stable. It has been explained that for stochastic stability certain properties similar to hyperbolicity are required. Thus if $f : \mathbb{R}^n \to \mathbb{R}^n$ is a linear automorphism, then stochastic stability is simply equivalent to hyperbolicity (Morimoto 1981). If $f : T^n \to T^n$ is an automorphism of an n-dimensional torus, then a similar assertion is true. An arbitrary stochastically stable trajectory-separating homeomorphism of T^n is topologically conjugate to a hyperbolic automorphism of the torus (Hiraide 1989).

On the other hand, in certain cases it has been proved that homeomorphisms whose properties essentially differ from hyperbolic behaviour of the trajectories are not stochastically stable. For example: isometries of Riemannian manifolds (see also the generalization in (Morimoto 1981)); homeomorphisms inducing minimal cascades; homeomorphisms of connected manifolds inducing distal cascades (Aoki 1982).

From now on we assume that the homeomorphism $f : M \to M$ separates trajectories.

By replacing conditions of hyperbolicity type by the condition of stochastic stability we are able to obtain a number of important results similar to those

---

[3] Axiom A and acyclicity guarantee the absence of an Ω-explosion (Chap. 1, Sect. 4.1). It remains to apply the theorem to a family of ϵ-trajectories (Chap. 1, Sect. 3.1) and use the separation of trajectories.

known for diffeomorphisms, primarily diffeomorphisms of Anosov type, in the non-smooth situation. Thus, if f is a stochastically stable homeomorphism, then by the natural introduction of the notion of local stable and unstable manifolds (more precisely, sets) of a point, one can show that in a sufficiently small neighbourhood of every point $x \in M$ one can introduce canonical coordinates similar to those introduced in the hyperbolic situation or, as one also says, the neighbourhoods of each point have a local structure of the direct product of its stable and unstable sets. This is a very important result which to a certain extent enables one to generalize to the case of stochastically stable homeomorphisms that part of the theory of Anosov diffeomorphisms for which smoothness is not essential. It turns out, for example, that a stochastically stable homeomorphism has a Markov partition of arbitrarily small diameter (Hiraide 1985). Furthermore, if we suppose that the restriction of a homeomorphism f to the set of nonwandering points of Ω separates trajectories and is stochastically stable, then the spectral decomposition theorem holds (Aoki 1983). In this connection we note that the restriction of the stochastically stable homeomorphism $f : M \to M$ to the set Ω is also stochastically stable. Furthermore, for homeomorphisms satisfying the property of tracing ϵ-trajectories, the following formula holds for the topological entropy (Shimomura 1987):

$$h_{\text{top}}(f) = \lim_{n\to\infty} \sup \frac{1}{n} \log N_n(f),$$

where $N_n(f)$ is the number of periodic points of period n.

As is well known, it is by no means the case that Anosov diffeomorphisms can be realized on every manifold. In this connection there arises the question on what spaces admit stochastically stable homeomorphisms. We mention some results in this direction. If $f : G \to G$ is a group automorphism of a compact connected Lie group that is stochastically stable with respect to some Riemannian metric, then G is isomorphic to a torus (Morimoto 1981). A stochastically stable homeomorphism on a compact surface is possible only in the case when this surface is a torus. Furthermore, if $f : M^2 \to M^2$ is a stochastically stable homeomorphism of a compact surface, then f is topologically conjugate to a hyperbolic automorphism of the torus. In the main this is connected with the property that a homeomorphism of such an M^2 that separates trajectories is topologically conjugate to a pseudo-Anosov diffeomorphism (Hiraide 1990), (Lewowicz 1989). We add that any orientable compact surface of positive genus admits a homeomorphism that separates trajectories but lacks the property of tracing ϵ-trajectories (Brien and Reddy 1970).

It is also known that any manifold whose universal cover is a rational homological sphere does not admit a stochastically stable homeomorphism. The same is true of the Lie groups $SO(n)$, $\text{Spin}(n)$, $SU(n)$ and certain others (Hiraide 1987a).

So far we have been dealing with cascades. But the property of tracing ϵ-trajectories is also of prime importance in the investigation of the properties of stable flows (in fact, ϵ-trajectories were first introduced in this connection (Anosov 1967)). However, in the present instance there are certain specific features.

In a metric space the smooth version of the notion of an ϵ-trajectory, introduced in (Anosov 1967) and recalled in (Anosov et al. 1985, Chap. 3, Sect. 2.2), drops out. The alternative version indicated in (Anosov et al. 1985) is suitable. However, the notion of (ϵ, T)-chain (*loc. cit.*) has become more widespread. It is clear that an (ϵ, T)-chain can be regarded as a piecewise continuous parametrized curve of special type and in accordance with this we talk about its δ-tracing by the (reparametrized) trajectory $\{f^t x\}$. (We also talk about the δ-tracing of the (ϵ, T)-chain by the point x.) The notion of chain recurrence is introduced by means of an (ϵ, T)-chain (Anosov et al. 1985, Chap. 3, Sect. 2.2). A study of the set of chain recurrent points enables one to obtain valuable information on the dynamical properties of a hyperbolic flow (Franke and Selgrade 1976), (Franke and Selgrade 1977), (Franke and Selgrade 1978), which in no small degree is due to the fact that for hyperbolic flows the theorem on ϵ-trajectories (which can be rephrased as the theorem on tracing of (ϵ, T)-chains with fixed ϵ) holds.

As in the case of homeomorphisms, we say that a flow $\{f^t\}$ is stochastically stable if it satisfies the condition that for any $\delta > 0$ there exists $\epsilon > 0$ and $T > 0$ such that every (ϵ, T)-chain is δ-traced by some point $x \in X$.

In fact there exist various kinds of stochastic stability of flows corresponding to the various properties of the reparametrization. If one allows reparametrizations h not satisfying the property $h(\mathbb{R}) = \mathbb{R}$, then one talks about *weak stochastic stability.* If, on the other hand, one imposes the extra condition

$$\left| \frac{h(s) - h(t)}{s - t} - 1 \right| \leq \epsilon$$

on the parametrization, then one talks about *strong stochastic stability.* Concerning the interrelationship between these notions and their relationship with the tracing of finite (ϵ, T)-chains see (Kato 1984), (Oka 1990). We merely note that for topological flows in compact metric spaces without fixed points all these notions are equivalent.

For flows satisfying certain assumptions there are analogues of most of the assertions dealing with stochastic stability that are true for homeomorphisms. In particular, every continuous trajectory-separating stochastically stable flow without fixed points is topologically stable. In turn, a topologically stable C^1-flow without fixed points on a compact manifold is stochastically stable (Thomas 1982).

We shall not dwell on the other results relating to stochastically stable flows, but merely refer the reader to (Kato 1985), (Kato and Morimoto 1979), (Thomas 1987).

The development of methods of investigation of dynamical systems relating to ϵ-trajectories is possible in various directions. First of all, it is clear that it is by no means essential to study ϵ-trajectories of a homeomorphism of a space onto itself. One can consider, for example, homeomorphisms of a space into some subset of itself, or the restriction of such a homeomorphism to some invariant subset. Properly speaking, this is what is done in the hyperbolic theory. Such an approach is equivalent to considering not all ϵ-trajectories but only those, for example, lying in a neighbourhood of a hyperbolic set. The following generalization of the notion of stochastic stability introduced in (Dateyama 1989) is also concerned with restricting the set of ϵ-trajectories to be considered. A homeomorphism $f : M \to M$ is *stochastically stable with respect to the partition* $\mathcal{A} = \{A_1, \ldots, A_n\}$ of the space M if for every $\delta > 0$ there exists $\epsilon > 0$ such that any ϵ-trajectory $\{x_n\}$, $n \in \mathbb{Z}$, with the property that fx_n and x_{n+1} belong to the same element of the partition for all $n \in \mathbb{Z}$ is δ-traced by some point $y \in M$. A stochastically stable homeomorphism is stochastically stable with respect to the partition consisting of a single element, namely the space M itself. An example of a stochastically stable homeomorphism with respect to some partition that is not stochastically stable is provided by a pseudo-Anosov diffeomorphism. The possibilities for this approach are clear from the following result: a homeomorphism that separates trajectories admits a Markov partition if and only if it is stochastically stable with respect to some partition (Dateyama 1989).

Another possible direction for generalizations is in connection with dropping the invertibility requirement for maps $f : M \to M$. Of course, the possibilities for such generalizations are fairly restricted since in order that the theorem on ϵ-trajectories should be true, to some extent one must preserve properties similar to hyperbolicity. One of the classes of non-invertible stochastically stable maps are the so-called Anosov maps (Sakai 1987), which are obtained as a certain generalization of Anosov diffeomorphisms. One can also consider maps having discontinuities, for example, piecewise-expanding maps. More precisely, let $M = \bigcup_{i=1}^{n} M_i$, and let $f : M_i \to M$ be a continuous expanding map for each $i = 1, 2, \ldots, n$. It is not difficult to see that in this case the theorem on ϵ-trajectories holds only when $f(M_i) = M$ for each i (see, for example, Komuro 1985). This corresponds to the fact that the restriction of a shift in the space of doubly infinite sequences from a finite alphabet to a closed invariant set enjoys the property of tracing ϵ-trajectories if and only if it is topologically conjugate to a topological Markov chain (this is equivalent to its being of finite type (see (Alekseev 1976, Theorem 3.1), (Walters 1978)).

The majority of results dealing with ϵ-trajectories of flows require assumptions enabling one to avoid the nuisance caused by the presence of singular points. For example, it is assumed that the flow satisfies Axiom A or, more often, that there is a complete absence of singular points. However, a flow corresponding to a Lorenz system has a non-isolated singular point in its non-wandering set. Therefore this flow does not come under the existing traditional ϵ-trajectory theory. Even so, there is a version of this theory that is valid even for flows of Lorenz type.

It is best to illustrate this version of the theorem on ϵ-trajectories by an example not of a flow, but rather a piecewise expanding map of an interval to itself. Let f be a piecewise continuous piecewise expanding map of an interval to itself satisfying certain innocuous conditions of general-position type. We will not give a precise statement of these conditions but merely note that they are satisfied by every map that has a sufficiently large coefficient of expansion in neighbourhoods of critical points, as well as by maps for which no critical point belongs to the set of inverse images of critical points. (By critical points we mean points of discontinuity and points at which there arises a change of direction of the monotonicity of the map.) For such maps the theorem on ϵ-trajectories holds in the following version: for any $\delta > 0$ there exist $\epsilon > 0$ and a piecewise expanding map f_δ such that $\rho_C(f_\delta, f) < \delta$ and every ϵ-trajectory is traced by some point under the map f_δ. In other words, close to every ϵ-trajectory of the map there is an actual trajectory of some map close to it. We note that in its standard formulation the theorem on ϵ-trajectories for this sort of map is not true (Shlyachkov 1985), (Shlyachkov 1988), (Komuro 1984).

Such a theorem on ϵ-trajectories is completely suitable for the study of properties of stability. Using it one can prove the following theorem on structural stability. Let f be a piecewise continuous piecewise expanding map of the interval I to itself with m critical points. There exists a $2m$-parameter family of maps $\{f_\mu\}$, $\mu = (\mu_1, \dots, \mu_{2m})$, $f_0 = f$, that is continuously dependent on the parameters and is structurally stable in the following sense: for any $\delta > 0$ there exists $\epsilon > 0$ such that for every map $\tilde{f}$ with the same properties and the same discontinuities as f for which $\rho_C(\tilde{f}, f) < \epsilon$, there exist values of the parameters $\mu^* = (\mu_1^*, \dots, \mu_{2m}^*)$ such that the maps $\tilde{f}$ and f_{μ^*} are topologically conjugate, that is, there exists a homeomorphism $h : I \to I$, $\rho_C(h, 1_I) < \delta$, such that $h \circ f = f_{\mu^*} \circ h$ with $\rho_C(f_{\mu^*}, f) < \delta$. Furthermore, the maps in the family $\{f_\mu\}$ are pairwise non-conjugate.

Associated with flows of Lorenz type are expanding maps with a single point of discontinuity, where the coefficient of local expansion tends to infinity as the point of discontinuity is approached. This enables one to prove for such a flow theorems similar to those stated above.

As it happens, the theorem on ϵ-trajectories in its new version is valid also for certain maps of an interval not having an expansion. Thus, in (Nusse and Yorke 1988) such a theorem is proved for quadratic maps of an interval satisfying conditions of a very general nature.

With regard to theorems on structural stability, apparently the existence of a finitely parametrized structurally stable family is a one-dimensional effect. Thus we can assume that maps inducing a Belykh attractor (Sect. 3) are not included in such a family, although the theorem on ϵ-trajectories in the version indicated above is true for such maps.

References*

Afraimovich, V.S., Bykov, V.V., Shil'nikov, L.P. (1977): The origin and structure of the Lorenz attractor. Dokl. Akad. Nauk SSSR *234*, 336–339. [English transl.: Sov. Phys., Dokl. *22*, 253–255 (1977)] Zbl. 451.76052

Afraimovich, V.S., Bykov, V.V., Shil'nikov, L.P. (1982): On structurally unstable attracting limit sets of Lorenz attractor type. Tr. Mosk. Mat. O.-va *44*, 150–212. [English transl.: Trans. Mosc. Math. Soc. 1983, No. 2, 153–216 (1983)] Zbl. 506.58023

Afraimovich, V.S., Pesin, Ya.B. (1987): Dimension of Lorenz type attractors. Sov. Sci. Rev., Sect. C, Math. Phys. Rev. *6*, 169–241. Zbl. 628.58031

Alekseev, V.M. (1976): Symbolic Dynamics (Eleventh Mathematical School). Akad. Nauk Ukr. SSR, Kiev (Russian)

Anosov, D.V. (1967): Geodesic flows on closed Riemannian manifolds of negative curvature. Tr. Mat. Inst. Steklova *90* (209 pp.) [English Transl.: Proc. Steklov Inst. Math. *90* (1969)] Zbl 163,436

Anosov, D.V., Aranson S. Kh., Bronshtein, I.U., Grines, V.Z. (1985): Dynamical Systems I: Smooth Dynamical Systems. Itogi Nauki Tekh., Ser. Sovrem. Probl. Mat., Fundam. Napravleniya *1*, 151–242. [English transl. in: Encycl. Math. Sci. *1*, 149–230. Springer-Verlag, Berlin Heidelberg New York 1988] Zbl. 605.58001

Aoki, N. (1982): Homeomorphisms without the pseudo-orbit tracing property. Nagoya Math. J. *88*, 155–160. Zbl. 466.54034

Aoki, N. (1983): Homeomorphism with the pseudo-orbit tracing property. Tokyo J. Math. *6*, No. 2, 329–334. Zbl. 537.58034

Arnol'd, V.I., Afraimovich, V.S., Il'yashenko, Yu.S., Shil'nikov, L.P. (1986): Dynamical Systems V: Bifurcation Theory. Itogi Nauki Tekh., Ser. Sovrem. Probl. Mat., Fundam. Napravleniya *5*, 5–218. [English transl. in: Encycl. Math. Sci. *5*. Springer-Verlag, Berlin Heidelberg New York 1993]

Barge, M. (1986): Horseshoe maps and inverse limits. Pac. J. Math. *121*, No. 1, 29–39. Zbl. 601.58049

Belykh, V.P. (1982): Models of discrete systems of phase synchronization. Chap. 10 in Shakhgil'dyan, V.V., Belyustinaya, L.N. (eds.): Radio and Communication, Moscow, 161–176 (Russian)

Bowen R. (1979): Methods of Symbolic Dynamics. Mir, Moscow (Russian)

Brindley, J., Moroz, I.M. (1980): Lorenz attractor behavior in a continuously stratified baroclinic fluid. Phys. Lett. *A77*, No. 6, 441–444

Bunimovich, L. (1983): Statistical properties of Lorenz attractors, in Barenblatt, G.I. (ed.): Nonlinear Dynamics and Turbulence. Pitman, Boston, 71–92. Zbl. 578.58025

Bunimovich, L.A., Sinai, Ya.G. (1980): Stochasticity of an attractor in the Lorenz model. In Gaponov-Grekhov, A.V. (ed.): Non-linear Waves, Nauka, Moscow, 212–226 (Russian)

Collet, P., Levy, Y. (1984): Ergodic properties of the Lozi mappings. Commun. Math. Phys. *93*, No. 4, 461–482. Zbl. 553.58019

Dateyama, M. (1989): Homeomorphisms with Markov partitions. Osaka J. Math. *26*, No. 2, 411–428. Zbl. 711.58026

Fedotov, A.G. (1980): On Williams solenoids and their realization in two-dimensional dynamical systems. Dokl. Akad. Nauk SSSR *252*, 801–804. [English transl: Sov. Math., Dokl. *21*, 835–839 (1980)] Zbl. 489.58019

Franke, J.E., Selgrade, J.F. (1976): Abstract ω-limit sets, chain recurrent sets, and basic sets for flows. Proc. Am. Math. Soc. *60*, 309–316. Zbl. 316.58014

* For the convenience of the reader, references to reviews in Zentralblatt für Mathematik (Zbl.), compiled using the MATH database, have, as far as possible, been included in this bibliography.

Franke, J.E., Selgrade, J.F. (1977): Hyperbolicity and chain recurrence. J. Differ. Equations *26*, No. 1, 27–36. Zbl. 329.58012

Franke, J.E., Selgrade, J.F. (1978): Hyperbolicity and cycles. Trans. Am. Math. Soc. *245*, 251–262. Zbl. 396.58023

Gibbon, J.D., McGuines, M.J. (1980): A derivation of the Lorenz equation for some unstable dispersive physical systems. Phys. Lett. *A77*, No. 5, 295–299

Guckenheimer, J., Williams, R.F. (1979): Structural stability of Lorenz attractors. Publ. Math., Inst. Hautes Etud. Sci. *50*, 59–72. Zbl. 436.58018

Haken, H. (1975): Analogy between higher instabilities in fluids and lasers. Phys. Lett. *53A*, No. 1, 77–79

Hiraide, K. (1985): On homeomorphisms with Markov partitions. Tokyo J. Math. *8*, No. 1, 219–229. Zbl. 639.54030

Hiraide, K. (1987a): Manifolds which do not admit expansive homeomorphisms with the pseudo-orbit tracing property. The theory of dynamical systems and its applications to nonlinear problems. Singapore World Sci. Publ., 32–34

Hiraide, K. (1987b): Expansive homeomorphisms of compact surfaces are pseudo-Anosov. Proc. Japan Acad., Ser. A. *63*, No. 9, 337–338. Zbl. 663.58024; Osaka J. Math. *27*, No. 1 (1990), 117–162. Zbl. 713.58042

Hiraide, K. (1988): Expansive homeomorphisms with the pseudo-orbit tracing property on compact surfaces. J. Math. Soc. Japan *40*, No. 1, 123–137. Zbl. 666.54023

Hiraide, K. (1989): Expansive homeomorphisms with the pseudo-orbit tracing property of n-tori. J. Math. Soc. Japan *41*, No. 3, 357–389. See also World Sci. Adv. Ser. Dyn. Syst. *5*, 35–41 (1987); Zbl. 678.54029

Hirsch, M., Pugh., C.C., Shub, M. (1977): Invariant Manifolds. Lect. Notes Math. *583*. Zbl. 355.58009

Hurley, M. (1984): Consequences of topological stability. J. Differ. Equations *54*, No. 1, 60–72. Zbl. 493.58013

Hurley, M. (1986): Fixed points of topologically stable flows. Trans. Am. Math. Soc. *294*, No. 4, 625–633. Zbl. 594.58030

Kamaev, D.A. (1977): On the topological indecomposability of certain invariant sets of A-diffeomorphisms. Usp. Mat. Nauk *32*, No. 1, 158

Kato. K. (1984): Pseudo-orbits and stabilities of flows, I, II. Mem. Fac. Sci., Kochi Univ., Ser. A *5*, 45–62. Zbl. 536.58017; *6* (1985), 33–43. Zbl. 571.58013

Kato, K., Morimoto, A. (1979): Topological Ω-stability of Axiom A flows with no Ω-explosions. J. Differ. Equations *34*, No. 3, 464–481. Zbl. 406.58025

Katok, A.B., Strelcyn, J.M. (1986): Invariant Manifolds, Entropy and Billiards; Smooth Maps with Singularities. Lect. Notes Math. *1222*. Zbl. 658.58001

Klinshpont, N.Eh. (1989): Instability of the cellular structure of Lorenz attractors. In Plykin, R.V. (ed.): Collection of Scientific Works of Students of Obninsk. OIATEh, Obninsk, 11–30 (Russian)

Knobloch, F. (1981): Chaos in a segmented disc dynamo. Phys. Lett. *A82*, No. 9, 439–440

Komuro, M. (1984): One parameter flows with the pseudo-orbit tracing property. Monatsh. Math. *98*, No. 3, 219–253. Zbl. 545.58037

Komuro, M. (1985): Lorenz attractors do not have the pseudo-orbit tracing property. J. Math. Soc. Japan *37*, No. 3, 489–514. Zbl. 552.58020

Lewowicz, J. (1989): Expansive homeomorphisms of surfaces. Bol. Soc. Bras. Mat., Nova Ser. *20*, No. 1, 113–133. Zbl. 753.58022

Lorenz, E.N. (1963): Deterministic nonperiodic flow. J. Atmosph. Sci. *20*, No. 2, 130–141

Lorenz, E.N. (1979): On the prevalence of aperiodicity in simple systems. In: Global Analysis. Proc. Semin. Calgary 1978, Lect. Notes Math. *755*, 53–75. Zbl. 438.34038

Lozi, R. (1978): Un attracteur étrange du type de Hénon. J. Phys. C. *39*, No. 5, 9–10

McLaughlin, J.B., Martin, P.C. (1974): Transition to turbulence of statically stressed fluids. Phys. Rev. Lett. *33*, No. 20, 1189–1192

McLaughlin, J.B., Martin, P.C. (1975): Transition to turbulence in statically stressed fluid systems. Phys. Rev. A *12*, No. 1, 186–203

Malkin, M.I. (1982): On the continuity of the entropy of discontinuous maps of the interval. In Leontovich-Andronova, E.A. (ed.): Metody Kachestvennoi Teorii Differents. Uravnenij: Mezhvuz. Temat. Sb. Nauchen. Tr., Gor'kij Gos. Univ., 1982, 35–47. [English transl.: Sel. Math. Sov. *8*, No. 2, 131–139 (1989)] Zbl. 572.54035

Milnor, J. (1985): On the concept of attractor. Commun. Math. Phys. *99*, No. 2, 177–195. Zbl. 595.58028

Morimoto, A. (1981): Some stabilities of group automorphisms. In: Manifolds and Lie groups, Prog. Math. *14*, 283–299. Zbl. 483.58007

Neimark, Yu.I., Landa, P.S. (1987): Stochastic and Chaotic Oscillations. Nauka, Moscow. [English transl.: Kluwer, Dordrecht 1992] Zbl. 644.58013

Nitecki, Z. (1971): On semistability for diffeomorphisms. Invent. Math. *14*, No. 2, 83–122. Zbl. 218.58007

Nusse, H.E., Yorke, J.A. (1988): Is every approximate trajectory of some process near an exact trajectory of a nearby process? Commun. Math. Phys. *114*, No. 3, 363–379. Zbl. 697.58036

Oka, M. (1990): Expansiveness of real flows. Tsukuba J. Math. *14*, No. 1, 1–8. Zbl. 733.54030

O'Brien, T., Reddy, W. (1970): Each compact orientable surface of positive genus admits an expansive homeomorphism. Pac. J. Math. *35*, No. 3, 737–742. Zbl. 206,524

Pedlosky, J., Frenzen, C. (1980): Chaotic and periodic behavior of finite amplitude baroclinic waves. J. Atmosph. Sci. *37*, No. 6, 1177–1196

Pesin, Ya.B. (1977): Characteristic Lyapunov exponents and smooth ergodic theory. Usp. Mat. Nauk *32*, No. 4, 55–111. [English transl.: Russ. Math. Surv. *32*, No. 4, 55–114 (1977)] Zbl. 359.58010

Pesin, Ya.B. (1988): Dimension type characteristics for invariant sets of dynamical systems. Usp. Mat. Nauk *43*, No. 4, 95–128. [English transl.: Russ. Math. Surv. *43*, No. 4, 111–151 (1988)] Zbl. 684.58024

Plykin, R.V. (1974): Sources and sinks of *A*-diffeomorphisms of surfaces. Mat. Sb., Nov. Ser. *94*, 243–264. [English transl.: Math. USSR, Sb. *23*, 233–253 (1975)] Zbl. 324.58013

Plykin, R.V. (1984): On the geometry of hyperbolic attractors of smooth cascades. Usp. Mat. Nauk *39*, No. 6, 75–113. [English transl.: Russ. Math. Surv. *39*, No. 6, 85–131 (1984)] Zbl. 584.58038

Plykin, R.V. (1990): Some problems of attractors of differentiable dynamical systems. Int. Conf. Topology, Abstracts, Math. Inst., Varna

Rand, D. (1978): The topological classification of Lorenz attractors. Math. Proc. Camb. Philos. Soc. *83*, No. 3, 451–460. Zbl. 375.58015

Robbins, K.A. (1977): A new approach to subcritical instability and turbulent transition in a simple dynamo. Math. Proc. Camb. Philos. Soc. *82*, No. 2, 309–325

Robinson, C. (1989): Homoclinic bifurcation to a transitive attractor of Lorenz type. Nonlinearity *2*, No. 4, 495–518. Zbl. 704.58031

Ruelle, D. (1976): The Lorenz attractor and the problem of turbulence. In Temam, R. (ed.): Turbulence and Navier Stokes Equations. Lect. Notes Math. *565*, 146–158. Zbl. 355.76036

Ruelle, D., Takens, F. (1971): On the nature of turbulence. Commun. Math. Phys. *20*, No. 3, 167–192. Zbl. 223.76041

Rychlik, M. (1990): Lorenz attractors through Shil'nikov type bifurcation. I. Ergodic Theory Dyn. Syst. *10*, No. 4, 793–821. Zbl. 715.58027

Sakai, K. (1987): Anosov maps on closed topological manifolds. J. Math. Soc. Japan *39*, No. 3, 505–519. Zbl. 647.58039

Saltzman, B. (1962): Finite amplitude free convection as an initial value problem. I. J. Atmosph. Sci. *19*, No. 2, 329–341

Sataev, E.A. (1992): Invariant measures for hyperbolic maps with singularities. Usp. Mat. Nauk *47*, No. 1, 147–202. [English transl.: Russ. Math. Surv. *47*, No. 1, 191–251 (1992)]

Shil'nikov, A.L. (1986): Bifurcations and chaos in the Marioka-Shimitsu model. In: Metody Kachestvennoi Teorii Differents. Uravnenij: Mezhvuz. Temat. Sb. Nauchen. Tr., Gor'kij Gos. Univ., 180–193 [English transl.: Sel. Math. Sov. *10*, No. 2, 105–117 (1991)] Zbl. 745.58036; II. Metody Kachestvennoi Teorii i Teorii Bifurkatsii: Mezhvuz. Temat. Sb. Nauchen. Tr., Gor'kij Gos. Univ. 1989, 130–138

Shil'nikov, L.P. (1963): On certain cases of the generation of periodic motions from singular trajectories. Mat. Sb., Nov. Ser. *61*, 433–466. Zbl. 121,75

Shil'nikov, L.P. (1981): Theory of bifurcations and quasihyperbolic attractors. Usp. Mat. Nauk *36*, No. 4, 240–241

Shimomura, T. (1987): Topological entropy and the pseudo-orbit tracing property. In: The Theory of Dynamical Systems and its Applications to Nonlinear Problems. Singapore World Sci. Publ., 35–41

Shlyachkov, S.V. (1985): A theorem on ϵ-trajectories for Lorenz mappings. Funkts. Anal. Prilozh. *19*, No. 3, 84–85. [English transl.: Funct. Anal. Appl. *19*, 236–238 (1985)] Zbl. 605.58029

Shlyachkov, S.V. (1988a): Structurally stable families of discontinuous maps of the interval into itself. OIATEh, Obninsk; Dep. VINITI, No. 4113-B88 (Russian)

Shlyachkov, S.V. (1988b): On the entropy of piecewise-monotone expanding maps of the interval. In Leontovich-Andronova, E.A. (ed.): Metody Kachestvennoi Teorii Differents. Uravnenij: Mezhvuz. Temat. Sb. Nauchen. Tr., Gor'kij Gos. Univ., 91–97

Sinai, Ya.G. (1972): Gibbs measures in ergodic theory. Usp. Mat. Nauk *27*, No. 4, 21–64. [English transl.: Russ. Math. Surv. *27*, No. 4, 21–69 (1972)] Zbl. 246.28008

Sinai, Ya.G., Shil'nikov, L.P. (eds.) (1981): Strange Attractors. Mir, Moscow (Russian)

Sinai. Ya.G., Vul, E.B. (1981): Hyperbolicity conditions for the Lorenz model. Physica *D2*, No. 1, 3–7

Smale, S. (1967): Differentiable dynamical systems. Bull. Am. Math. Soc. *73*, No. 6, 747–817. Zbl. 202,552

Sparrow, C. (1982): The Lorenz equations: Bifurcations, chaos and strange attractors. Appl. Math. Sci. *41*. Springer-Verlag, Berlin Heidelberg New York. Zbl. 504.58001

Thomas, R.F. (1982): Stability properties of one-parameter flows. Proc. Lond. Math. Soc., III Ser. *45*, No. 3, 479–505. Zbl. 449.28019

Thomas, R.F. (1987): Entropy of expansive flows. Ergodic Theory Dyn. Syst. *7*, No. 4, 611–625. Zbl. 612.28015

Ustinov, Yu.I. (1987): Algebraic invariants of topological conjugacy of solenoids. Mat. Zametki *42*, No. 1, 132–144. [English transl.: Math. Notes *42*, 583–590 (1987)] Zbl. 713.58043

Walters, P. (1978): On the pseudo-orbit tracing property and its relationship to stability. The structure of attractors in dynamical systems, Proc. North Dakota 1977, Lect. Notes Math. *668*, 231–244. Zbl. 403.58019

Welander, P. (1967): On the oscillatory instability of a differentially heated fluid loop. J. Fluid. Mech. *29*, No. 1, 17–30. Zbl. 163,207

Williams, R.F. (1967): One-dimensional non-wandering sets. Topology *6*, No. 4, 473–487. Zbl. 159,537

Williams, R.F. (1970a): Classification of one dimensional attractors. Global Analysis. Proc. Symp. Pure Math.*14*, 341–361. Zbl. 213,504

Williams, R.F. (1970b): The *DA*-maps of Smale and structural stability. Global Analysis. Proc. Symp. Pure Math. *14*, 329–334. Zbl. 213,503

Williams, R.F. (1974): Expanding attractors. Publ. Math., Inst. Hautes Etud. Sci. *43*, 169–204. Zbl. 279.58013

Williams, R.F. (1977): The structure of Lorenz attractors. Turbulence Seminar, Berkeley 1976/77. Lect. Notes. Math. *615*, 94–116. Zbl. 363.58005

Williams, R.F. (1979): The structure of Lorenz attractors. Publ. Math., Inst. Hautes. Etud. Sci. *50*, 73–100. Zbl. 484.58021

Yorke, J., Yorke, E. (1979): Metastable chaos: the transition to sustained chaotic behaviour in the Lorenz model. J. Stat. Phys. *21*, No. 3, 263–278

Young, L.-S. (1985): Bowen-Ruelle measures for certain piecewise hyperbolic maps. Trans. Am. Math. Soc. *287*, No. 1, 41–48. Zbl. 552.58022

Yuri, M. (1983): A construction of an invariant stable foliation by the shadowing lemma. Tokyo J. Math. *6*, No. 2, 291–296. Zbl. 541.58041

Zhirov, A.Yu. (1989): Matrices of intersections of one-dimensional hyperbolic attractors on a two-dimensional sphere. Mat. Zametki *45*, No. 6, 44–56. [English transl.: Math. Notes *45*, No. 6, 461–469 (1989)] Zbl. 687.58019

Chapter 3 Cascades on Surfaces

S.Kh. Aranson, V.Z. Grines

In this chapter we give a survey of modern results relating to the qualitative investigation of *cascades* on surfaces (two-dimensional manifolds); the main attention is devoted to questions of the topological classification of cascades satisfying *Axiom* A and its various modifications.

The recently published survey (Aranson and Grines 1990) by the authors is also devoted to these questions, therefore we have tried to include in the present survey information not contained in (Aranson and Grines 1990), although we have kept some of the material in (Aranson and Grines 1990) for the sake of completeness of exposition. Questions relating to topological classification for flows are fairly comprehensively explained in (Anosov 1985), (Aranson and Grines 1985), (Aranson and Grines 1986), (Palis and de Melo 1982).[1]

§1. Morse-Smale Diffeomorphisms

The simplest structurally stable diffeomorphisms on closed orientable surfaces M of genus $p \geq 0$ are the Morse-Smale diffeomorphisms (Smale 1967). These are diffeomorphisms satisfying *Axiom* A, the *strict transversality condition* and having a nonwandering set that is the union of a finite number of periodic points (see (Plykin 1984), (Liao 1980), (Mañé 1982), (Nitecki 1971), (Sannami 1981), (Smale 1967) concerning these concepts).

Whereas Morse-Smale flows on M have been completely studied, the qualitative investigation of Morse-Smale cascades on M are considerably more complex and many questions still remain open. In particular, this is connected with the fact that, in general, Morse-Smale diffeomorphisms on M induce a non-trivial action both on the fundamental group and on the group of one-dimensional homologies of M, which imposes certain restrictions on the homotopy classes of homeomorphisms on M allowing the existence of Morse-Smale diffeomorphisms in them. The consequence of this is that there arise additional algebraic relations on the nonwandering set (consisting of a finite number of periodic points in the present instance) by comparison with the Lefschetz formula (1967).[2] Apart from this, Morse-Smale diffeomorphisms on

---

[1] By a *cascade* (or *flow*) defined on a manifold M we mean a dynamical system f^t (a one-parameter group of homeomorphisms acting on M), where $t \in \mathbb{Z}$ (or $t \in \mathbb{R}$), $\mathbb{Z}$ is the group of integers, $\mathbb{R}$ is the group of real numbers.

[2] The initiative in posing and solving the problems arising in this connection is due in the main to overseas mathematicians (Roche, Narasimkhan, Batterson, Smillie, and others). The reader can obtain an account of the results along these lines in the survey (Aranson and Grines 1990, Sect. 1.1).

M can have heteroclinic points which in turn lead to a complex structure of the nonwandering set.

At present a complete topological invariant of Morse-Smale diffeomorphisms on M has been found in classes of varying generality: 1) the class of *gradient-like* diffeomorphisms (Bezdenezhnykh and Grines 1985a), (Bezdenezhnykh and Grines 1985c); 2) the class of diffeomorphisms with *orientable heteroclinic sets* (Bezdenezhnykh and Grines 1985b); 3) the class of diffeomorphisms with a finite set of *heteroclinic trajectories.*

The class 3), introduced and studied by Grines, is the most general of the above classes, therefore we will go into this in more detail (we denote this class by $\Phi(M)$).[3]

We say that a Morse-Smale diffeomorphism f is of class $\Phi(M)$ if:

1) f preserves the orientation of M;

2) the set of heteroclinic trajectories of f is finite.

We recall that a trajectory $\Gamma = \bigcup_{n=-\infty}^{+\infty} f^n(x)$ of the cascade generated by a diffeomorphism $f : M \to M$ satisfying Axiom A is said to be *heteroclinic* if $x \in W^{\mathrm{s}}(p_1) \cap W^{\mathrm{u}}(p_2)$, where p_1, p_2 are saddle-type periodic points; $W^{\mathrm{s}}(p_1), W^{\mathrm{u}}(p_2)$ are the stable and unstable manifolds of the points p_1, p_2 respectively. The point x itself is called a *heteroclinic point.* If $p_1 = p_2$, then Γ is a *homoclinic trajectory.*

In (Umanskij 1990) a complete topological invariant is found for Morse-Smale flows with a finite number of singular trajectories on closed orientable three-dimensional manifolds M^3. By the *suspension* over a Morse -Smale diffeomorphism $f \in \Phi(M)$ we mean the Morse-Smale flow f^t ($t \in \mathbb{R}$) belonging to the class in (Umanskij 1990), where M is a global transversal for f^t. However, it does not generally follow from the topological equivalence of flows on M^3 with a global transversal M that the diffeomorphisms on M induced by them are topologically conjugate. For this reason, the problem of finding topological invariants of diffeomorphisms $f \in \Phi(M)$ is of interest in its own right.

We denote by $\Omega(f)$ ($f \in \Phi(M)$) the set of periodic points of f, and by $\Omega_1(f)$ the subset of all saddle-type periodic points in $\Omega(f)$.

Let W^{s} (W^{u}) be the union of the stable (unstable) manifolds of all points of $\Omega_1(f)$, and let $\hat{W}^{\sigma}$ be the set consisting of all the connected components of $W^{\sigma} \backslash \Omega_1(f)$ containing heteroclinic points, where $\sigma \in \{s, u\}$.

We set $W = W^{\mathrm{s}} \cup W^{\mathrm{u}}$, $\hat{W} = \hat{W}^{\mathrm{s}} \cup \hat{W}^{\mathrm{u}}$, $\bar{\sigma} = s$, if $\sigma = u$ and $\bar{\sigma} = u$, if $\sigma = s$. We denote by $\bar{X}$ the closure of a set X belonging to M.

Let $L^{\sigma}(p)$ be one of the connected components of $W^{\sigma}(p) \backslash p$, where $p \in \Omega_1(f)$.

[3] See also the survey (Aranson and Grines 1990, Sect. 1.2) concerning the classes 1), 2). Some particular cases of Morse-Smale diffeomorphisms with a finite or countable set of heteroclinic trajectories on M are also considered in (Borevich 1980), (Borevich 1986), (Borevich 1984), (Borevich 1989). However the results on topological conjugacy of Morse-Smale diffeomorphisms stated in (Borevich 1980), (Borevich 1984), (Borevich 1989) require substantial refinements.

Lemma 1.1. *If $L^\sigma(p) \notin \hat{W}^\sigma$, $p \in \Omega_1(f)$, $\sigma \in \{s,u\}$, then $\bar{L}^\sigma(p)\backslash(L^\sigma(p) \cup p)$ consists either of a sink ω (if $\sigma = u$), or a source α (if $\sigma = s$); if $L^\sigma(p) \subset \hat{W}^\sigma$, then $\bar{L}^\sigma(p)\backslash(L^\sigma(p) \cup p)$ is the connected one-dimensional complex $\bigcup_q W^{\mathrm{u}}(q)$, where the union is taken over all points $q \in \Omega_1(f)$ such that $W^{\bar{\sigma}}(q) \cap L^\sigma(p) \neq \emptyset$, in addition $W^\sigma(q)\backslash q \notin \hat{W}^\sigma$.*

Lemma 1.2. *Suppose that $\Omega_1(f) \neq \emptyset$. Then the boundary ∂Q of the connected component Q of $M\backslash\overline{(W\backslash\hat{W})}$ consists of two connected one-dimensional complexes $\partial^{\mathrm{s}}Q \subset \overline{W^{\mathrm{s}}\backslash\hat{W}^{\mathrm{s}}}$ and $\partial^{\mathrm{u}}Q \subset \overline{W^{\mathrm{u}}\backslash\hat{W}^{\mathrm{u}}}$. In addition, if $\partial^{\mathrm{s}}Q \cap \partial^{\mathrm{u}}Q = \emptyset$, then Q is homeomorphic to an open disc, while if $\partial^{\mathrm{s}}Q \cap \partial^{\mathrm{u}}Q = \emptyset$, then Q is homeomorphic to an open annulus.*

Definition 1.1. A connected component of the set $M\backslash\overline{(W\backslash\hat{W})}$ containing heteroclinic points is called a *heteroclinic domain.*

Examples of Heteroclinic Domains (see Fig. 1). Let x be a heteroclinic point of the diffeomorphism f, and $L^{\mathrm{s}}(p)$ $(L^{\mathrm{u}}(p))$ the connected components of $W^{\mathrm{s}}(p)\backslash p$ $(W^{\mathrm{u}}(q)\backslash q)$ such that $x \in L^{\mathrm{s}}(p) \cap L^{\mathrm{u}}(q)$. Let $m \in \mathbb{Z}^+$ be the smallest number for which $f^m(L^{\mathrm{s}}(p)) = L^{\mathrm{s}}(p)$.

We define the orientation on the curve $L^{\mathrm{s}}(p)$ $(L^{\mathrm{u}}(q))$ towards the point p (from the point q) and two numberings of the set of points $[x, f^m(x))^{\mathrm{s}} \cap L^{\mathrm{u}}(p)$; the first is in accordance with the orientation on $L^{\mathrm{s}}(p)$, and the second on $L^{\mathrm{u}}(q)$, where $[x, f^m(x))^{\mathrm{s}}$ is the arc of the curve $L^{\mathrm{s}}(p)$.

We denote by $S(x)$ the permutation that takes the first numbering to the second. We fix an orientation of the surface M and choose the tangent vector ξ^{s}_x (ξ^{u}_x) to the curve $L^{\mathrm{s}}(p)$ $(L^{\mathrm{u}}(p))$ at x inducing the direction chosen on it. We ascribe the *weight* $+1$ (-1) to the permutation $S(x)$ if the frame $(\xi^{\mathrm{s}}_x, \xi^{\mathrm{u}}_x)$ is compatible (incompatible) with the orientation of M.

Lemmas 1.1 and 1.2 enable us to associate with the diffeomorphism f the *oriented graph* $G(f)$ whose vertices correspond to sinks, sources, saddle-type periodic points and heteroclinic domains, while the edges correspond to the connected components of $W\backslash\Omega_1(f)$. Here if the edge l of $G(f)$ is incident to vertices a, b and l corresponds to the component $L^\sigma(p)$ $p \in \Omega_1(f)$, then one of the vertices, for example, a, corresponds to the point p, while there are two logical possibilities for the other vertex b: 1) if $L^\sigma(p) \notin \hat{W}^\sigma$, then b corresponds to a source (sink) when $\sigma = s$ $(\sigma = u)$; 2) if $L^\sigma(p) \subset \hat{W}^\sigma$, then b corresponds to a heteroclinic domain.

The orientation on the edge l is from b to a (from a to b) if $\sigma = s$ $(\sigma = u)$. We ascribe the weight $+1$ (-1) to a vertex a corresponding to a saddle-type periodic point p of period m if the map $Df^m_p : E^{\mathrm{u}}_p \to E^{\mathrm{u}}_p$ preserves (changes) the orientation of E^{u}_p. Here E^{u}_p is a one-dimensional expanding subspace of the tangent space to M at p which is invariant with respect to Df^m_p.

For a vertex γ corresponding to a heteroclinic domain Q we denote by $Q^{\mathrm{s}}(\gamma)$ $(Q^{\mathrm{u}}(\gamma))$ the set of all edges for which γ is the initial (end) point.

We number the edges in $Q^\sigma(\gamma)$ as follows. We choose a simple closed curve $L \subset Q$ such that:

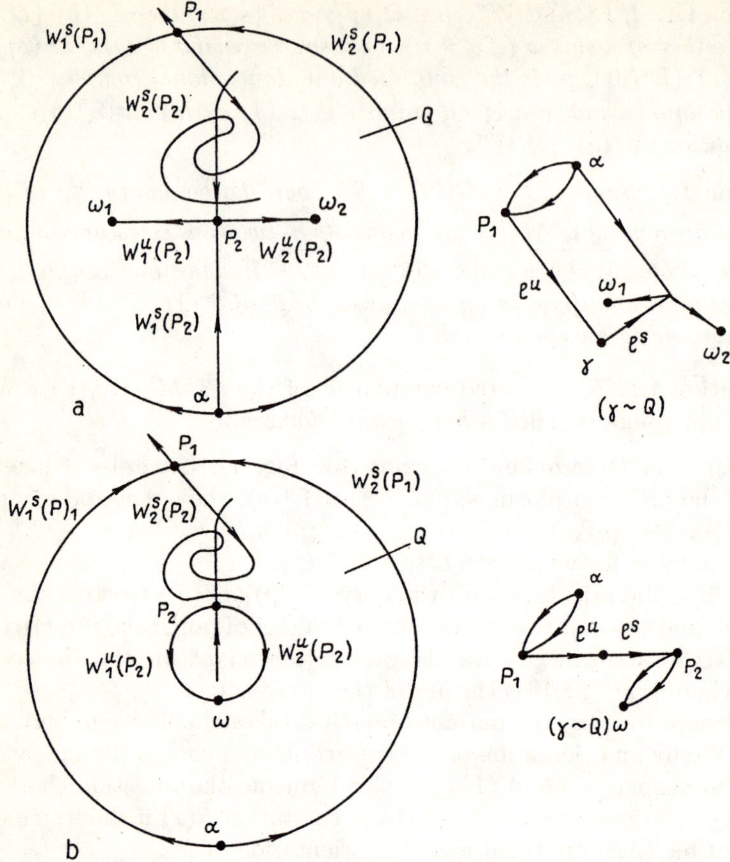

Fig. 1. Examples of heteroclinic domains Q and their corresponding discriminating sets

a) the domain Q homeomorphic to the disc

$$\partial^s Q = \overline{W_1^s(P_1) \cup W_1^s(P_2) \cup W_2^s(P_1)},\ \partial^u Q = \overline{W_1^u(P_2) \cup W_2^u(P_2)},\ \partial^s Q \cap \partial^u Q = P_2;$$

b) the domain Q homeomorphic to the annulus

$$\partial^s Q = \overline{W_1^s(P_1) \cup W_2^s(P_1)},\ \partial^u Q = \overline{W_1^u(P_2) \cup W_2^u(P_2)},\ \partial^s Q \cap \partial^u Q = \emptyset.$$

1) L bounds a disc D containing no points of ∂Q;

2) for each component $L^\sigma(p) \subset Q$, $p \in \partial Q$, there exists a neighbourhood $U(p)$ of p such that the intersection $L \cap (L^\sigma(p) \cap U(p))$ consists of a single point.

We fix the direction round the curve L so that D is to the left. We number the edges entering $Q^\sigma(\gamma)$ in the order in which one encounters the points of the components in the neighbourhoods $U(p)$, $p \in \partial Q$, corresponding to them in going round L starting with some point on L. We call this numbering the *orientation numbering.*

Let $N^\sigma(\gamma)$ be the number of edges in $Q^\sigma(\gamma)$. For each edge l_i^σ, $i \in \{1,2,\ldots,N^\sigma(\gamma)\}$, we denote by L_i^σ the components in $\hat{W}^\sigma$ corresponding to l_i^σ, and by $N_i^\sigma(\gamma)$ the number of heteroclinic trajectories that intersect the curve L_i^σ.

Let $x_1^\sigma,\ldots,x_{N_i^\sigma(\gamma)}$ be heteroclinic points on L_i^σ numbered in accordance with the orientation on L_i^σ, belonging to different heteroclinic trajectories and chosen so that there are no heteroclinic points distinct from them on the arc $[x_1^\sigma, x_{N_i^\sigma(\gamma)}^\sigma]^\sigma \subset L_i^\sigma$.

We define the map $\phi_\gamma^{i\sigma}$ from the set $\{1,2,\ldots,N_i^\sigma(\gamma)\}$ to $\{1,2,\ldots,N^{\bar\sigma}(\gamma)\}$ by setting $\phi_\gamma^{i\sigma}(j)=k$, where k is the number of the edge $l_k^{\bar\sigma} \in Q^{\bar\sigma}(\gamma)$ for which $L_k^{\bar\sigma} \cap L_i^\sigma = x_j$.

We define the map $\psi_\gamma^{i\sigma}$ from the set $\{1,2,\ldots,N_i^\sigma(\gamma)\}$ to the set of permutations by setting $\psi_\gamma^{i\sigma}(j)=S(x_j)$.

For the edge $l_k^\sigma \in Q^\sigma\mathrm{u}(\gamma)$ we denote by $K_\gamma^\sigma(l_k^\sigma)$ the subcomplex of the graph $G(f)$ consisting of the two edges corresponding to the connected components of $W^{\bar\sigma}(q_k)\backslash q_k$, where the point $q_k \in \Omega_1(f)$ corresponds to one of the vertices incident to the edge l_k^σ.

We choose the tangent vectors $\xi^\sigma(q_k)$, $\xi^{\bar\sigma}(q_k)$ to the curves $W^\sigma(q_k), W^{\bar\sigma}(q_k)$, respectively, so that $\xi^\sigma(q_k)$ induces on $W^\sigma(q_k)$ the direction corresponding to that on L_k^σ and the frame $(\xi^\sigma(q_k),\ \xi^{\bar\sigma}(q_k))$ is compatible with the orientation of M.

The vector $\xi^{\bar\sigma}(q_x)$ induces an orientation on $W^{\bar\sigma}(q_k)$. We number a pair of edges in $K_\gamma^\sigma(l_k^\sigma)$ in accordance with this orientation and call it the *orientation numbering.*

The boundaries of the connected components of $M\backslash\overline{(W\backslash\hat{W})}$ correspond to subcomplexes on the graph $G(f)$ which, by analogy with (Peixoto 1973), we call the *discriminating sets of the graph* $G(f)$.

Definition 1.2. By the *discriminating graph* $G^*(f)$ we mean the graph $G(f)$ endowed with:

1) the collection of discriminating sets;

2) the weights of the saddle vertices;

3) the orientation numberings in the sets $Q^\sigma(\gamma)$, $K_\gamma^\sigma(l_k^\sigma)$, $l_k^\sigma \in Q^\sigma(\gamma)$, the numbers $N^\sigma(\gamma)$, $N_i^\sigma(\gamma)$, the maps $\phi_\gamma^{i\sigma}$, $\psi_\gamma^{i\sigma}$ and the weights of the permutations $\psi_\gamma^{i\sigma}(j)$, $i \in \{1,2,\ldots,N^\sigma(\gamma)\}$, $j \in \{1,2,\ldots,N_i^\sigma(\gamma)\}$.

Definition 1.3. The discriminating graphs $G^*(f)$, $G'^*(f')$ of the diffeomorphisms f, f' are said to *isomorphic* if there exists an orientation numbering in the sets $Q^\sigma(\gamma)$, $Q'^\sigma(\gamma')$, $K_\gamma^\sigma(l_k^\sigma)$, $K'^\sigma_{\gamma'}(l'^\sigma_k)$, maps $\phi_\gamma^{i\sigma}$, $\psi_\gamma^{i\sigma}$, $\phi'^{i\sigma}_{\gamma'}$, $\psi'^{i\sigma}_{\gamma'}$ and an isomorphism ϕ of the graphs $G^*(f)$, $G'^*(f')$ such that:

1) ϕ takes vertices corresponding to sinks, sources, saddle-type periodic points and heteroclinic domains of $G^*(f)$ to the corresponding vertices of $G'^*(f')$;

2) ϕ preserves the weights of the vertices corresponding to the saddle-type periodic points;

3) if $\gamma' = \phi(\gamma)$, then $\phi(l_k^\sigma) = l_k^{'\sigma}$, where $l_k^\sigma \in Q^\sigma(\gamma)$, $l_k^{'\sigma} \in Q'^\sigma(\gamma')$, $N^\sigma(\gamma) = N'^\sigma(\gamma')$, $N_i^\sigma(\gamma) = N_i'^\sigma(\gamma')$, $\phi_\gamma^{i\sigma}(j) = \phi_{\gamma'}'^{i\sigma}(j)$, $\psi_\gamma^{i\sigma}(j) = \psi_{\gamma'}'^{i\sigma}(j)$ and the weights of the permutations $\psi_\gamma^{i\sigma}(j)$, $\psi_{\gamma'}'^{i\sigma}(j)$ are the same (different) if the numbers of isomorphic edges in $K_\gamma^\sigma(l_k^\sigma)$, $K_{\gamma'}'^\sigma(l_k'^\sigma)$ under ϕ are the same (different), where $k = \phi_\gamma^{i\sigma}(j)$, $i \in \{1, \dots, N^\sigma(\gamma)\}$, $j \in \{1, \dots, N_1^\sigma(\gamma)\}, \sigma \in \{s, u\}$.

The diffeomorphisms f, f' induce the permutations $P(f')$, $P'(f')$ on the sets of vertices of the graphs $G^*(f)$, $G'^*(f')$.

Theorem 1.1 (Grines). *In order that the Morse-Smale diffeomorphisms $f, f' \in \Phi(M)$ be topologically conjugate, it is necessary and sufficient that there exist an isomorphism ϕ of the graphs $G^*(f)$, $G'^*(f')$ such that $P' = \phi P \phi^{-1}$.*

§2. Cascades with a Countable Set of Periodic Points

2.1. Topological Classification of Basic Sets. In the problem of the topological classification of diffeomorphisms satisfying Axiom A (*A-diffeomorphisms*) with a countable set of periodic points, the centrepiece is the study of the topological structure of the restriction of the diffeomorphisms to *non-trivial* (distinct from periodic orbits) *basic sets* and also finding topological invariants of embedding the basic sets in the manifold.

The next step in this direction is to single out and find topological invariants of the basic classes of A-diffeomorphisms with non-trivial basic sets and also homeomorphisms not satisfying Axiom A which are the simplest representatives of those classes of topological conjugacy that either contain diffeomorphisms with Axiom A but are more general (they do not require smoothness or that estimates on the compression and expansion should be fulfilled, and so on), or do not contain diffeomorphisms with Axiom A, but which naturally arise in various important situations (for example, in the Teichmüller and Nielsen-Thurston theories).

We note that, according to (Shub and Sullivan 1975), there exists an A-diffeomorphism in each homotopy class of homeomorphisms on M; however, this does not generally hold for every topological conjugacy class, for example, in any topological conjugacy class of homeomorphisms on M containing a *pseudo-Anosov* homeomorphism (see below for the definition) there are no A-diffeomorphisms.

The fundamental results on the description of A-diffeomorphisms with a countable set of periodic points are due to Anosov and Smale (Anosov 1967), (Smale 1967).

In the case when f is an Anosov diffeomorphism, $M = T^2$ and the entire torus T^2 is a nonwandering set, the set Per$f$ being dense on T^2. This was proved for U-diffeomorphisms with an invariant measure in (Anosov 1967), while in the general case it follows from the topological conjugacy of a U-diffeomorphism on T^2 to an algebraic automorphism (Sinai 1968). Thus when f is a U-diffeomorphism, the nonwandering set $\Omega(f)$ is a basic set of dimension two. Conversely, if $f : M \to M$ has a basic set Ω of dimension two, then $\Omega = T^2$ and f is a U-diffeomorphism.

An example is given in (Smale 1967) of an A-diffeomorphism on S^2 whose nonwandering set contains a non-trivial zero-dimensional basic set (*Smale horseshoe*). Based on the Anosov-diffeomorphism of the torus T^2 Smale in (Smale 1967) presented a construction (*surgical operation*) of an A-diffeomorphism on T^2 (which he called a DA-diffeomorphism) whose nonwandering set contains a one-dimensional basic set (see also (Plykin 1984), (Williams 1970)).

Using a DA-diffeomorphism on T^2 one can construct A-diffeomorphisms with one-dimensional basic sets on any closed orientable surface M of genus $p \geq 1$. For example, in (Robinson and Williams 1971) an A-diffeomorphism is constructed on an orientable closed surface M of genus $p = 2$ (on a pretzel) with nonwandering set consisting of two one-dimensional basic sets.

An essential fact in the understanding of the structure of A-diffeomorphisms on surfaces was the construction by Plykin of examples of A-diffeomorphisms on S^2 with one-dimensional basic sets (see the survey (Plykin 1984) and the bibliography cited therein).

In this connection we note the paper (Fedotov 1980) in which sufficient conditions are given for the realization of a one-dimensional solenoid as an attractor of a diffeomorphism of the two-dimensional sphere.

Topological invariants describing the structure of the restriction of an A-diffeomorphism f to a basic set are based on the notion of a *Markov partition*, introduced by Sinai and improved by Bowen.

It was established in (Sinai 1968), (Bowen 1970) that the restriction of an A-diffeomorphism f to a basic set Ω is the continuous (and in the case when Ω is zero-dimensional, homeomorphic) image of a topological Markov chain.

At the same time, as was established in (Williams 1967), the restriction of f to a one-dimensional basic set is topologically conjugate to a shift of a solenoid.

Alekseev (Alekseev 1969) and Batterson (Batterson 1980) considered the problem of the realization of *Markov chains* by zero-dimensional basic sets.

The problem of the topological equivalence of basic sets with regard to their embeddings in a manifold is posed in the following way. Let Ω and Ω' be basic sets of the diffeomorphisms f and f' on M. Find necessary and sufficient conditions for the existence of a homeomorphism $\phi : M \to M$ such that $\phi(\Omega) = \Omega'$ and $f'|_{\Omega'} = \phi f \phi^{-1}|_{\Omega'}$.

One can extract from (Bowen 1970), (Williams 1967) conditions for the existence of a homeomorphism $\psi : \Omega \to \Omega'$ that conjugates the maps $f|_{\Omega}$, $f'|_{\Omega'}$. However, in the case $\dim \Omega = \dim \Omega' = 1$ the homeomorphism ψ cannot always be extended to the whole of M. In the case $\dim \Omega = \dim \Omega' = 0$ the homeomorphism ψ can always be extended to the whole of M, but the problem of the topological conjugacy of zero-dimensional basic sets still remains open, since the problem of actually finding matrices defining the Markov chains corresponding to the given basic sets Ω and Ω' has so far not been solved.

The first classification results were obtained by Grines for one-dimensional basic sets satisfying the *orientability* condition (Grines 1974), (Grines 1975). These results were generalized by Plykin to the case of one-dimensional *exteriorly situated* basic sets and one-dimensional basic sets with *connections of degree not greater than two* (Plykin 1980a), (Plykin 1980b), (Plykin 1984).[4]

Recently in (Grines and Kalay 1985), (Grines and Kalay 1988a), (Grines and Kalay 1988b) the problem of topological equivalence on a closed orientable surface M of genus $p \geq 0$ was finally solved for *basic sets without pairs of conjugate points* (see (Aranson and Grines 1990) for the notion of basic sets without pairs of conjugate points). This class of basic sets includes arbitrary one-dimensional basic sets. Furthermore, using constructions of the authors of this chapter, the problem of the topological classification of such basic sets has been solved (see the survey (Aranson and Grines 1990, Sect. 2) for the details).

The basic concept for this is that of the *carrier of a basic set.* According to (Anosov 1970), (Bowen 1971), any basic set Ω of an A-diffeomorphism $f : M \to M$ can be represented as a finite union of m $(m \geq 1)$ *C-dense components* $\Omega_1, \ldots, \Omega_m$, that is, closed subsets of Ω such that $f^m(\Omega_i) = \Omega_i$ and for any point $x \in \Omega_i$ the following relation holds:

$$\overline{W^\sigma(x) \cap \Omega_i} = \Omega_i, \ \sigma \in \{s; u\}.$$

To keep the exposition simple we suppose that Ω consists of a single C-dense component. In (Grines and Kalay 1988a), (Grines and Kalay 1988b) Grines and Kalay construct for a basic set Ω without pairs of conjugate points

---

[4] It was proved in (Plykin 1971) that any one-dimensional basic set Ω of a diffeomorphism $f : M \to M$ $(\dim M = 2)$ is either an *attractor* or a *repeller*, that is, either $\Omega = \bigcap_{n\geq 0} f^n(U)$ or $\Omega = \bigcap_{n\leq 0} f^n(U)$ respectively, where U is an open set containing Ω.

The notion of an *orientable* basic set can be found in (Grines 1974), (Grines 1975), while the notion of an *exteriorly situated* basic set can be found in (Plykin 1974) (these notions are also given in the survey (Aranson and Grines 1990)). We note that exteriorly situated basic sets do not exist on the sphere S^2, the projective plane P^2 or the Klein bottle K^2 (Plykin 1977). On the torus T^2 and a non-orientable manifold M of genus $p = 3$ exteriorly situated basic sets can only be orientable. On all the other closed surfaces there can be both orientable and non-orientable exteriorly situated basic sets. In (Plykin 1974) an estimate is given of the number N of one-dimensional exteriorly situated basic sets on a connected closed orientable surface of genus $p \geq 0: \ N \leq p + \max\{[p/2] - 1; \ 0\}$, where $[\cdot]$ denotes the integer part.

a submanifold $N_\Omega \subset M$ containing Ω, of genus $q \geq 0$ and with k ($k > 0$) components of the boundary such that:

1) the Euler characteristic $\chi(N_\Omega) = 2 - 2q - k$ is negative;
2) $N_\Omega \backslash \Omega$ contains no points of the nonwandering set $\Omega(f)$;
3) the numbers q and k are uniquely defined by Ω.

The submanifold N_Ω is called the *carrier of the basic set* Ω. The invariants of the topological equivalence of basic sets are expressed in terms of the actions of the diffeomorphisms in the fundamental group of the carriers of the basic sets (see Aranson and Grines 1990, Theorems 1–3).

As was pointed out above, the condition of topological conjugacy of one-dimensional basic sets was first obtained for orientable (Grines 1974), (Grines 1975, Part 1) and for exteriorly situated basic sets (Plykin 1984, Theorem 3.3) on closed orientable surfaces M of genus $p \geq 2$. (In (Plykin 1984, Theorem 3.3) M can even be a non-orientable closed surface of genus $p \geq 3$.)

The selection of the carrier N_Ω for an arbitrary basic set enabled the authors of (Grines and Kalay 1985), (Grines and Kalay 1988a), (Grines and Kalay 1988b) to extend the technique of (Grines 1975, Part 1) to a manifold N_Ω with boundary on which Ω is exteriorly situated in a certain sense.

In (Plykin 1980a), (Plykin 1984, Theorems 2.5–2.7) the invariants of topological conjugacy obtained in (Grines 1975, Part 2) for one-dimensional exteriorly situated basic sets on a two-dimensional torus T^2 (such basic sets of T^2 are always orientable) were generalized in a strengthened version to the case of orientable and non-orientable basic sets of A-diffeomorphisms on closed orientable surfaces M of genus $p \geq 0$ under the hypothesis that these basic sets have connections of degree one or two. A detailed account of problems related to the investigation of basic sets on T^2, as well as the topological structure of attractors on the sphere S^2 is given in the survey (Plykin 1984).

2.2. Topological Classification of *A*-diffeomorphisms with Non-trivial Basic Sets and Homeomorphisms with Two Invariant Transversal Foliations. We indicate some fundamental classes of cascades on a closed orientable surface M of genus $p \geq 0$ for which a complete topological invariance has been found.

A wide class, the introduction of which is based on the results described in Sect. 1 and Sect. 2.1 has been studied in (Grines and Kalay 1988a), (Kalay 1988). Representatives of this class are A-diffeomorphisms satisfying a strict transversality condition (such diffeomorphisms are *structurally stable*) and their nonwandering set can contain non-trivial one-dimensional and zero-dimensional basic sets without pairs of conjugate points.

A complete topological invariant of diffeomorphisms of this class is the discriminating graph analogous to the graph of the Morse-Smale diffeomorphism (see Sect. 1) but supplemented by the set of vertices corresponding to non-trivial basic sets and endowed with the group invariants described in (Aranson and Grines 1990, Sect. 2).

It seems likely that such A-diffeomorphisms exist in each homotopy class of homeomorphisms on M.

We go into this class, which we denote by $\Gamma(M)$, in more detail. Prior to introducing this class we give certain definitions and notions.

Given an A-diffeomorphism $f: M \to M$, let Ω be a basic set, Λ a nonwandering set, Λ' the union of all saddle-type periodic points in the trivial basic sets, and Λ_0 the union of all periodic points that are sinks or sources.

A region $G \subset M$ is said to be *wandering* for $f: M \to M$ if the relation $f^n(G) \cap G = \emptyset$ holds for all $n \in \mathbb{Z}$, $n \neq 0$.

For a saddle-type periodic point $p \in \Lambda'$ we denote by $W_i^\sigma(p)$ $(i = 1, 2)$ the connected components of $W^\sigma(p) \backslash p$. We set $W_{i,\alpha}^\sigma(p) = \{x \in W_i^\sigma(p) | l(x,p) < \alpha\}$, where $l(x,p)$ is the distance in the metric of the curve $W^\sigma(p)$. Let $p, q \in \Lambda'$ be saddle-type periodic points for which $W_i^\sigma(p) \cap W_j^{\bar\sigma}(q) \neq \emptyset$, $i, j \in \{1, 2\}$. We say that the *intersection is orientable* if for any $\alpha > 0$, $\beta > 0$ the index of intersection of the curves $W_{i,\alpha}^\sigma(p), W_{j,\beta}^{\bar\sigma}(q)$ is the same for all points of intersection.

We say that the A-diffeomorphism $f: M \to M$ is of *class* $\Gamma(M)$ if

1) f satisfies the strict transversality condition;

2) there does not exist a wandering region (cell) bounded by a simple closed curve consisting of an arc of the manifold $W^\sigma(p)$ and an arc of the manifold $W^{\bar\sigma}(q)$ of saddle-type periodic points $p, q \in \Lambda'$;

3) if for some component $W_i^\sigma(p)$, where $p \in \Lambda'$, the intersection $W^\sigma(p) \cap W^{\bar\sigma}(\Omega) = \emptyset$ for any non-trivial basic set Ω, then for any component $W_j^\sigma(q)$, $q \in \Lambda'$, $j \in \{1, 2\}$, the intersection $W_i^\sigma(p) \cap W_j^{\bar\sigma}(q)$ is either empty or orientable.

It follows from condition 2) that all non-trivial basic sets of a diffeomorphism $f \in \Gamma(M)$ contain no pairs of conjugate points.

Our problem is to establish necessary and sufficient conditions for topological conjugacy of diffeomorphisms in this class. For this we need some further concepts.

A periodic point $p \in \Omega$ is called a *boundary periodic point of type* σ if one of the connected components of $W^\sigma(p) \backslash p$ is disjoint from Ω, while both the connected components of $W^{\bar\sigma}(p) \backslash p$ intersect Ω; it is called a *boundary periodic point of type* (s, u) if one of the connected components of the sets $W^{\mathrm{s}}(p) \backslash p$, $W^{\mathrm{u}}(p) \backslash p$ is disjoint from Ω.

If one of the connected components of $W^\sigma(p) \backslash p$ is disjoint from Ω, then we denote it by $W_\emptyset^\sigma(p)$, the other connected component intersecting Ω being denoted by $W_\infty^\sigma(p)$.

We denote by W^{s} (W^{u}) the union of the stable (unstable) manifolds of all the points of Λ' and of the manifolds $W_\emptyset^{\mathrm{s}}(p)$, $W_\emptyset^{\mathrm{u}}(p)$ of all the boundary periodic points. We set $W = W^{\mathrm{s}} \cup W^{\mathrm{u}}$.

Let $W_i^{\mathrm{s}}(p) \cap W_j^{\mathrm{u}}(q) \neq \emptyset$, $j \in \{1, 2\}$, where $p, q \in \Lambda'$ and $W_i^{\mathrm{s}}(p) \cap W^{\mathrm{u}}(\Omega) = \emptyset$ or $W_j^{\mathrm{u}}(q) \cap W^{\mathrm{s}}(\Omega) = \emptyset$ for any non-trivial basic set Ω. It can then be established that there exists a set k_{pq} on M such that:

1) k_{pq} is homeomorphic to an open annulus;

2) k_{pq} contains $W_i^{\mathrm{s}}(p) \cup W_j^{\mathrm{u}}(q)$;

3) the boundary of k_{pq} is the union $\partial W_i^{\mathrm{s}}(p) \cup \partial W_j^{\mathrm{u}}(q)$, where $\partial W_i^{\mathrm{s}}(p) \backslash p$, $(\partial W_j^{\mathrm{u}}(q) \backslash q)$ are connected one-dimensional complexes consisting of certain

points of Λ_0 that are sources (sinks) and of stable (unstable) manifolds of certain points of Λ'.

A set k_{pq} satisfying the above properties is called a *heteroclinic annulus* of the diffeomorphism f.

The union $N_\Omega = \bigcup_{i=1}^m N_{\Omega_i}$ of the non-intersecting carriers of the C-dense components N_{Ω_i} of the basic set Ω is called the *carrier* of Ω.

A pair of carriers $(N_{\Omega_0}, n_{\Omega_0}^m)$ of a C-dense component Ω_0 of a basic set Ω is called a *compatible pair* if $n_{\Omega_0}^m$ and $f^m(n_{\Omega_0}^m)$ belong to N_{Ω_0}.

It can be established that for any C-dense component Ω_0 of a basic set Ω there exists a compatible pair of carriers $(N_{\Omega_0}, n_{\Omega_0}^m)$.

We consider a compatible pair of carriers $(N_{\Omega_0}, n_{\Omega_0}^m)$ of a C-dense component Ω_0 of the basic set Ω. Since the Euler characteristic of N_{Ω_0} is negative, there exists a subset H_F of the Lobachevskij plane U in the Poincaré realization on the interior of the disc $|z| < 1$ of the complex z-plane and a discrete group of motions F on U leaving subset H_F invariant such that H_F/F is homeomorphic to N_{Ω_0}.

We denote by π the natural projection $H_F \to N_{\Omega_0}$, and by $\bar{n}_{\Omega_0}^m$ the inverse image of $n_{\Omega_0}^m$ on H_F. The diffeomorphism $\bar{f}_m : \bar{n}_{\Omega_0}^m \to H_F$, which covers the diffeomorphism $f^m : n_{\Omega_0}^m \to N_{\Omega_0}$ $(\pi \bar{f}_m = f^m \pi)$, induces an automorphism $\bar{f}_{m^*}$ of the group F in accordance with the formula $\bar{f}_{m^*}(\gamma) = \bar{f}_m \gamma \bar{f}_m^{-1}$, $\gamma \in F$.

We set $B_F(f^m) = \bigcup_{\gamma \in F} A_\gamma \bar{f}_{m^*}$, where A_γ is the inner automorphism of F $(A_\gamma(\alpha) = \gamma\alpha\gamma^{-1};\ \alpha, \gamma \in F)$. We associate with the basic set Ω the pair $(F, B_F(f^m))_\Omega$.

We denote by $\bar{\Omega}_0$ the inverse image of Ω_0 on H_F (Ω_0 is a C-dense component of the basic set Ω). Let $p \in \Omega_0$ be a boundary periodic point of period k. There exists a map $\bar{f}_k : \bar{\Omega}_0 \to \bar{\Omega}_0$ covering the map $f^k|_{\Omega_0} (\pi \bar{f}_k|_{\Omega_0} = f^k \pi|_{\bar{\Omega}_0})$ such that $\bar{f}_k(\bar{p}) = \bar{p}$, where $\bar{p}$ is a point of the inverse image of p on H.

We set $\alpha = \sigma$ if p is of type σ, and $\alpha = s$ if p is of type (s, u). We denote by L^α the connected component of $(W_\infty^\alpha(p) \cup p) \cap N_{\Omega_0}$ containing p, and by $\bar{L}^\alpha$ the inverse image of L^α on H_F containing the point $\bar{p}$.

It can be established that for any $\bar{x} \in \bar{L}^\alpha \cap \bar{\Omega}_0$ the sequence of points $\{\bar{f}_k^n(\bar{x})\}$ converges to a unique point $z_{\bar{p}}$ on the absolute $E : |z| = 1$, where $n \in \mathbb{Z}^+$ if $\alpha = u$ and $n \in \mathbb{Z}^-$ if $\alpha = s$.

The set $\mu_F(W_\infty^\alpha(p)) = \bigcup_{\gamma \in F} \gamma(z_{\bar{p}})$ is called the *homotopy rotation class* of $W_\infty^\alpha(p)$.

We associate with each diffeomorphism $f \in \Gamma(M)$ the *discriminating* graph $G^*(f)$. The vertices of the graph correspond to the periodic points of the trivial basic sets, the C-dense components of the non-trivial basic sets and the heteroclinic annuli. The edges of the graph correspond to the connected components of the set $W \backslash \Lambda$ and are oriented in accordance with the stability of the components corresponding to them.

Furthermore, certain vertices and edges of the discriminating graph are equipped as follows. The sets of vertices corresponding to one basic set Ω are equipped with the pair $(F, B_F(f^m))_\Omega$, while edges incident to vertices corresponding to C-dense components are equipped with homotopy rotation classes.

The diffeomorphism f induces a permutation $P(f)$ on the vertices of the graph. As in Sect. 1 we introduce the notion of an isomorphism ϕ of the discriminating graphs $G^*(f_1), G^*(f_2)$ of the diffeomorphisms $f_1, f_2 \in \Gamma(M)$.

Theorem 2.1 (Grines and Kalay). *The diffeomorphisms $f_1, f_2 \in \Gamma(M)$ are topologically conjugate if and only if there exists an isomorphism ϕ of the graphs $G^*(f_1), G^*(f_2)$ such that $P(f_2) = \phi P(f_1)\phi^{-1}$.*

The reader is referred to (Grines and Kalay 1988a), (Kalay 1988) for a precise statement and proofs.

Other important classes are classes of cascades on M containing homeomorphisms with two invariant foliations with singularities (or without them) that are *transversal in the topological sense.*

We denote three such classes by $Q_i^{(m)}(M)$ $(i = 0, 1, 2)$. The homeomorphisms of these classes can, in particular, be Anosov diffeomorphisms, pseudo-Anosov homeomorphisms, and *generalized pseudo-Anosov homeomorphisms* (see below for the definition). Such classes have been considered in (Aranson 1986), (Aranson 1988, Part 2), (Grines 1982), and also in part, in the survey (Aranson and Grines 1990, Sect. 4).[5]

Before introducing these classes we give the requisite concepts.

By a foliation F *with singularities* (the set of singularities can be empty) defined on M we mean a partition of M into curves that are homeomorphic in a neighbourhood of non-singular points to a family of parallel lines, while for sets of singularities, for each class of foliations we have to make separate stipulations (from this point of view, each flow on M can also be treated as a foliation with singularities, where the singularities of this foliation are the equilibrium points of the flow).

We introduce on M two classes of foliations $T_0^{(m)}(M)$, $T_1^{(m)}(M)$, the first class being called the class of *transitive foliations* and the second the *class of foliations with non-trivial minimal sets*, where by a *minimal set* Ω of a foliation F we mean a non-empty closed set that is invariant with respect to F and contains no proper subsets with these properties.[6]

A foliation F is of *class* $T_0^{(m)}(M)$ if:

1) F contains an everywhere dense leaf on M;

2) F contains a non-empty finite set of singularities, each singularity being a topological saddle of integral or half-integral index;[7]

3) there are no separatrices going from saddle to saddle;

---

[5] The study of such homeomorphisms began with (Anosov 1967). More recently there has been a renewed interest in the study of such homeomorphisms, much of which is due to the work of Thurston in (Thurston 1988), (Poénaru 1979).

[6] A set Ω is called *invariant* with respect to F if for any point $x \in \Omega$ the leaf L_x passing through x lies entirely in Ω.

[7] By the *index* of a topological saddle with ν separatrices we mean the number $1 - \nu/2$.

4) if F contains no spines (saddles with a single separatrix), then $m = 0$, while if there are spines, then m is the number of singularities of half-integral index.

A foliation F is of *class* $T_1^{(m)}(M)$ if:

1) F contains a non-trivial minimal set Ω (Ω is not a singularity or a compact leaf);

2) F contains no singularities and is nowhere dense on M;

3) F contains a non-empty finite set of singularities and each singularity is a topological saddle of non-zero integral or half-integral index; in each connected component of $M \setminus \Omega$ there is exactly one singularity and there are no separatrices going from saddle to saddle;

4) if F contains no spines, then $m = 0$, while if there are spines, then m is the number of singularities of half-integral index.

Since the sum of the indices of the singularities of the foliation F is equal to $2-2p$, where p ($p \geq 0$) is the genus of the closed orientable surface M, then the number m occurring in the definition of the classes $T_0^{(m)}(M)$, $T_1^{(m)}(M)$ is always even, and if $M = S^2$, then $m \geq 4$, while if $M \neq S^2$, then $m \geq 0$.[8]

We denote by $Q_0^{(m)}(M)$ the class of homeomorphisms on M such that any homeomorphism $f \in Q_0^{(m)}(M)$ satisfies the following properties:

1) f has two invariant transitive rectifiable foliations F^{u}, $F^{\mathrm{s}} \in T_0^{(m)}(M)$ having the same non-empty finite set of singularities $A_1, \ldots, A_r$, these singularities being topological saddles of non-zero integral or half-integral index;

2) F^{u} (F^{s}) is *expanding* (*contracting*), that is, for any points a, b of the leaf L^{u} (L^{s}) of the foliation F^{u} (F^{s}) the following relation holds:

$$\rho_{\mathrm{u}}(f^{-n}(a),\ f^{-n}(b)) \to 0 \quad (\rho_{\mathrm{s}}(f^{n}(a),\ f^{n}(b)) \to 0) \quad \text{as } n \to +\infty,$$

where ρ_{u} (ρ_{s}) is the intrinsic metric on L^{u} (L^{s});

3) $F^{\mathrm{u}}, F^{\mathrm{s}}$ are *transversal in the topological sense* everywhere except at the points $A_1, \ldots, A_r$ (see (Solodov 1982) for the definition of transversality in the topological sense).

In particular, the class $Q_0^{(m)}(M)$ contains the pseudo-Anosov and generalized pseudo-Anosov homeomorphisms.

---

[8] In (Aranson 1986), (Aranson 1988, Part 1), (Aranson and Zhuzhoma 1984) a topological classification of such foliations on M is given in the language of the *orbit of a homotopy rotation class* of the semi-leaves. This invariant is a generalization of the *homotopy rotation class* of the survey introduced by the authors for the classification of flows on M (Aranson and Grines 1973), (Aranson and Grines 1984), (Aranson and Grines 1985), (Aranson and Grines 1986). In (Aranson 1990) the technique of foliations developed in (Aranson 1986), (Aranson 1988, Part 1) is used to give a classification of vector fields on the plane having *strange attractors* consisting entirely of equilibrium states and which are asymptotically attracting limit sets that are locally homeomorphic to the direct product of a Cantor set and an interval. See also the survey (Aranson and Grines 1990, Sect. 4) with regard to the classification of such foliations.

We introduce alongside the class $Q_0^{(m)}(M)$ two further classes $Q_1^{(m)}(M)$ and $Q_2^{(m)}(M)$ for which we need some preliminary notions.

We note that on M the set of points $\Sigma^{(m)} = \{s_1, \ldots, s_m\}$ (m is even, $m \geq 4$ if $M = S^2$; $m \geq 0$ if $M \neq S^2$; here if $m = 0$, then $\Sigma^{(m)} = \emptyset$).

By the Klein-Poincaré uniformization theorem (Krushkal', Apanasov, and Gusevskij 1981), M can be represented as the quotient space $\overline{M}/G$ where $\overline{M}$ is the disc $x^2 + y^2 < 1$, G is the discrete group of motions on $\overline{M}$ (G is called the *covering group*), and the cover $\pi : \overline{M} \to M$ is the universal non-branched cover in the case $m = 0$, while in the case $m \neq 0$ it is the universal branched cover with branchings of order two over the points $s_i \in \Sigma^{(m)}$ ($i = \overline{1, m}$). The circle $E : x^2 + y^2 = 1$ is called the *absolute*.

An automorphism τ of the group G is called *hyperbolic* if for any two elements $g, \gamma \in G$ ($\gamma^2 \neq \mathrm{id}$) and any integer $n \neq 0$ we have $g\tau^n(\gamma)g^{-1} \neq \gamma$.

We say that a homeomorphism $f : M \to M$ is of class $Q_i^{(m)}(M)$ ($i = 1; 2$) if:

1) f contains two invariant rectifiable foliations $F^{\mathrm{u}}, F^{\mathrm{s}}$ such that $F^{\mathrm{u}} \in T_1^{(m)}(M)$, $F^{\mathrm{s}} \in T_{i-1}^{(m)}(M)$; F^{u} and F^{s} have the same non-empty finite set of singularities $A_1, \ldots, A_r$ and these singularities are topological saddles of non-zero integral or half-integral index;

2) in the case $i = 1$, the foliation F^{u} has the property of being expanding by at least one of its separatrices;

3) $F^{\mathrm{u}}, F^{\mathrm{s}}$ are transversal in the topological sense everywhere except at the points $A_1, \ldots, A_r$;

4) f induces a hyperbolic automorphism τ of G, that is, there exists a homeomorphism $f_1 : M \to M$, $f_1(\Sigma^{(m)}) = \Sigma^{(m)}$, topologically conjugate to f such that at least one lifting of it $\bar{f}_1 : \overline{M} \to \overline{M}$ induces the hyperbolic automorphism τ of the group G ($\tau(\xi) = \bar{f}_1 \xi \bar{f}_1^{-1}$, $\xi \in G$).[9]

[9] Since $f_1(\Sigma^{(m)}) = \Sigma^{(m)}$, it follows that f_1 is lifted to $\overline{M}$. However, f_1 cannot always be lifted to a closed orientable surface $\hat{M}$ of genus $2p - 1 + \frac{m}{2}$ that is a two-sheeted branched cover for M with branchings of order two over the points of $\Sigma^{(m)}$. It is shown in (Aranson 1988, Part 2) that if $M = S^2$, then $f_1 : S^2 \to S^2$ ($f_1(\Sigma^{(m)}) = \Sigma^{(m)}$) is always lifted to $\hat{M}$. If, on the other hand, $M \neq S^2$, then we have the following proposition (O'Brien 1970). Let $\pi_1(\hat{M} \backslash \hat{\Sigma}^{(m)}, a)$, $\pi_1(\hat{M} \backslash \hat{\Sigma}^{(m)}, a')$ be the fundamental groups for $\hat{M} \backslash \hat{\Sigma}^{(m)}$ with base points $a, a' \in \hat{M}$, $q : \hat{M} \to M$ the natural projection, and $\hat{\Sigma}^{(m)}$ the inverse image of $\Sigma^{(m)}$ on $\hat{M}$ ($\Sigma^{(m)} = q(\hat{\Sigma}^{(m)})$), where $q\big|_{\hat{\Sigma}^{(m)}}$ is a one-to-one map. Then in order that the homeomorphism $f_1 : M \to M$ ($f_1(\Sigma^{(m)}) = \Sigma^{(m)}$) can be lifted to $\hat{M}$, realizing the homeomorphism $\hat{f}_1 : \hat{M} \to \hat{M}$ of it, where $\hat{f}_1(\hat{\Sigma}^{(m)}) = \hat{\Sigma}^{(m)}$, $\hat{f}_1(a) = a'$, it is necessary and sufficient that the group $f_{1_*} q_* \pi_1(\hat{M} \backslash \hat{\Sigma}^{(m)}, a)$ be conjugate to the group $q_* \pi_1(\hat{M} \backslash \hat{\Sigma}^{(m)}, a')$, where f_{1_*} and q_* are the automorphism and homomorphism induced by f_1 and q. In particular, if f_1 is homotopic to the identity via a homotopy $f_{1_\mu} : M \to M$ such that $f_{1_\mu}(\Sigma^{(m)}) = \Sigma^{(m)}$, then f_1 can be lifted to $\hat{M}$.

It follows from the conditions defining the classes $Q_i^{(m)}(M)$ $(i = 0, 1, 2)$, that each leaf of the foliations $F^{\mathrm{u}}, F^{\mathrm{s}}$ contains at most one singularity and that the indices of the singularities of $F^{\mathrm{u}}, F^{\mathrm{s}}$ are the same at each point A_i $(i = \overline{1, r})$.

Definition 2.1. The homeomorphisms $f, f' : M \to M$ are called $\pi_1^{(m)}$-*conjugate* if there exist homeomorphisms $f_1 f_1' : M \to M$ topologically conjugate to them that leave the set $\Sigma^{(m)}$ invariant and are such that there exists an automorphism τ of G and liftings $\bar{f}_1, \bar{f}_1' : \overline{M} \to \overline{M}$ for f_1, f_1' such that $\bar{f}_{1*}' = \tau \bar{f}_{1*} \tau^{-1}$, where the automorphisms $\bar{f}_{1*}, \bar{f}_{1*}'$ are induced by $\bar{f}_1, \bar{f}_1'$.

Remark. For the case $m = 0$ the notion of $\pi_1^{(m)}$-conjugacy (Aranson 1986) is the same as that of π_1-conjugacy (Grines 1982).

We say that the homeomorphisms $f, f' : M \to M$ are *topologically conjugate on the invariant sets* Ω_0, Ω_0' *(with respect to* f, f'*)* if there exists a homeomorphism $\phi : M \to M$ such that $\Omega_0' = \phi(\Omega_0)$ and $f'|_{\Omega_0'} = \phi f \phi^{-1}|_{\Omega_0'}$.

Let $f, f' \in Q_i^{(m)}(M)$ $(i = 0, 1, 2)$; $F^{\mathrm{u}}, F^{\mathrm{s}}$ and $F'^{\mathrm{u}}, F'^{\mathrm{s}}$ are the invariant foliations with respect to f, f' involved in the definition of the classes $Q_i^{(m)}(M)$; $\Omega^{\mathrm{u}}, \Omega'^{\mathrm{u}}$ and $\Omega^{\mathrm{s}}, \Omega'^{\mathrm{s}}$ are non-trivial minimal sets of the foliations $F^{\mathrm{u}}, F'^{\mathrm{u}}$ and $F^{\mathrm{s}}, F'^{\mathrm{s}}$ respectively.

It follows from the properties of these classes that in the case $i = 0$ the minimal sets $\Omega^{\mathrm{u}}, \Omega'^{\mathrm{u}}, \Omega^{\mathrm{s}}, \Omega'^{\mathrm{s}}$ coincide with M, that is, they have dimension two; in the case $i = 1$ the sets $\Omega^{\mathrm{u}}, \Omega'^{\mathrm{u}}$ are one-dimensional and each connected component of the complement of these sets with respect to M is simply connected and $\Omega^{\mathrm{s}}, \Omega'^{\mathrm{s}}$ coincide with M; in the case $i = 2$, $\Omega^{\mathrm{u}}, \Omega'^{\mathrm{u}}, \Omega^{\mathrm{s}}, \Omega'^{\mathrm{s}}$ are one-dimensional and each connected component of the complement of these sets with respect to M is simply connected.

Theorem 2.2 (Aranson). *The following three statements hold:*

I. *Let* $f, f' \in Q_0^{(m)}(M)$. *Then in order that* f, f' *be topologically conjugate on* M *it is necessary and sufficient that* f, f' *be* $\pi_1^{(m)}$*-conjugate;*

II. *Let* $f, f' \in Q_1^{(m)}(M)$ *and let* M *be distinct from* S^2 *with four distinguished points. Then in order that* f, f' *be topologically conjugate on the sets* $\Omega^{\mathrm{u}}, \Omega'^{\mathrm{u}}$, *it is necessary and sufficient that* f, f' *be* $\pi_1^{(m)}$*-conjugate;*

III. *Let* $f, f' \in Q_2^{(m)}(M)$ *and let* M *be distinct from* S^2 *with four distinguished points. Then:*

1) *if* f, f' *are topologically conjugate on the sets* $\Omega_0 = \Omega^{\mathrm{u}} \cap \Omega^{\mathrm{s}}$, $\Omega_0' = \Omega'^{\mathrm{u}} \cap \Omega'^{\mathrm{s}}$ *and the conjugating homeomorphism* ϕ: $M \to M$ *takes* $\Omega^{\mathrm{u}}, \Omega^{\mathrm{s}}$ *to* $\Omega'^{\mathrm{u}}, \Omega'^{\mathrm{s}}$, *then* f, f' *are* $\pi_1^{(m)}$*-conjugate;*

2) *if* f, f' *are* $\pi_1^{(m)}$*-conjugate, then* f, f' *are topologically conjugate on the sets* $\Omega_0 = \Omega^{\mathrm{u}} \cap \Omega^{\mathrm{s}}$, $\Omega_0' = \Omega'^{\mathrm{u}} \cap \Omega'^{\mathrm{s}}$.[10]

[10] In the situation of Statements II and III of Theorem 2.2 it is not difficult to give examples of homeomorphisms $f, f' \in Q_i^{(m)}(M)$ $(i = 1, 2)$, such that f, f' are

Remark 1. Statement I of Theorem 2.2 was considered by Grines in (Grines 1982) for $m = 0$ in the presence of two invariant orientable transversal foliations for the diffeomorphisms f, f'; it was considered by Aranson in (Aranson 1986), (Aranson 1988, Part 2) for $m = 0$ in the presence of non-orientable foliations for the homeomorphisms f, f' and also for $m \neq 0$. Statements II and III of Theorem 2.2 were considered by Aranson in (Aranson 1988, Part 2).

Remark 2. In (Aranson 1988, Part 2) the analogous assertion for Statements II and III of Theorem 2.2 was considered for S^2 with four distinguished points and for T^2 without distinguished points. Here it is possible that there exist connected components that belong to the complements with respect to S^2 or T^2 of nowhere dense non-trivial minimal sets of invariant foliations with respect to f, f' such that these components contain no singularities of the foliations.

Theorem 2.3 (Aranson). *The following assertions hold:*

I. 1) *Let* $f \in Q_0^{(m)}(M)$ *and in the case* $m \neq 0$, *suppose that the set of singularities of half-integral index of the foliation* F^{u} *(or* F^{s}*) coincides with* $\Sigma^{(m)}$. *Then the automorphism* $\tau : G \to G$ *induced by any lifting* $\bar{f} : \overline{M} \to \overline{M}$ *of* f *is hyperbolic.*

2) *For any hyperbolic automorphism* $\tau : G \to G$ *there exists a homeomorphism* $f \in Q_0^{(n)}(M)$ *(*n *may not be the same as* m*) such that for some lifting* $\bar{f} : \overline{M} \to \overline{M}$ *of* f *we have* $\bar{f}_* = \tau$, *where* $\bar{f}_*$ *is the automorphism of* G *induced by* $\bar{f}$.

II. *Let* M *be distinct from* S^2 *with four distinguished points, and* $\tau : G \to G$ *a hyperbolic automorphism. Then there exists a homeomorphism* $f \in Q_i^{(n)}(M)$ $(i = 1, 2)$ *(*n *may not be the same as* m*) such that for some lifting* $\bar{f} : \overline{M} \to \overline{M}$ *of* f *we have* $\bar{f}_* = \tau$.

Remark. This theorem is proved in (Aranson 1988, Part 2). In this same work the author considers the analogous assertion to II of Theorem 2.3 for S^2 with four distinguished points and for T^2 without distinguished points, where it is possible that there exist connected components that belong to complements of nowhere dense non-trivial minimal f-invariant foliations with respect to S^2 or T^2 such that these components contain no singularities of the foliations.

topologically conjugate on the sets $\Omega^{\mathrm{u}}, \Omega'^{\mathrm{u}}$ (for $i = 1$) or on the sets $\Omega_0 = \Omega^{\mathrm{u}} \cap \Omega^{\mathrm{s}}$ and $\Omega'_0 = \Omega'^{\mathrm{u}} \cap \Omega'^{\mathrm{s}}$ (for $i = 2$), but f, f' are not topologically conjugate on the whole of M.

§3. Various Approaches to the Problem of the Realization of Homotopy Classes of Homeomorphisms with Given Topological Properties

In this section we give results relating to the various approaches to the problem of realizing homotopy classes of homeomorphisms on closed orientable surfaces M of genus $p \geq 2$. The first highly exhaustive results on this question were obtained by Nielsen in the twenties and thirties (Nielsen 1927), (Nielsen 1937). However in recent times, starting from the seventies, Nielsen's theory has been amplified by new results in principle which are at the interface of various methods and theories (Anosov and Smale's hyperbolic theory of diffeomorphisms, the Teichmüller theory of quasiconformal maps, Thurston's theory on the geometry and dynamics of maps of surfaces).

A fairly detailed bibliography on these questions is given in the surveys (Aranson and Grines 1990), (Plykin 1984).

3.1. Geodesic Laminations and Their Role in the Construction of Representatives of Homotopy Classes of Homeomorphisms. In his papers (Nielsen 1927), (Nielsen 1937), Nielsen called attention to the special role of geodesics in the solution of the problem of constructing homeomorphisms in a given homotopy class, although to the very end this role was never explained by him. A new impetus in this direction was given by Thurston (Thurston 1976) who rediscovered the fundamental aspects of Nielsen's theory, stating them in another language and amplifying them with new results of principle, the principal one of which is the discovery of a new representative in one of the types of homotopy classes of homeomorphisms of surfaces, namely, the pseudo-Anosov homeomorphism.

As a preliminary, we give a description of the properties of *geodesic laminations* which play a key role for an understanding of the connection between the geometry of surfaces and the topological properties of foliations, flows and cascades defined on it (Aranson and Grines 1978).

We introduce a metric of constant negative curvature on M. Following (Thurston 1976), we call a set Λ on M a *geodesic lamination* if Λ is non-empty, closed in M and is the union of disjoint geodesics without self-intersections.

We represent M as the quotient space $\overline{M}/G$, where $\overline{M}$ is the Lobachevskij plane in the Poincaré realization in the disc $x^2 + y^2 < 1$ and G is the discrete group of hyperbolic elements isomorphic to the fundamental group M. Each element $g \in G$ ($g \neq \mathrm{id}$) has exactly two fixed points which lie on the absolute $E : x^2 + y^2 = 1$; one of these points is an attractor and the other a repeller.

Those points of the absolute that are fixed points of non-trivial elements of G are called *rational points.* The remaining points of the absolute are called *irrational points.*

Let L be a directed geodesic on M and l its connected inverse image on $\overline{M}$. Then l has exactly one limit point on E under the motion along l corresponding to the direction chosen on this geodesic.

We denote this limit point by $\sigma(l)$ and call it the *asymptotic direction* given by the directed geodesic l.

The following definition is taken from (Aranson and Grines 1973).

Definition 3.1. By the *homotopy rotation class* of a directed geodesic L defined on M we mean the orbit of the action of the group G on the point $\sigma(l)$, that is, the set $\mu(L) = \bigcup_{g \in G} g(\sigma(l))$.

It has been established (Aranson and Grines 1978) that if L is either a simple closed geodesic or a non-closed geodesic without self-intersections and Ω is the limit set of the geodesic L under the motion along L in accordance with the chosen direction, then for the case when the homotopy rotation class $\mu(L)$ of L is rational (consists of rational points) Ω is a simple closed geodesic, while if $\mu(L)$ is irrational (consists of irrational points), then Ω is a geodesic lamination consisting of recurrent geodesics.[11]

Throughout the rest of this subsection we assume, unless otherwise stated, that Ω is a geodesic lamination consisting of non-closed recurrent geodesics without self-intersections.

Theorem 3.1 (Aranson and Grines). *In order that two geodesic laminations Ω and Ω' of the above type coincide on M, it is necessary and sufficient that there exist two directed geodesics $L \subset \Omega$ and $L' \subset \Omega'$ such that their homotopy rotation classes are the same ($\mu(L) = \mu(L')$).*

It is known (Nielsen 1927) that every automorphism τ of the group G can be uniquely extended to the absolute E by means of the homeomorphism τ^* and this homeomorphism takes rational points to rational ones and irrational points to irrational ones.

We denote by H^* the set of all homeomorphisms of the absolute induced by all the automorphisms of G.

Definition 3.2. The set $H^*(\mu(L)) = \bigcup_{\tau^* \in H^*} \tau^*(\mu(L))$ is called the orbit $O(L)$ of the homotopy rotation class $\mu(L)$ of a directed geodesic L defined on M.

Remark. Geometrically, for a non-closed recurrent geodesic L the orbit $O(L)$ of the homotopy rotation class is a countable everywhere dense set of irrational points on the absolute.

We say that the *geodesic laminations* Ω and Ω' defined on M are *topologically equivalent* if there exists a homeomorphism $\phi : M \to M$ such that $\Omega' = \phi(\Omega)$ and ϕ takes geodesics in Ω to geodesics in Ω'.

It has been established by Aranson and Grines that in order that two geodesic laminations Ω and Ω' of the above type be topologically equivalent

[11] A directed geodesic L with parametrization t introduced on L is said to be *recurrent* if for any $\epsilon > 0$ there exists $T > 0$ such that for any arc on L corresponding to the values $t \in [T, -T]$, the geodesic L lies entirely in the ϵ-neighbourhood of this arc.

on M it is necessary and sufficient that there exist two directed geodesics $L \subset \Omega$, $L' \subset \Omega'$ such that the orbits of their homotopy rotation classes are the same ($O(L) = O(L')$).

We say that a point $\sigma \in E$ is *admissible* for a geodesic lamination $\Omega \subset M$ if there exists at least one directed geodesic $L \subset \Omega$ having the point σ as the limit set of at least one of its inverse images on $\overline{M}$.

It has been established by Aranson and Grines that the set of admissible points on the absolute for geodesic laminations on M is everywhere dense on the absolute, has power of the continuum and is a set of zero Lebesgue measure.

In (Aranson and Grines 1982) an effective method is given for constructing geodesics without self-intersections that are recurrent; this method is based on the properties of an automorphism of the fundamental group. These geodesics are analogous to geodesics on T^2 in a metric of zero curvature whose inverse images on the universal covering space $\mathbb{R}^2$ are parallel to the eigen-directions of integer-valued unimodular 2×2 matrices the moduli of the eigenvalues of which are distinct from unity.

It is natural to call these geodesics *algebraic geodesics* defined on M.

We consider an automorphism τ of the group G satisfying the following properties:

1) $\tau(g) \neq g$ for any $g \in G$ distinct from the identity;

2) the homeomorphism τ^* of the absolute induced by τ contains at least four fixed points;

3) τ^* preserves the orientation of the absolute.

By virtue of the results of (Nielsen 1927) the automorphism τ enjoys the following properties:

1) τ^* has an even number μ of fixed points on the absolute ($\mu \leq 8p-4$, $p \geq 2$ is the genus of M); they are all irrational, half of them being attracting and half repelling;

2) the image $L = \pi(l)$ on M of a geodesic $l \subset \overline{M}$ joining two neighbouring attracting (or repelling) fixed points of the homeomorphism τ^* is a non-closed curve without self-intersections.

We consider on the absolute E any two attracting fixed points σ_1^+, σ_2^+ of a homeomorphism τ^* of the above type such that on one of the arcs of the absolute partitioned by these two points there are no attracting fixed points other than σ_1^+, σ_2^+.

We denote by l the geodesic on $\overline{M}$ for which σ_1^+, σ_2^+ are the boundary points, and by $L = \pi(l)$ its image on M, where $\pi : \overline{M} \to M$ is the natural projection. In view of (Aranson and Grines 1981), (Aranson and Grines 1982), L is a recurrent geodesic. A similar property holds for images of geodesics joining any two repelling neighbouring fixed points of a homeomorphism τ^* of the above type.

In (Aranson and Grines 1978), (Aranson and Grines 1984), (Aranson and Grines 1985), (Aranson and Grines 1986) it is shown what role is played by geodesic laminations in the description of flows on M, and in (Aranson 1988,

Part 1) in the description of foliations with singularities defined on M (in the latter case, the genus p of the surface can be greater than or equal to zero).

It follows in particular from these papers that if a geodesic lamination Ω has the property of *Whitney orientability*[12] then there exists on M a Lipschitz vector field v inducing a flow f^t on M with the following properties:

1) the geodesics in Ω are trajectories for f^t and Ω is an invariant nowhere-dense set;

2) if Ω consists of recurrent geodesics, then Ω is a minimal set of the flow f^t, while if there are non-recurrent geodesics in Ω, then Ω cannot contain more than p minimal sets consisting of non-closed recurrent geodesics (p being the genus of M).

We also indicate the role played by geodesics in the asymptotic behaviour of trajectories of flows defined on M or, in other words, how far the trajectories of the covering flows on $\overline{M}$ oscillate (deviate at a finite or infinite distance) from geodesics of the same asymptotic direction.

It turns out that in this setting the question whether the flow under consideration contains a finite or an infinite set of equilibrium points plays an essential role.

The most studied class which is sufficiently wide for applications is the class of so-called Maier flows; these are flows with a finite number of equilibrium points and a finite number of separatrices. Maier flows can contain both closed trajectories and non-closed Poisson-stable trajectories, so that they possess many of the fundamental properties of three-dimensional and multi-dimensional dynamical systems.

We suppose that M is a closed orientable surface of genus $p \geq 1$ and we again represent M as a quotient space $\overline{M}/G$, where $\overline{M} : x^2 + y^2 < 1$ is the universal cover, G is the discrete group of motions on $\overline{M}$ and d is a metric of constant curvature K on M (in the case $p = 1$, the curvature $K = 0$, while in the case $p > 1$ the curvature $K < 0$). Let $\bar{d}$ be a metric on $\overline{M}$ such that $\bar{d}$ induces d under the projection $\pi : \overline{M} \to M$.

Theorem 3.2 (Aranson and Grines). *Let f^t be a Maier flow on M. Then any half-trajectory l of the covering flow $\bar{f}^t$ on $\overline{M}$ such that l has an asymptotic direction has the property of bounded deviation from a geodesic l_0 (in the $\bar{d}$ metric) of the same asymptotic direction.*

Remark 1. By the *asymptotic direction* of a half-trajectory l of a flow $\bar{f}^t$ on $\overline{M}$ covering the flow f^t on M, we mean the point σ on the absolute $E : x^2 + y^2 = 1$ that is a limit set for l (cf. the notion of asymptotic direction for a directed geodesic). If f^t is a Maier flow and any half-trajectory has a limit set that is not contractible (does not lie in a disc), then any connected inverse image $l \subset \overline{M}$ of a half-trajectory $L \subset M$ has an asymptotic direction.

[12] In the case of a closed orientable surface M of genus p Whitney orientability of a geodesic lamination Ω means that the *index* $J(C, \Omega)$ of any simple closed curve C with respect to geodesics in Ω is an integer (see (Aranson 1988, Part 1) for details).

Remark 2. On a manifold M of genus $p \geq 2$ there exists a C^∞-flow f^t with a non-contractible continuum of equilibrium states for which there is a half-trajectory L with the properties: 1) any connected inverse image l of it has an asymptotic direction on $\overline{M}$; 2) l moves off unboundedly from a geodesic l_0 of the same asymptotic direction.

A construction of such an example was carried out by the authors of the survey by using an A-diffeomorphism on M of genus 2 (on a pretzel) with two orientable basic sets, namely an attractor Ω_1 and a repeller Ω_2 (Robinson and Williams 1973). As noted by Grines, the connected inverse image of each connected component of the set $W^u(p)\backslash p$ ($W^s(p)\backslash p$), $p \in \Omega_1$ ($p \in \Omega_2$) moves off boundedly from a geodesic of the same asymptotic direction, so that this inverse image and the inverse image of each connected component of $W^s(p)\backslash p$ ($W^u(p)\backslash p$), where $p \in \Omega_1$ ($p \in \Omega_2$) that is not a boundary periodic point, deviates unboundedly from the corresponding geodesic.

We consider the set of curves on M that are unions of sets $W^s(\Omega_1) \cup W^s(\Omega_2)$ of all stable manifolds in Ω_1 and Ω_2. By Whitney's theorem, there exists a C^0-flow f_0^t on M such that the curves in $W^s(\Omega_1)$ are one-dimensional trajectories of it and the set $W^s(\Omega_2)$ consists entirely of equilibrium points. By Guttierrez's theorem, there exists on M a C^∞-flow f^t that is topologically equivalent to the flow f_0^t and for which the assertions 1), 2) of Remark 2 to Theorem 3.2 hold.

Algebraic recurrent geodesics were used by the authors of the survey for the construction of so-called *hyperbolic homeomorphisms*, for which it has been proved that Axiom A* holds and the topological entropy is minimal among the entropies of all homeomorphisms homotopic to a given hyperbolic homeomorphism (see (Aranson and Grines 1980), (Aranson and Grines 1981), (Aranson and Grines 1984) as well as the survey (Aranson and Grines 1990) in which the appropriate definitions are given and the construction of hyperbolic homeomorphisms is explained).[13]

These homeomorphisms are similar to pseudo-Anosov homeomorphisms (furthermore, they are semiconjugate to them), but by comparison with them, they have the advantage that the invariant foliations of hyperbolic homeomorphisms consist of two foliations each of which contains a geodesic lamination as a minimal set.

Definition 3.3 A homeomorphism $f : M \to M$ is said to be *hyperbolic* if: 1) f satisfies Axiom A*; 2) there exist two f-invariant transversal geodesic

[13] The construction of hyperbolic homeomorphisms was suggested by the authors in 1979 and in a weaker version (without investigating questions relating to estimates on the contracting and expanding of points lying on geodesic invariant laminations) was repeated by Miller in (Miller 1982). A visual representation of the invariant foliations of a hyperbolic homeomorphism is given in Figure 2, where they are depicted on the universal cover $\overline{M}$, where $+$ and $-$ denote attracting and repelling fixed points of the homeomorphism τ^* induced by the automorphism τ. The picture is then projected onto M via the projection $\pi : \overline{M} \to M$.

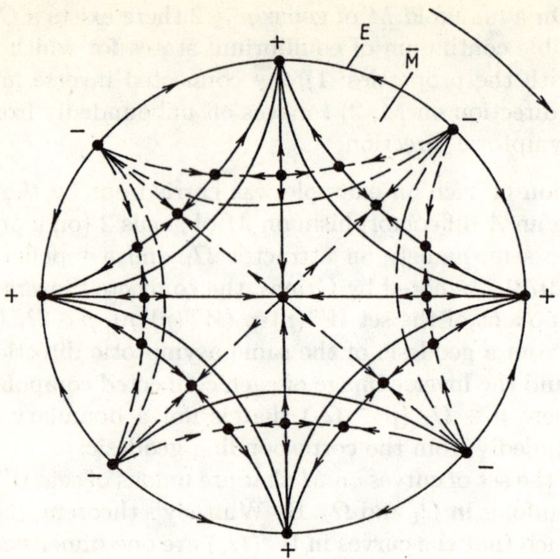

Fig. 2. Construction of hyperbolic homeomorphisms

laminations (in a metric of constant negative curvature on M of genus $p \geq 2$) Ω^{u}, Ω^{s} such that $M \backslash \Omega^{\mathrm{u}}$ ($M \backslash \Omega^{\mathrm{s}}$) is a union of a finite number of regions homeomorphic to a disc; 3) the nonwandering set $\Omega(f)$ of the diffeomorphism f is a union of finitely many isolated periodic points and one zero-dimensional local maximal set $\Omega_0 = \Omega^{\mathrm{u}} \cap \Omega^{\mathrm{s}}$, the stable and unstable manifolds of points of which lie on the leaves of the geodesic laminations $\Omega^{\mathrm{u}}, \Omega^{\mathrm{s}}$, respectively.

In the smooth case, Axiom A* (see the survey (Aranson and Grines 1990) for the precise definition) is a consequence of Axiom A. In (Bowen 1971) Bowen noted quite a number of important properties of diffeomorphisms on manifolds satisfying Axiom A which can be proved by methods of topological dynamics without invoking smoothness properties but merely relying on Axiom A*. Such properties include, for example, the existence of a Markov partition, the formula for calculating the entropy in terms of the growth of the periodic points (see also (Alekseev 1976), (Alekseev and Yakobson 1979)).

An f-invariant set Ω_0 is said to be *locally maximal* if there exists a neighbourhood U of Ω_0 such that any f-invariant set contained in U is in fact contained in Ω_0 (Anosov 1970). An automorphism τ of a group G is called *hyperbolic* (Aranson and Grines 1980), (Aranson and Grines 1981), (Aranson and Grines 1984), if for any g, γ ($\gamma \neq \mathrm{id}$), $n \neq 0$ the relation $g\tau^n(\gamma)g^{-1} \neq \gamma$ holds. (Here and in what follows M has genus $p \geq 2$.)

Two fixed points x, y over continuous transformation $f : M \to M$ belong to the same *fixed-point class* if they can be joined by a path w such that the loop $w^{-1}fw$ is null homotopic.

By the *Nielsen number* $\tilde{N}(f)$ of a transformation $f : M \to M$ we mean the number of fixed point classes having non-zero index.[14] Nielsen numbers are homotopy invariants (Nielsen 1927).

Theorem 3.3 (Aranson and Grines 1980), (Aranson and Grines 1981), (Aranson and Grines 1984), (Aranson and Grines 1990). *Let $M = \overline{M}/G$ be a closed Riemann surface of genus $p \geq 2$ endowed with a metric of constant negative curvature. Then for each hyperbolic automorphism τ of G:*

1) a pair of transversal geodesic laminations $\Omega^{\mathrm{u}}, \Omega^{\mathrm{s}}$ can be uniquely constructed consisting of non-closed recurrent geodesics and a homeomorphism $\phi : \Omega_0 \to \Omega_0$, where $\Omega_0 = \Omega^{\mathrm{u}} \cap \Omega^{\mathrm{s}}$;

2) ϕ can be extended to a hyperbolic homeomorphism $\tilde{f}_0 : M \to M$ such that:

a$_1$) $\tilde{f}_0$ induces the automorphism τ;

a$_2$) $\Omega^{\mathrm{u}}, \Omega^{\mathrm{s}}$ are invariant geodesic laminations for $\tilde{f}_0$;

3) $\tilde{f}_0$ has the minimal topological entropy[15] *$h(\tilde{f}_0)$ among all entropies of homeomorphisms that are homotopic to $\tilde{f}_0$ and is calculated from the formula $h(\tilde{f}_0) = \limsup_{n\to+\infty} \frac{1}{n} \ln \tilde{N}_n(\tilde{f}_0)$, where $\tilde{N}_n(\tilde{f}_0)$ is the Nielsen number of the homeomorphism $\tilde{f}_0^n$.*

Remark. Assertion 3) of Theorem 3.3 has a simple proof. For homeomorphisms satisfying Axiom A* and, in particular, for the homeomorphism $\tilde{f}_0 : M \to M$, the topological entropy, by virtue of (Bowen 1971) is calculated from the formula: $h(\tilde{f}_0) = \limsup_{n\to+\infty} \frac{1}{n} \ln N_n(\tilde{f}_0)$, where $N_n(\tilde{f}_0)$ is the number of fixed points of the transformation $\tilde{f}_0^n$. On the other hand, it follows from the construction of $\tilde{f}_0$ that there exists $K > 0$ such that for all $n > 0$ we have the estimate $|N_n(\tilde{f}_0) - \tilde{N}_n(\tilde{f}_0)| < K$, where $\tilde{N}_n(\tilde{f}_0)$ is the Nielsen number of the map $\tilde{f}_0^n$. Hence it follows that $h(\tilde{f}_0) = \limsup_{n\to+\infty} \frac{1}{n} \ln \tilde{N}_n(\tilde{f}_0)$. Since for any homeomorphism $f : M \to M$ homotopic to $\tilde{f}_0$ the entropy $h(f) \geq \limsup_{n\to+\infty} \frac{1}{n} \ln \tilde{N}_n(f)$, it follows that $h(f) \geq h(\tilde{f}_0)$. The last estimate was announced by the authors of the present survey in 1981 at the ninth international conference on nonlinear oscillations in Kiev (Aranson and Grines 1984) and was also obtained by Ivanov in (Ivanov 1982b).[16]

---

[14] In the case when the number of fixed points in the fixed-point class is finite, the index of this class is just the sum of the indices of the fixed points in this class. See (Jiang 1981), (Nielsen 1927) for the definition of the index of a fixed-point class and Nielsen numbers in the general case.

[15] See (Adler, Konheim and Andrew 1965) and (Aranson and Grines 1990) for the definition of *topological entropy*.

[16] In connection with this estimate there arises the question posed by Nielsen: does there exist in each homotopy class of homeomorphisms on M a representative f such that $N_n(f) = \tilde{N}_n(f)$ $(n = 1, 2, \dots)$? For connected closed surfaces this problem was affirmatively solved by Jiang (Jiang 1981) and later by Ivanov (Ivanov 1982a). We note that for pseudo-Anosov maps the relation $N_n(\tilde{f}) = \tilde{N}_n(f)$ does not always hold since they can have fixed points of zero index (if these points are removed, then such maps may cease to be pseudo-Anosov: see, for example, (Jiang 1981)).

3.2. The Nielsen-Thurston Theory on the Homotopy Classification of Homeomorphisms of Surfaces. The results of Nielsen's theory on the homotopy classification of homeomorphisms of surfaces were written down by him in a whole series of papers (see, for example, (Nielsen 1927), (Nielsen 1937), (Nielsen 1944)) spread over a long period of time, and were little known to most mathematicians. Recently two articles by Gilman (Gilman 1981), (Gilman 1982) have appeared in which a concise account of Nielsen's theory is given and it is explicitly indicated how a number of Thurston's results (Thurston 1976) (see also (Poénaru et al. 1979)), including homotopy classification, follow from Nielsen's results. In these same articles (Gilman 1981), (Gilman 1982) it is also shown that Nielsen's homotopy classification is equivalent to Bers's classification (Bers 1975) obtained by him in the language of Teichmüller spaces.

In this subsection we shall, in the main, follow the account of the theory of Nielsen and Thurston in accordance with the papers (Gilman 1981), (Gilman 1982). In addition we recommend that the reader acquaint himself with the appendix by Ivanov in the book (Zieschang, Vogt, and Coldewey 1980), which also refers to this question, and with the authors' survey (Aranson and Grines 1990, Sect. 3).[17]

As before, we assume M to be a closed orientable surface of genus $p \geq 2$. Again we introduce on M an analytic structure converting M into a Riemann surface and denote by π the conformal map of the universal cover $\overline{M}$ on M where $\overline{M}$ is the Lobachevskij plane in the Poincaré realization in the disc $x^2 + y^2 < 1$ with metric of constant negative curvature. Again, as before, (see Sect. 3.1) we represent M as a quotient space $\overline{M}/G$, where G is a discrete group of hyperbolic elements isomorphic to the fundamental group of M and we denote by $E : x^2 + y^2 = 1$ the absolute of the plane $\overline{M}$.

We begin by considering Nielsen's theory on the homotopy classification of homeomorphisms of surfaces.

Let $f : M \to M$ be a homeomorphism, $\{f\}$ the homotopy class containing f, and $T = \langle f \rangle$ the cyclic group generated by f. (The elements of T are transformations f^n, $n \in \mathbb{Z}$.) Let $L(T)$ be the group of liftings to $\overline{M}$ of elements of T (the elements of $L(T)$ are homeomorphisms of $\overline{M}$ of the form $h = \gamma \bar{f}^n$, where $\gamma \in G$, $n \in \mathbb{Z}$, and $\bar{f}$ is a homeomorphism covering f, that is, $f\pi = \pi\bar{f}$). For each $h \in L(T)$ we let h^* denote the homeomorphism of the absolute induced by h (h^* is the extension of h to the absolute).

The *Nielsen type* of $h \in L(T)$ is the pair of integers (v_h, u_h), defined as follows. Let $F_h = \{\gamma \in G | h\gamma h^{-1} = \gamma\}$; then v_h denotes the minimal number of generators of F_h, where we set $v_h = 0$ if F_h consists only of the identity element of G.

We define u_h by the rule:

[17] The account given below is an expanded account of the Nielsen-Thurston theory on the homotopy classification of homeomorphisms of surfaces in the survey (Aranson and Grines 1990, Sect. 3).

1) if $v_h = 0$, then u_h is the number of isolated attracting fixed points of h^*;

2) if $v_h = 1$, that is, if F_h is a free cyclic group with generator $\gamma \in G$, then u_h is the number of F_h-orbits of isolated attracting fixed points of h^* distinct from γ^+, γ^- (here γ^+, γ^- are respectively the attracting and repelling fixed points of the element $\gamma \in G$ lying on the absolute), where the type (1,0) splits into two subtypes: $(1,0)^+$ for which γ^+, γ^- are neutral fixed points of h^* (an isolated fixed point of $h^* : E \to E$ is said to be *neutral* if locally it is neither attracting nor repelling); $(1,0)^{++}$ for which one of the points γ^+, γ^- is attracting for h^* and the other repelling;

3) if $v_h \neq 1$, then u_h is the number of F_h-orbits of isolated attracting fixed points of h^*.

Theorem 3.4 (Nielsen 1927). *The set of all homotopy classes $\{f\}$ can be represented as the union of four disjoint classes N_1, N_2, N_3, N_4 distinguished by the following properties:*

1) *if $\{f\} \in N_1$, then any $h \in L(T)$, $h \neq \mathrm{id}$, is either of type $(0,0)$ or of type $(1,0)$;*[18]

2) *if $\{f\} \in N_2$, then any $h \in L(T)$ is either of type $(0,1)$ or of type $(v_h, 0)$. In the latter case there exists $h' \in L(T)$, $h' \neq \mathrm{id}$, such that either* a$_1$) $v_{h'} \geq 2$, *or* a$_2$) *h' is of type $(1,0)^+$ with $h' = h''^2$, where $h'' \in L(T)$ is of type $(0,0)$;*

3) *if $\{f\} \in N_3$, then there exist $h, h' \in L(T)$ $(h \notin G)$ of types (v_h, u_h), $(v_{h'}, u_{h'})$ such that $v_h \neq 0$, $u_{h'} \neq 0$; here if $h \neq h'$, then $u_{h'} \geq 2$;*

4) *if $\{f\} \in N_4$, then any $h \in L(T)$, $h \notin G$, is of type $(0, u_h)$ and there exists $h' \in L(T)$ of type $(0, u_{h'})$, where $u_{h'} \geq 0_2$.*[19]

The classes N_i $(i = \overline{1,4})$ in Theorem 3.4 are called *Nielsen classes.* The classes N_1, N_2 were studied in detail by Nielsen himself in (Nielsen 1937), (Nielsen 1944). He proved that each homotopy class $\{f\}$ belongs to N_1 (N_2) if and only if $\{f\}$ contains a periodic homeomorphism (a non-periodic homeomorphism of algebraically finite type).

We recall that a homeomorphism $f : M \to M$ is said to be *algebraically of finite type* if there exists a finite invariant family of disjoint cylinders $\sigma_1, \dots, \sigma_k$ embedded in M such that the restriction of f to the set $\Delta = M \setminus \bigcup_{i=1}^{k} \mathrm{int}\, \sigma_i$ is a *periodic homeomorphism*, that is, there exists an integer $n_0 > 0$ such that $f^{n_0}\big|_{\Delta} = \mathrm{id}$, where id is the identity transformation (Nielsen 1944).

As regards the classes N_3, N_4, although Nielsen had examples of homeomorphisms whose homotopy classes belonged to N_3, N_4 he did not study them as exhaustively as the classes N_1, N_2. Nevertheless, he indicated the complicated

[18] If $h \in L(T)$ is of type (1,0), then this type is either $(1,0)^{++}$ or $(1,0)^+$, depending on whether h preserves or reverses the orientation of the universal cover.

[19] In fact, the existence of an element $h' \in L(T)$ of type $(0, u_{h'})$, where $u_{h'} \geq 2$, is superfluous for distinguishing the class N_4, since in view of the survey by the authors (Aranson and Grines 1980, Lemma 1) it follows merely from the condition that any $h \in L(T)$, $h \notin G$, is of type $(0, u_{h'})$. This result was proved subsequently in (Handel and Thurston 1985), (Miller 1982). We note that Nielsen himself did not prove this fact.

dynamical properties of homeomorphisms whose homotopy classes belong to N_3, N_4.

The first examples that shed light on the complex dynamics of homeomorphisms whose homotopy classes belong to N_4 (and, with slight modifications, to N_3) were constructed in 1970 by O'Brien and Reddy (O'Brien 1970), (O'Brien and Reddy 1970) in connection with the solution of the problem of the existence of expansive homeomorphisms on surfaces. These homeomorphisms have positive topological entropy and a pair of contracting and expanding transitive foliations with singularities and are semiconjugate to Anosov diffeomorphisms of the torus T^2.

We recall that a homeomorphism $f : M \to M$ is said to be *expansive* if there exists $\epsilon > 0$ such that for any $x, y \in M$ $(x \neq y)$ there exists a number $n = n(x, y)$ such that $d(f^n(x), f^n(y)) > \epsilon$, where d is a metric on M.

Recently Hiraide (Hiraide 1990) announced the following result: every expansive homeomorphism on a closed surface is pseudo-Anosov (Anosov for T^2). There are no expansive homeomorphisms on the sphere, the projective plane, or the Klein bottle.

It immediately follows from Grines's papers (Grines 1974), (Grines 1975, Corollary 4.1) that any A-diffeomorphism whose nonwandering set contains an orientable (and, as it later turned out, exteriorly situated) basic set of necessity belongs to a homotopy class in N_3 or N_4. In particular, the homotopy class containing the A-diffeomorphism on M of genus $p = 2$ with two orientable basic sets, constructed in (Robinson and Williams 1973), belongs to N_3.

We note further that the notion of a hyperbolic automorphism τ of a group G given in Sect. 3.1 in front of the statement of Theorem 3.3 is equivalent to the class N_4, since any hyperbolic automorphism τ defines a homotopy class $\{f\} \in N_4$ and conversely: for any $f \in \{f\} \in N_4$ the automorphism τ defined by the formula $\tau(\gamma) = \bar{f}\gamma\bar{f}^{-1}$ $(\gamma \in G)$ is hyperbolic. Thus Theorem 3.3, proved by the authors of the survey in 1979 is in point of fact the bridge linking Nielsen's algebraic theory of the thirties on homotopy classification of homeomorphisms of surfaces with Thurston's geometric theory given below, which was devoted to the same questions and developed by him in 1976 (Thurston 1976).

We now turn to Thurston's theory on the homotopy classification of homeomorphisms of surfaces and their interpretation in the language of Teichmüller spaces, developed by Bers (Bers 1978).

Prior to this we give some requisite notions.

A homeomorphism $f : M \to M$ is called a *pseudo-Anosov diffeomorphism* (Thurston 1976) if:

1) there exist two f-invariant foliations $F^{\mathrm{u}}, F^{\mathrm{s}}$ with a common set of singularities $A_1, \dots, A_r$ which are topological saddles, the number of separatrices of each of which is at least three, such that:

a$_1$) $F^{\mathrm{u}}, F^{\mathrm{s}}$ are transversal everywhere except at $A_1, \dots, A_r$;

a$_2$) $F^{\mathrm{u}}, F^{\mathrm{s}}$ are *measurable foliations with transversal measures* $\mu_{\mathrm{u}}, \mu_{\mathrm{s}}$;

2) there exists a number $\lambda > 1$, called the *stretch factor* such that for any arc $\alpha \subset M$ we have:

$$\mu_{\mathrm{u}}(f(\alpha)) = \lambda^{-1}\mu_{\mathrm{u}}(\alpha), \quad \mu_{\mathrm{s}}(f(\alpha)) = \lambda\mu_{\mathrm{s}}(\alpha);$$

3) f is a diffeomorphism everywhere except at $A_1, \dots, A_r$.

At the points $A_1, \dots, A_r$ the pseudo-Anosov diffeomorphism is everywhere non-smooth, otherwise the estimates in condition 2) would not hold; therefore the terminology "*pseudo-Anosov diffeomorphism*" introduced by Poénaru in (Poénaru et al. 1979, pp. 159–180) is more natural. If a homeomorphism f satisfies conditions 1), 2) but not necessarily 3), then f is called a *pseudo-Anosov homeomorphism* (see, for example, the appendix in (Zieschang, Hogt, and Coldewey 1980), where this question is discussed).

If the set of singularities of the foliations of $F^{\mathrm{u}}, F^{\mathrm{s}}$ is empty and conditions 1), 2) hold, then it is natural to call f an *Anosov homeomorphism.* In this case $M = T^2$. If the singularities $A_1, \dots, A_r$ of $F^{\mathrm{u}}, F^{\mathrm{s}}$ include spines (saddles with a single separatrix) and conditions 1), 2) hold, then f is called a *generalized pseudo-Anosov homeomorphism.* A pseudo-Anosov homeomorphism can exist on orientable M of genus $p \geq 2$ and non-orientable M of genus $p \geq 3$, while a generalized pseudo-Anosov homeomorphism can exist on orientable M of genus $p \geq 0$ and non-orientable M of genus $p \geq 1$.

For an acquaintance with the questions of realization of pseudo-Anosov, generalized pseudo-Anosov homeomorphisms and homeomorphisms topologically conjugate to them, as well as smooth and even analytic diffeomorphisms, we refer the reader to the following sources: (Aranson 1986), (Aranson 1988, Part 2), (Aranson and Grines 1990), (Grines 1982), (Zieschang, Hogt, and Coldewey 1980) (Abikoff 1980), (Arnoux and Yoccoz 1981), (O'Brien 1970), (O'Brien and Reddy 1970), (Gerber and Katok 1982), (Katok 1979), (Poénaru et al. 1979).

Let F be a C^k ($k \geq 0$) foliation on M with set of singularities $A_1, \dots, A_r$. Then F is said to be *measurable* (Thurston 1976) if there exists a covering of $M \backslash \bigcup_{j=1}^{r} A_j$ by simply connected regions U_i ($i \in \mathbb{Z}^+$) and a family of homeomorphisms $\Phi_i : U_i \to \mathbb{R}^2$ ($\mathbb{R}^2$ is the Euclidean plane with coordinates x, y) such that:

1) Φ_i takes the leaves of $F|_{U_i}$ to the straight lines $y = \text{const}$;

2) the evaluation map $\Phi_{ij} = \Phi_i \Phi_j^{-1} : \Phi_j(U_{ij}) \to \Phi_i(U_{ij})$ ($U_{ij} = U_i \cap U_j \neq \emptyset$) has the form $\Phi_{ij}(x, y) = f_{ij}(x, y), c_{ij} \pm y)$, where f_{ij} is a C^k-smooth or analytic function and the c_{ij} are constants which depend on i, j.

Let β be a simple arc on M lying in U_i, and $\bar{\beta} = \Phi_i(\beta) : x = \phi_i(t)$, $y = \psi_i(t)$ ($t_0 \leq t \leq t_1$) its image on $\mathbb{R}^2$.

By the *transversal measure* $\mu(\bar{\beta})$ of the arc β in U_i we mean the variation $\mathrm{Var}_{[t_0,t_1]}\Psi_i(t)$ (Thurston 1976). Suppose now that $\alpha \subset M$ is any simple arc (possibly a closed curve). We represent the set $\alpha \backslash \bigcup_{j=1}^{r} A_j$ as a union $\bigcup_k \beta_k$ of simple arcs with disjoint interiors such that each β_k lies entirely in at least one of the regions U_i. Then we call the sum $\mu(\alpha) = \sum_k \mu(\beta_k)$ the *transversal measure* $\mu(\alpha)$ of the arc (or closed curve) α (Thurston 1976).

We point out certain properties of pseudo-Anosov homeomorphisms $f : M \to M$ (Gerber and Katok 1982), (Katok 1979), (Thurston 1976), (Poénaru et al. 1979):

1) f is topologically transitive and the set of periodic points is everywhere dense on M;

2) the stretch factor λ and its reciprocal are algebraic numbers;

3) the topological entropy $h(f)$ of f is equal to $\ln \lambda$ and is minimal among all entropies of homeomorphisms homotopic to f;

4) two pseudo-Anosov homeomorphisms are topologically conjugate if and only if they are π_1-conjugate;

5) the foliations $F^{\mathrm{u}}, F^{\mathrm{s}}$ are transitive and have no separatrices going from one singularity to another or to the same singularity;

6) f induces a hyperbolic automorphism of G;

7) f preserves the natural absolutely continuous measure which is C^∞-dense and positive with the exception of a finite number of periodic points $A_1, \ldots, A_r$ of f at which it vanishes; f is a Bernoulli homeomorphism with respect to this measure. (With regard to ergodic properties of central limit theorem type, no such results are known to the authors).

A homeomorphism $f : M \to M$ is called *reducible* by a system Σ of disjoint simple closed curves C_i $(i = \overline{1,l})$ which are not null-homotopic and not homotopic to one another if Σ is invariant with respect to f.

Theorem 3.5 (Thurston 1976). *The set of all homotopy classes $\{f\}$ of homeomorphisms of a closed orientable surface M of genus $p \geq 2$ can be represented as a union of four disjoint classes T_1, T_2, T_3, T_4 distinguished by the following conditions:*

1) *if $\{f\} \in T_1$, then $\{f\}$ contains a periodic homeomorphism f_0;*

2) *if $\{f\} \in T_2$, then $\{f\}$ contains a reducible aperiodic homeomorphism f_0 of algebraically finite type;*

3) *if $\{f\} \in T_3$, then $\{f\}$ contains a reducible homeomorphism f_0 that is not a homeomorphism of algebraically finite type;*

4) *if $\{f\} \in T_4$, then $\{f\}$ contains a pseudo-Anosov homeomorphism f_0.*

The classes T_i $(i = \overline{1,4})$ are called *Thurston classes.*

The representatives $f_0 \in \{f\}$ of the classes T_i $(i = \overline{1,4})$ in Theorem 3.5 are called the *canonical Thurston forms.* We now go into the geometrical description of them in more detail (see, for example, (Handel and Thurston 1985).

In the cases 1) and 4), when f_0 is either periodic or a pseudo-Anosov homeomorphism the definition has been given above.

In the case 2) M contains a finite invariant family of disjoint cylinders $\sigma_1, \ldots, \sigma_k$ embedded in M such that the generators of these cylinders are not null-homotopic, are not homotopic to one another, and the following properties hold:

a_1) the restriction of f_0 to $\Delta = M\setminus\bigcup_{i=1}^{k} \operatorname{int}\sigma_i$ is a periodic homeomorphism;

a_2) the restriction of $f_0^{s_i} : \sigma_i \to \sigma_i$ to each of the cylinders σ_i is a generalized rotation,[20] where $s_i = s_i(\sigma_i)$ are integers such that σ_i is invariant with respect to $f_0^{s_i}$.

In the case 3), f_0 satisfies the same properties as in case 2) with the exception that in some of the connected components α_ν of Δ the homeomorphism $f_0^{p_\nu} : \alpha_\nu \to \alpha_\nu$ is periodic, while in the remaining connected components β_λ of Δ the homeomorphism $f_0^{q_\lambda} : \beta_\lambda \to \beta_\lambda$ is pseudo-Anosov, where $p_\nu = p_\nu(\alpha_\nu)$, $q_\lambda = q_\lambda(\beta_\lambda)$ are integers such that α_ν and β_λ are invariant with respect to $f_0^{p_\nu}, f_0^{q_\lambda}$ respectively.

We now turn to the Bers interpretation (Bers 1978). We denote by $T(M)$ the *Teichmüller space*,[21] where we consider as a model of this space one whose points of this space are distinguished Riemannian manifolds (M, α) oriented the same way, where α is the equivalence class of the standard systems of generators.

By the *Teichmüller distance* between two distinguished similarly oriented Riemannian manifolds (M, α) and (M, α') of genus $p \geq 2$ we mean the number

$$\rho[(M, \alpha), (M, \alpha')] = \inf_f \ln K[f],$$

where the infimum is taken over all maps $f : (M, \alpha) \to (M, \alpha')$ and $K[f]$ is the maximal deviation of f from a conformal one.

It is known that $T(M)$ is homeomorphic to the space $\mathbb{R}^{6p-6}$, where $\mathbb{R}^n$ is n-dimensional Euclidean space.

The class $\{f\}$ of homotopic homeomorphisms on M induces a map $\hat{f}$ in $T(M)$ according to the rule: each point $(M, \alpha) \in T(M)$ is associated with a point $(M, \alpha') \in T(M)$ such that $\alpha' = f_*(\alpha)$, where f_* is the isomorphism of the fundamental groups of M induced by any $f \in \{f\}$ (this does not depend on the choice of the representative $f \in \{f\}$).

We associate with each homotopy class $\{f\}$ of homeomorphisms on M the number

$$\eta(\hat{f}) = \inf_{(M,\alpha)\in T(M)} \rho((M, \alpha), (M, \alpha')),$$

where $\hat{f}$ is the map in $T(M)$ induced by $\{f\}$ and taking (M, α) to (M, α').

Let $\{f\}$ be a homotopy class of homeomorphisms of the surface M, and $\hat{f}$ the map in $T(M)$ induced by $\{f\}$. Following (Bers 1978) (see also (Gilman 1981), (Gilman 1982)) we say that:

---

[20] Let C_i be the generator of the cylinder σ_i and C_i^-, C_i^+ simple closed curves bounding σ_i. We introduce on σ_i the parametrization x, y ($-1 \leq x \leq 1$; $-\infty < y < +\infty$) such that $(x, y), (x, y+n)$ $(n \in \mathbb{Z})$ corresponds to the same point in σ_i and C_i^-, C_i, C_i^+ are given by the equations: $x = -1$, $x = 0$, $x = 1$. Then by a *generalized rotation* of the homeomorphism $\phi : \sigma_i \to \sigma_i$ on the cylinder σ_i we mean a homeomorphism ϕ of the form: $\phi(x, y) = (\zeta x, \lambda x + \zeta y + a)$, where λ, a are constants, $\zeta = \pm 1$.

[21] See (Abikoff 1980) for details.

1) $\{f\}$ is an *elliptic class* if $\hat{f}$ has a fixed point in $T(M)$ and, consequently, $\eta(\hat{f}) = 0$ (we denote this class by B_1);

2) $\{f\}$ is a *parabolic class* if $\hat{f}$ has no fixed points in $T(M)$, but $\eta(\hat{f}) = 0$ (we denote this class by B_2);

3) $\{f\}$ is a *pseudo-hyperbolic class* if $\eta(\hat{f}) > 0$ and for any point $(M, \alpha) \in T(M)$ we have $\rho((M,\alpha),(M,\alpha')) > \eta(\hat{f})$ where $\hat{f}$ takes (M,α) to (M,α') (we denote this class by B_3);

4) $\{f\}$ is a *hyperbolic class* if $\eta(\hat{f}) > 0$ and there exists a point $(M,\alpha) \in T(M)$ such that $\eta(\hat{f}) = \rho((M,\alpha),(M,\alpha'))$, where $\hat{f}$ takes (M,α) to (M,α') (we denote this class by B_4).

Theorem 3.6 (Bers 1978). *We have the relations:* $B_i = T_i$ $(i = \overline{1,4})$, *where* T_1, T_2, T_3, T_4 *are the Thurston classes.*

Theorem 3.7 (Gilman 1981), (Gilman 1982). *The following relations hold:* $T_i = N_i$ $(i = \overline{1,4})$, *where* N_1, N_2, N_3, N_4 *are the Nielsen classes.*[22]

The canonical Thurston forms $f_0 \in \{f\}$ described above and appearing in Theorem 3.5 are not *structurally stable* on a closed orientable surface M of genus $p \geq 2$. At the same time, according to (Smale 1967), $\{f\}$ contains a structurally stable diffeomorphism with zero-dimensional basic sets (the so-called *Smale diffeomorphism*). However, *a priori* this Smale diffeomorphism can have a very complicated topological structure. It is therefore of interest to indicate the simplest structurally stable representatives in these classes. Such classes are discussed in (Aranson and Grines 1990, Sect. 3).

Namely, the simplest representatives in each homotopy class $\{f\} \in N_i$ $(i = 1, 2)$ can be taken to be Morse-Smale diffeomorphisms $\tilde{\tilde{f}}_0$ such that for N_1 the diffeomorphism is gradient-like, while for N_2 it has an orientable heteroclinic set and M is representable as a union of two compact invariant sets. The first set is the union of k $(k \geq 2)$ compact surfaces of genus greater than zero and contains no heteroclinic trajectories; the second set is a union of subsets homeomorphic to a closed annulus whose interiors consist of wandering points and contain an orientable heteroclinic set of $\tilde{\tilde{f}}_0$.

We recall that a Morse-Smale diffeomorphism is called *gradient-like* if it contains no heteroclinic trajectories. See Sect. 1 of this chapter for the definition of a heteroclinic point and a heteroclinic trajectory.

[22] We again emphasize that it is the hyperbolic homeomorphism $\tilde{f}_0$ constructed in Theorem 3.3 for each homotopy class $\{f\} \in N_4$ $(N_4 = T_4)$ that brings about the connection between Nielsen's theory and Thurston's theory. Note that $\tilde{f}_0$ is not only homotopic to a pseudo-Anosov homeomorphism $f_0 \in \{f\}$, but also $\tilde{f}_0$ is semi-conjugate to f_0 via a continuous transformation that is homotopic to the identity. Note also that, in view of (Abikoff 1980), the "pseudo-Anosov diffeomorphism" is a *quasiconformal map* on M of smallest deviation from a conformal map in the class $\{f\}$. The existence of such an extremal quasiconformal map follows from the Teichmüller theory.

A set of heteroclinic points of $\tilde{\tilde{f}}_0$ is called an *orientable heteroclinic set* if for each pair of points $x, y \in \Omega'$ and any $\alpha > 0$, $\beta > 0$ the index of the intersection of the curves $W^{\mathrm{u}}_{j,\alpha}(y)$ and $W^{\mathrm{s}}_{i,\beta}(x)$ is the same at all points of the intersection $W^{\mathrm{u}}_j(y)$ with $W^{\mathrm{s}}_i(x)$, where the definition of the curve $W^{\mathrm{u}}_{j,\alpha}(y)$ and $W^{\mathrm{s}}_{i,\beta}(x)$ is given in Sect. 2.2 and Ω' is the set of saddle-type periodic points of $\tilde{\tilde{f}}_0$.

In each homotopy class $\{f\} \in N_3$ it is natural to regard as the simplest structurally stable diffeomorphism a diffeomorphism $\tilde{\tilde{f}}_0$ of the following type. The closed orientable surface M of genus $p \geq 2$ is representable as a union of three compact sets A_1, A_2, A_3. The sets A_1 and A_2 are unions of compact surfaces of genus greater than zero, where A_1 contains finitely many periodic points and contains no heteroclinic trajectories, while A_2 contains finitely many exteriorly situated one-dimensional attractors and there are no other nonwandering points in the interior of this set. The set A_3 is a union of subsets homeomorphic to a closed annulus whose interiors contain an orientable heteroclinic set. Here A_1 can be empty, but A_2 and A_3 are not empty.

The simplest structurally stable diffeomorphism in each homotopy class $\{f\} \in N_4$ can be taken to be a diffeomorphism whose nonwandering set consists of finitely many isolated periodic points and a single one-dimensional exteriorly situated attractor Ω such that $M \backslash \Omega$ consists of the union of finitely many regions homeomorphic to a disc. Moreover, $M \backslash \Omega$ contains no heteroclinic trajectories.

See (Aranson and Grines 1990, Sect. 1) for the construction of gradient-like diffeomorphisms in the class N_1. With regard to the classes N_2, N_3, N_4, no rigorous construction of the structurally stable diffeomorphisms described above is known to the authors.

References*

Abikoff, W. (1980): The Real Analytic Theory of Teichmüller Space. Lect. Notes Math. *820*. Zbl. 452.32015

Adler, R.L, Konheim, A.G., Andrew, M.N. (1965): Topological entropy. Trans. Am. Math. Soc. *114*, No. 2, 309–319

Alekseev, V.M. (1969): Perron sets and topological Markov measures. Usp. Mat. Nauk *24*, No. 5, 227–228. Zbl. 201,566

Alekseev, V.M. (1976): Symbolic Dynamics (Eleventh Mathematical School). Inst. Mat. Akad. Nauk Ukr. SSR, pp. 6–206

Alekseev, V.M., Yakobson, M.V. (1979): Symbolic dynamics and hyperbolic dynamical systems. In Bowen, R. (ed.): Methods of Symbolic Dynamics, Mir, Moscow, 196–240 (Russian)

* For the convenience of the reader, references to reviews in Zentralblatt für Mathematik (Zbl.), compiled using the MATH database, have, as far as possible, been included in this bibliography.

Anosov, D.V. (1967): Geodesic flows on closed Riemannian manifolds of negative curvature. Tr. Mat. Inst. Steklova *90* (209 pp.) [English transl.: Proc. Steklov Inst. Math. *90* (1969)] Zbl. 163,436

Anosov, D.V. (1970): On a class of invariant sets of smooth dynamical systems. Teor. Nelin. Koleb. *2*, Tr. 5 Konf. 1969, 39–45. Zbl. 243.34085

Anosov, D.V. (1985): Structurally stable systems. Tr. Mat. Inst. Steklova *169*, 59–93. [English transl.: Proc. Steklov Inst. Mat. *169*, 61–95 (1985)] Zbl. 619.58027

Aranson, S.Kh. (1986): On topological equivalence of foliations with singularities and homeomorphisms with invariant foliations on two-dimensional manifolds. Usp. Mat. Nauk *41*, No. 3, 167–168. [English transl.: Russ. Math. Surv. *41*, No. 3, 197–198 (1986)] Zbl. 619.57013

Aranson, S.Kh. (1988): Topological classification of foliations with singularities and homeomorphisms with invariant foliations on closed surfaces. Ch. 1 Foliations. Gor'kij Gos. Univ., 1–195. Dep. VINITI, No. 6887–B 88; (1989): Ch. 2 Homeomorphisms. 1–137. Dep. VINITI, No. 1043–B 89

Aranson, S.Kh. (1990): Topological invariants of vector fields in a disc on the plane with limit sets of Cantor type. Usp. Mat. Nauk *45*, No. 4, 139–140. [English transl.: Russ. Math. Surv. *45*, No. 4, 157–158 (1990)] Zbl. 715.57014

Aranson, S.Kh., Grines, V.Z. (1973): On some invariants of dynamical systems on two-dimensional manifolds (necessary and sufficient conditions for topological equivalence of transitive dynamical systems). Mat. Sb., Nov. Ser. *90*, No. 3, 372–402. [English transl.: Math. USSR, Sb. *19*. 365–393 (1973)] Zbl. 252.54027

Aranson, S.Kh., Grines, V.Z. (1978): On the representation of minimal sets of flows on two-dimensional manifolds by geodesic lines. Izv. Akad. Nauk SSSR, Ser. Mat. *42*, No. 1, 104–129. [English transl.: Math. USSR, Izv. *12*, No. 1, 103–124 (1978)] Zbl. 384.58013

Aranson, S.Kh., Grines, V.Z. (1980): Dynamical systems with minimal entropy on two-dimensional manifolds. Gor'kij Gos. Univ. 1–71. [English transl.: Sel. Math. Sov. *2*, No. 2, 123–158 (1982)] Zbl. 528.58026

Aranson, S.Kh., Grines, V.Z. (1981): Homeomorphisms with minimal entropy on two-dimensional manifolds. Usp. Mat. Nauk *36*, No. 2, 175–176. [English transl.: Russ. Math. Surv. *36*, No. 2. 163–164 (1981)] Zbl. 471.58013

Aranson, S.Kh., Grines, V.Z. (1982): On recurrent geodesics without self-intersections on two-dimensional manifolds of negative curvature. In Otrokov, N.F. (ed.): Differ. Integral'nye Uravn., Gor'kij Gos. Univ., 21–24

Aranson, S.Kh., Grines, V.Z. (1984): Classification of dynamical systems on two-dimensional manifolds. Tr. IX Mezhdunar. Konf. po Nelineinym Kolebaniyam. Kachestvennye Metody (1981). Inst. Mat. Akad. Nauk Ukr. SSR *2*. Naukova Dumka, Kiev, 23–25

Aranson, S.Kh., Grines, V.Z. (1985): Dynamical Systems I: Flows on Two-Dimensional Manifolds. Itogi Nauki Tekh., Ser. Sovrem. Probl. Mat., Fundam. Napravleniya *1*, 229–240. [English transl. in: Encycl. Math. Sci. *1*, 219–229. Springer-Verlag, Berlin Heidelberg New York 1988] Zbl. 605.58001

Aranson, S.Kh., Grines, V.Z. (1986): Topological classification of flows on closed two-dimensional manifolds. Usp. Mat. Nauk *41*, No. 1, 149–169. [English transl.: Russ. Math. Surv. *41*, No. 1, 183–208 (1986)] Zbl. 615.58015

Aranson, S.Kh., Grines, V.Z. (1990): The topological classification of cascades on closed two-dimensional manifolds. Usp. Mat. Nauk *45*, No. 1, 3–32. [English transl.: Russ. Math. Surv. *45*, No. 1, 1–35 (1990)] Zbl. 705.58038

Aranson, S.Kh., Zhuzhoma, E.V. (1984): Classification of transitive foliations on the sphere with four singularities of "spine" type. In Leontovich-Andronov, E.A. (ed.): Tr. Mezhvuz. Temat. Sb. Nauchn. Metody Kachestvennoj Teorii Differents. Uravn., Gor'kij Gos. Univ. 3–10. [English transl.: Sel. Math. Sov. *9*, No. 2, 117–121 (1990)] Zbl. 742.57014

Arnoux, P., Yoccoz, C. (1981): Construction de difféomorphismes pseudo-Anosov. C.R. Acad. Sci., Paris, Ser. I *292*, No. 1, 75–78. Zbl. 478.58023

Batterson, S. (1980): Constructing Smale diffeomorphisms on compact surfaces. Trans. Am. Math. Soc. *257*, 237–245. Zbl. 426.58019

Bers, L. (1978): An extremal problem for quasiconformal mappings and a theorem of Thurston. Acta Math. *141*, 73–98. Zbl. 389.30018

Bezdenezhnykh A.N., Grines, V.Z. (1985a): Realization of gradient-like diffeomorphisms of two-dimensional manifolds. In Otrokov, N.F. (ed.): Differ. Integral'nye Uravn., Gor'kij Gos. Univ., 33–37

Bezdenezhnykh A.N., Grines, V.Z. (1985b): Diffeomorphisms with orientable heteroclinic sets on two-dimensional manifolds. In Leontovich-Andronov, E.A. (ed.): Tr. Mezhvuz. Temat. Sb. Nauchn. Metody Kachestvennoj Teorii Differents. Uravn., Gor'kij Gos. Univ., 139–152

Bezdenezhnykh A.N., Grines, V.Z. (1985c): Dynamical properties and topological classification of gradient-like diffeomorphisms on two-dimensional manifolds. Part 1, in Leontovich-Andronov, E.A. (ed.): Tr. Mezhvuz. Temat. Sb. Nauchn. Metody Kachestvennoj Teorii Differents. Uravn., Gor'kij Gos. Univ., 22–38; Part 2 (1987): (*ibid*), 24–31

Borevich, E.A. (1980): Conditions for the topological equivalence of two-dimensional Morse-Smale diffeomorphisms. Izv. Vyssh. Uchebn. Zaved., Mat. 1980, No. 11, 12–17. [English transl.: Sov. Math. *24*, No. 11, 11–17 (1980)] Zbl. 463.58017

Borevich, E.A. (1981): Conditions for the topological equivalence of two-dimensional Morse-Smale diffeomorphisms. Differ. Uravn. *17*, No. 8, 1481–1482. Zbl. 489.58017; Topological equivalence of two-dimensional Morse-Smale diffeomorphisms without heteroclinic points (1986): Izv. Vyssh. Uchebn. Zaved., Mat. 1986, No. 9, 6–9. [English transl.: Sov. Math. *30*, No. 9, 6–11 (1986)] Zbl. 641.58018

Borevich, E.A. (1984): Topological equivalence of two-dimensional Morse-Smale diffeomorphisms. Izv. Vyssh. Uchebn. Zaved., Mat. 1984, No. 4, 3–6. [English transl.: Sov. Math. *28*, No. 4, 1–15 (1984)] Zbl. 545.58030

Borevich, E.A. (1989): Two-dimensional Morse-Smale diffeomorphisms with oriented heteroclinic connections. Izv. Vyssh. Uchebn. Zaved., Mat. 1989, No. 9, 77–79. [English transl.: Sov. Math. *33*, No. 9, 76–78 (1989)] Zbl. 693.58008

Bowen, R. (1970): Topological entropy and axiom A. Global analysis. Proc. Symp. Pure Math. *14*, 23–41. Zbl. 207,544

Bowen, R. (1971): Periodic points and measures for axiom A-diffeomorphisms. Trans. Am. Math. Soc. *154*, 377–397. Zbl. 212,291

Fedotov, A.G. (1980): On Williams solenoids and their realization in two-dimensional dynamical systems. Dokl. Akad. Nauk SSSR *252*, 801–804. [English transl.: Sov. Math., Dokl. *21*, 835–839 (1980)] Zbl. 489.58019

Gerber, M., Katok, A. (1982): Smooth models of Thurston's pseudo-Anosov maps. Ann. Sci. Ec. Norm. Super., IV. Ser. *15*, No. 1, 173–204. Zbl. 580.58029. See also Gerber, M. (1985): Conditional stability and real analytic pseudo-Anosov maps. Mem. Am. Math. Soc. 321. Zbl. 572.58018.

Gilman, J. (1981): On the Nielsen type and the classification for the mapping-class group. Adv. Math. *40*, No. 1, 68–96. Zbl. 474.57005

Gilman, J. (1982): Determining Thurston classes using Nielsen types. Trans. Am. Math. Soc. *272*, No. 2, 669–675. Zbl. 507.57005

Grines, V.Z. (1974): On the topological equivalence of one-dimensional basic sets of diffeomorphisms on two-dimensional manifolds. Usp. Mat. Nauk *29*, No. 6, 163–164. Zbl. 304.58009

Grines, V.Z. (1975): On the topological conjugacy of diffeomorphisms of a two-dimensional manifold onto one-dimensional basic sets, Part 1. Tr. Mosk. Mat. O.-va *32*, 35–61. Part 2 (*ibid*) *34*, 243–252 (1977). [English transl.: Trans. Mosc. Math. Soc. *32*, 31–56. Part 2 (*ibid*) *34*, 237–245 (1978)] Zbl. 355.58011; Zbl. 417.58014

Grines, V.Z. (1982): Diffeomorphisms of two-dimensional manifolds with transitive foliations. In Leontovich-Andronov, E.A. (ed.): Tr. Mezhvuz. Temat. Sb. Nauchn. Metody Kachestvennoj Teorii Differents. Uravn., Gor'kij Gos. Univ. 62–72. [English transl.: Transl., Ser. 2, Am. Math. Soc. *149*, 193–199 (1991)] Zbl. 598.58034

Grines, V.Z., Kalaj, Kh.Kh. (1985): Topological equivalence of attractors on two-dimensional manifolds. In Otrokov, N.F. (ed.): Differ. Integral'nye Uravn., Gor'kij Gos. Univ., 104–105

Grines, V.Z., Kalaj, Kh.Kh. (1988a): On the topological equivalence of diffeomorphisms with non-trivial basic sets on two-dimensional manifolds. In Leontovich-Andronov, E.A. (ed.): Tr. Mezhvuz. Temat. Sb. Nauchn. Metody Kachestvennoj Teorii Differents. Uravn., i Teorii Bifurkatsij. Gor'kij Gos. Univ., 40–48

Grines, V.Z., Kalaj, Kh.Kh. (1988b): The topological classification of basic sets without pairs of conjugate points of *A*-diffeomorphisms on surfaces. Gor'kij Gos. Univ., 1–95. Dep. VINITI, No. 1137–B 88

Handel, M., Thurston, W.P. (1985): New proofs of some results of Nielsen. Adv. Math. *56*, No. 2, 173–191. Zbl. 584.57007

Hiraide, K. (1987): Expansive homeomorphisms of compact surfaces are pseudo-Anosov. Proc. Japan Acad., Ser. A *63*, No. 9, 337–338. Zbl. 663.58024; Osaka J. Math. *27* (1990), No. 1, 117–162. Zbl. 713.58042

Ivanov, N.V. (1982a): Nielsen numbers of mappings of surfaces. Zap. Nauchn. Semin. LOMI *122*, No. 4, 56–65. Nauka, Leningrad. [English transl.: J. Sov. Math. *26*, 1636–1671 (1984)] Zbl. 492.55001

Ivanov, N.V. (1982b): Entropy and Nielsen numbers. Dokl. Akad. Nauk SSSR *265*, 284–287. [English transl.: Sov. Math., Dokl. *26*, 63–66 (1982)] Zbl. 515.54016

Jiang, B. (1981): Fixed points of surface homeomorphisms. Bull. Am. Math. Soc., New Ser. *5*, No. 2, 176–178. Zbl. 479.55003

Jiang, B. (1983): Lectures on Nielsen fixed point theory. Am. Math. Soc., Contemp. Math. *14*. Providence. Zbl. 512.55003

Kalay, Kh.Kh. (1988): On the topological equivalence of *A*-diffeomorphisms with basic sets without pairs of conjugate points on two-dimensional manifolds. Gor'kij Gos. Univ., 1–62. Dep. VINITI, No. 8822–B 88

Katok, A. (1979): Bernoulli diffeomorphisms on surfaces. Ann. Math., II. Ser. *110*, No. 3, 529–547. Zbl. 435.58021

Krushkal', S.L., Apanasov, B.N., Gusevskij, N.A. (1981): Kleinian Groups and Uniformization in Examples and Problems. Nauka, Novosibirsk. [English transl.: Transl. Math. Monogr., Vol. 62, Providence. (1986)] Zbl. 485.30001

Liao, S.T. (1980): On the stability conjecture. Chin. Ann. Math. *1*, No. 1, 9–30. Zbl. 449.58013

Mañé, R. (1982): An ergodic closing lemma. Ann. Math., II. Ser. *116*, No. 2, 503–540. Zbl. 511.58029

Miller, R.T. (1982): Geodesic laminations from Nielsen's viewpoint. Adv. Math. *45*, No. 2, 189–212. Zbl. 496.57003

Nielsen, J. (1927): Untersuchungen zur Topologie der geschlossenen zweiseitigen Flächen. I. Acta Math. *50*, 189–358. FdM 53,545; II. Acta Math. *53* (1929), 1–76. FdM 55,971; III. Acta Math. *58* (1932), 87–167. Zbl. 4,275

Nielsen, J. (1937): Die Struktur periodischer Transformationen von Flächen. Det. Kgl. Danske Vid. Selsk., Mat.-Fys. Medd. *15*, 1–77. Zbl. 17,133

Nielsen, J. (1944): Surface transformation classes of algebraically finite type. Det. Kgl. Danske Vid. Selsk., Mat.-Fys. Medd. *21*, 1–89

Nitecki, Z. (1971): Differentiable Dynamics. An Introduction to the Orbit Structure of Diffeomorphisms. MIT Press, Cambridge (Mass.) London. Zbl. 246.58012

O'Brien, T. (1970): Expansive homeomorphisms on compact manifolds. Proc. Am. Math. Soc. *24*, No. 2, 767–771. Zbl. 195,526

O'Brien, T, Reddy, W. (1970): Each compact orientable surface of positive genus admits an expansive homeomorphism. Pac. J. Math. *35*, No. 3, 737–742. Zbl. 206,524

Palis, J., de Melo, W. (1982): Geometric Theory of Dynamical Systems. An Introduction. Springer-Verlag, Berlin Heidelberg New York. Zbl. 491.58001

Peixoto, M. (1973): On the classification of flows on two-manifolds. Dynamical Systems. Proc. Symp., Univ. Bahia, Salvador, Brazil, 1971, 389–419. Zbl. 299.58011

Plykin, R.V. (1971): On the topology of basic sets of Smale diffeomorphisms. Mat. Sb., Nov. Ser. *84*, 301–312. [English transl.: Math. USSR, Sb. *13*, 297–307 (1971)] Zbl. 211,563

Plykin, R.V. (1974): Sources and sinks of A-diffeomorphisms of surfaces. Mat. Sb., Nov. Ser. *94*, 243–264. [English transl.: Math. USSR, Sb. *23*, 233-253 (1975)] Zbl. 324.58013

Plykin, R.V. (1977): On the existence of attracting (repelling) periodic points of A-diffeomorphisms of the projective plane and the Klein bottle. Usp. Mat. Nauk *32*, No. 3, 179. Zbl. 363.58010

Plykin, R.V. (1980a): On hyperbolic attractors of diffeomorphisms. Usp. Mat. Nauk *35*, No. 3, 94–104. [English transl.: Russ. Math. Surv. *35*, No. 3, 109–121 (1980)] Zbl. 445.58016

Plykin, R.V. (1980b): On hyperbolic attractors of diffeomorphisms (non-orientable case). Usp. Mat. Nauk *35*, No. 4, 205–206. [English transl.: Russ. Math. Surv. *35*, No. 4, 186–187 (1980)] Zbl. 448.58011

Plykin, R.V. (1984): On the geometry of hyperbolic attractors of smooth cascades. Usp. Mat. Nauk *39*, No. 6, 75–113. [English transl.: Russ. Math. Surv. *39*, No. 6, 85–131 (1984)] Zbl. 584.58038

Poénaru, V. et al. (1979): Travaux de Thurston sur les surfaces. Astérisque *66–67*. Zbl. 446.57005–008, 011, 017, 019

Robinson, R.C., Williams, R.F. (1973): Finite stability is not generic. Dynamical Systems. Proc. Symp., Univ. Bahia, Salvador, Brazil, 1971, 451–462. Zbl. 305.58009

Sanami, A. (1981): The stability theorems for discrete dynamical systems on two-dimensional manifolds. Proc. Japan Acad., Ser. A *57*, No. 8, 403–407. Zbl. 509.58028; Nagoya Math. J. *90* (1983), 1–55. Zbl. 515.58022

Shub, M., Sullivan, D. (1975): Homology theory and dynamical systems. Topology *14*, No. 2, 109–132. Zbl. 408.58023

Sinai, Ya.G. (1968): Markov partitions and C-diffeomorphisms. Funkts. Anal. Prilozh. *2*, No. 1, 64–69. [English transl.: Funct. Anal. Appl. *2*, 61–82 (1968)] Zbl. 182,550

Solodov, V.V. (1982): Components of topological foliations. Mat. Sb., Nov. Ser. *119*, 340–354. [English transl.: Math. USSR, Sb. *47*, 329–343 (1984)] Zbl. 518.57014

Smale, S. (1967): Differentiable dynamical systems. Bull. Am. Math. Soc. *73*, No. 6, 747–817. Zbl. 202,552

Thurston, W. (1976): On geometry and dynamics of diffeomorphisms of surface. [Preprint, 1–18.] Bull. Am. Math. Soc. *19* (1988), No. 2, 417–431. Zbl. 674.57008

Umanskij, Ya.L. (1990): Necessary and sufficient conditions for the topological equivalence of three-dimensional Morse-Smale dynamical systems with a finite number of singular trajectories. Mat. Sb. *181*, 212–239. [English transl.: Math. USSR, Sb. *69*, 227–253 (1991)] Zbl. 717.58033

Williams, R.F. (1967): One-dimensional nonwandering sets. Topology *6*, No. 4, 473–487. Zbl. 159,537

Williams, R.F. (1970): The DA-maps of Smale and structural stability. Global Analysis. Proc. Symp. Pure Math. *14*, 329–334. Zbl. 213,503

Zieschang, H., Vogt, E., Coldewey, H.-D. (1980): Surfaces and Planar Discontinuous Groups. Lect. Notes Math. *835*. Zbl. 438.57001, Zbl. 204,240

Chapter 4
Dynamical Systems with Transitive Symmetry Group.
Geometric and Statistical Properties

A.V. Safonov, A.N. Starkov, A.M. Stepin

Introduction

The main topic of discussion in this survey is the geometric and metric properties of geodesic and horocycle flows on surfaces of constant curvature and their multidimensional analogues, the so-called G-induced actions (that is, restrictions to a suitable subgroup F of the natural action of G on a homogeneous space of it). One usually takes as F one-parameter or cyclic subgroups of G and the horospherical subgroups corresponding to them (the details are given in Sect. 1). However, in the course of the exposition of our main topic we sometimes mention results relating to the case when F is an n-parameter subgroup of G or a lattice. This class of dynamical systems consists precisely of group actions admitting an extension to a transitive action of a Lie group. We shall not be considering in our survey (or, at any rate, not systematically) the following themes which are close to our main subject matter:

Topological dynamics and metric properties of general group actions (in particular, the work of Zimmer on semisimple group actions (Zimmer 1984);

Group automorphisms;

Affine transformations of homogeneous spaces;

G-induced flows on spaces with an infinite invariant measure. The papers (Hopf 1939), (Sullivan 1981) and the recent book (Nicholls 1989) are devoted to the connections between G-induced flows (and also geodesic flows) and the theory of discrete groups of motions of a hyperbolic space;

The actions of subgroups on the double cosets related to geodesic (for $n > 2$) and horocycle (for $n > 3$) flows on an n-dimensional manifold of constant negative curvature (see (Gel'fand and Fomin 1952) and, in particular, (Mautner 1957)). Flows of this type in somewhat more general form differing from geodesic and horocycle flows have not been investigated.

The following separate questions have been considered: the algebraic structure of Anosov flows (Tomter 1975) and flows of frames on manifolds of negative curvature (which is not necessarily constant, so that this area of activity is only partly related to DS's of algebraic origin) (see the survey (Pesin 1981)).

In the study of homogeneous flows a fundamental role is played by the algebraic results on the structure of spaces of finite volume (see Appendix A), as well as results of the theory of unitary Lie group representations (Mautner phenomenon, see Sect. 1).

A certain acquaintance with ergodic theory is required for reading this article (which is necessary for the subsequent account of measure theory; these include: metric entropy, the various statistical properties from ergodicity to the Bernoulli condition, including spectral properties and unique ergodicity) as well as the theory of Lie groups and algebras (as far as the Levi-Mal'tsev decomposition). Other key words characterizing the boundary of the material used are: the adjoint action and Jordan decomposition for an element of Lie algebra, various types of Lie groups and algebras, from nilpotent to reductive, and the nil-radical. Here and there algebraic groups and the Zariski topology are used. We draw attention to one change in the terminology that has occurred in recent years and is adopted by us. Those solvable groups that were previously called groups of type (R) (from "rigid" or "rotation", see (Auslander, Green, and Hahn 1963) are now called groups of type (I) (from "imaginary"), while inside another class (E) (from "exponential") we now select groups of type (R) (from "real", see (Vinberg, Gorbatsevich and Shvartsman 1988, Chap. 2, Sect. 5).

We give a few words on the history of the earlier research, roughly over a twenty year period. Hadamard was the first to draw attention to the behaviour of geodesic lines on (not necessarily closed) surfaces of negative curvature. The work of Morse, with occasional participation of other well known mathematicians, played an important role in the earlier history of the subject relating to the formation of symbolic dynamics; see the beginning of (Alekseev 1976). While only topological properties of geodesic flows were first investigated, subsequently Hedlund drew attention to horocycle flows (Hedlund 1936), while Hedlund and Hopf also started to investigate metric properties of geodesic and horocycle flows. This period is concluded by the article (Hopf 1939) and the survey (Hedlund 1939).

The algebraic and functional-analytic approach to the study of these flows was first proposed by Gel'fand and Fomin (Gel'fand and Fomin 1952). This line of approach was further developed in the paper by Mautner (Mautner 1957) and especially in the book (Auslander, Green, and Hahn 1963). Apparently, the works of the last three authors were started independently of (Gel'fand and Fomin 1952) and mark the passage to the investigation of homogeneous flows differing from geodesic and horocycle ones (leaving out of consideration the irrational winding of the torus, known in the time of Kronecker and Poincaré). In fact this area of research started with the investigation of nil-flows in which a group of American mathematicians took an interest in connection with the search for examples of minimal flows. Finally, the combination of geometric and probability-theoretic approaches with the use of ergodic theory in the work of Anosov and Sinai (Anosov 1967), (Anosov and Sinai 1967), (Sinai 1960), (Sinai 1961), (Sinai 1966), although not specifically concerned with homogeneous flows, enriched the methods of investigation of subsequent research and stimulated the interest of a large number of mathematicians in this subject.

§1. Basic Concepts and Constructions. Examples

In this survey we consider dynamical systems of algebraic origin. The general construction occurs from the left action of a subgroup F of a Lie group G on the homogeneous space G/D of right cosets of G with respect to another subgroup $D \subset G$. We denote such a dynamical system by $(G/D, F)$. The standard (and best studied) situation is when it is required of the subgroup $D \subset G$ that there exist on G/D a finite G-invariant measure μ (in other words, the space G/D must have a finite volume). Such a measure is unique to within a constant factor and it is usually normalized by the condition $\mu(G/D) = 1$. If in addition to the finiteness of the volume of G/D the subgroup $D \subset G$ is discrete, then it is called a *lattice* in G. A subgroup $D \subset G$ is considered to be *uniform* or *non-uniform* depending on whether or not G/D is compact. An example of a uniform lattice is the integer lattice $\mathbb{Z}^n$ in the abelian group $\mathbb{R}^n$. The classical example of a non-uniform lattice is the discrete subgroup of integer-valued matrices $SL(n, \mathbb{Z})$ of the semisimple group $SL(n, \mathbb{R})$ of all $n \times n$ matrices with determinant 1. A summary of results on the structure of spaces of finite volume is given in Appendix A.

The subgroup $F \subset G$ is usually taken to be some one-parameter subgroup $\exp(\mathbb{R}x)$. In this case the dynamical system $(G/D, \exp(\mathbb{R}x))$ is called the *G-induced flow.* (Sometimes the action of any connected subgroup $F \subset G$ is called the flow. With the exception of the so-called horospherical flows, we shall be considering one-parameter flows.) The homogeneous space G/D is always assumed to be of finite volume.

Remark 1.1. If the subgroup $D \subset G$ contains a connected normal subgroup H of G, then the flow $(G/D, F)$ is equivalent to the flow $(G'/D', F')$, where $G' = G/H$, $D' = D/H$ and $F' = FH/H$. It is therefore natural to restrict oneself to the case when the subgroup $D \subset G$ contains no non-trivial connected normal subgroups of G. Here we call the space G/D *non-cancellable* (sometimes these spaces are called *locally faithful* or *presentations*).

Remark 1.2. The adjoint elements x and $\mathrm{Ad}_g x$ induce equivalent flows. This equivalence is given by the homeomorphism $L_g(hD) = ghD$ of G/D, which is equivariant with respect to the action of the flows since $L_g(\exp(tx)hD) = g\exp(tx)g^{-1}ghD = \exp(t\,\mathrm{Ad}_g x)L_g(hD)$.

The class of homogeneous flows is a good proving ground for the study of smooth dynamical systems considered in (Bunimovich et al. 1985). Such concepts of the smooth theory as expanding, contracting and neutral foliations, partial hyperbolicity, and so on, are also naturally applicable to them.

However, in the homogeneous case these notions have a precise algebraic interpretation. For example, it is not difficult to construct in explicit form a contracting leaf of the flow $(G/D, \exp(\mathbb{R}x))$ at any point $gD \in G/D$. It is the orbit $G^- gD$ of the so-called contracting horospherical subgroup $G^- \subset G$ for the element $\exp(x) \in G$.

Definition 1.3. The subgroup G^- of all elements $h \in G$ for which $a^n h a^{-n} \to 1$ as $n \to +\infty$ is called the *contracting horospherical subgroup* for the element $a \in G$. Similarly, the *expanding horospherical subgroup* G^+ consists of all $h \in G$ for which $a^{-n} h a^n \to 1$ as $n \to +\infty$.

It is known that any horospherical subgroup is a connected unipotent[1] subgroup of G (but not conversely (Dani 1976a)).

It is clear from the definition of a contracting horosphere that the orbit of a point $hgD \in G/D$ for $h \in G^-$ is exponentially approached by the orbit of the point $gD \in G/D$, since $\exp(tx)hgD = \exp(tx)h\exp(-tx)\exp(tx)gD \to \exp(tx)gD$ in the metric on G/D induced by the right-invariant metric on G. Furthermore, the orbits of the action of the subgroups G^+, G^- have constant dimension and therefore form smooth foliations. This is obvious for a discrete lattice $D \subset G$, while for spaces of finite volume it is proved in (Starkov 1991)[2]. Thus we have the following result.

Proposition 1.4. *The orbits of the horospherical flows* $(G/D, G^+)$ *and* $(G/D, G^-)$ *form smooth expanding and contracting foliations, respectively, for the flow* $(G/D, \exp(\mathbb{R}x))$.

We note that, in general, horospherical flows are induced by the action of multidimensional subgroups G^+ and G^-. They play an important role in the study of the measure theoretic and topological properties of the original flow. We note also that either both the horospherical foliations of a given homogeneous flow are trivial (each leaf consists of a single point) or both are non-trivial in view of the existence of a finite invariant measure on G/D.

Flows on Surfaces of Constant Negative Curvature. The classical examples of homogeneous flows are geodesic, horocycle and periodic flows on a surface M^2 of constant negative curvature. The phase space of these flows (that is, the space of unit tangent vectors of M^2) is the homogeneous space $SL(2,\mathbb{R})/\Gamma$, where Γ is a discrete subset of $SL(2,\mathbb{R})$ containing the centre $\mathbb{Z}_2 \subset SL(2,\mathbb{R})$. This follows from the observation that $SL(2,\mathbb{R})/\mathbb{Z}_2$ is the group of conformal motions of the upper half-plane, which is a Lobachevskij space in the Poincaré model. The *geodesic, horocycle* and *periodic* flows are induced by the following three elements of the Lie algebra $\mathfrak{sl}(2,\mathbb{R})$:

---

[1] In what follows, a subgroup $F \subset G$ is called unipotent if for each $f \in F$ the adjoint action of $\mathrm{Ad}\, f$ is a unipotent operator on the Lie algebra $\mathfrak{g}$ of G.

[2] Stepin and Chumak have established the general fact that partial hyperbolicity is preserved under factoring (see Chumak and Stepin 1989).

$$a = \begin{pmatrix} 1 & 0 \\ 0 & -1 \end{pmatrix}, \quad n = \begin{pmatrix} 0 & 1 \\ 0 & 0 \end{pmatrix}, \quad c = \begin{pmatrix} 0 & 1 \\ -1 & 0 \end{pmatrix}.$$

We note that a horocycle subgroup is an expanding horospherical subgroup for a geodesic flow, since $[a, n] = 2n$. This fact (in a more geometric form) is known to have been used back in the 30s by Hopf, Hedlund and others. The subsequent generalization of the corresponding geometric situation led to Anosov flows. We give a brief résumé of the results on the measure theoretic and topological properties of these flows, taking the area of M^2 to be finite.

A geodesic flow is a Bernoulli flow (Ornstein and Veiss 1973). Earlier a number of weaker properties of this flow had been established which, in fact, were used in the proof of the Bernoulli property. In order of strengthening they are: ergodicity and mixing (Hedlund 1939), (Hopf 1936), (Hopf 1939), the countability of the Lebesgue spectrum (Gel'fand and Fomin 1952), and the K-property (Sinai 1960). All these are special cases of results on Anosov flows (Anosov 1967), (Anosov and Sinai 1967), (Bowen 1979), (Sinai 1966). But in the theory of Anosov systems the phase manifold is assumed to be compact, while the results given above are valid in the more general situation when M^2 has a finite area. (Hopf 1939) and (Gel'fand and Fomin 1952) refer directly to the general case, while the K-property and the Bernoulli property follow from results of Dani (see Sect. 5). As is necessary for a smooth ergodic flow, almost all its orbits are dense, however there are at least two other classes of orbits forming a dense set of measure 0 in the phase manifold:

a) periodic orbits,

b) *exotic orbits*, the closures of which are not manifolds (but which, for example, are constructed locally as a product of a Cantor set and an interval). Their discovery is naturally associated with the name of Morse (Morse 1921) who was interested in minimal sets (see (Alekseev 1976) concerning the history of this). Nowadays the existence of exotic orbits is obvious. For example, it follows from the existence of transversal homoclinic points, while the existence of the latter is ensured by the well known properties of transversal foliations for Anosov systems.

In the non-compact case one more class should be added:

c) orbits, one half of which goes off to infinity.

A horocycle flow also is ergodic and mixing (Hedlund 1936), where any degree of mixing is possible (Marcus 1978) and has a Lebesgue spectrum (Parasyuk 1953). By contrast with a geodesic flow, the rate of scattering of nearby orbits is not exponential but polynomial (the precise statement is given in infinitesimal terms; here it is a question of the rate of increase of solutions of the variational equations). Therefore it has zero entropy, which immediately follows from the Millionshchikov-Pesin formula for entropy (Millionshchikov 1976), (Pesin 1977), (Mañé 1981), (Mañé 1987), as well as from the weaker estimate due to Kushnirenko (Kushnirenko 1965). However, it was Gurevich who first calculated the entropy of a horocycle flow (Gurevich 1961) (in the non-compact case, see also (Dani 1977a)).

In the compact case a horocycle flow is *minimal* (that is, all its orbits are dense, (Hedlund 1936)), and even uniquely ergodic (Furstenberg 1963). In the non-compact case, apart from the dense orbits there is just one additional class, namely, the periodic orbits (Hedlund 1936).

We also note a helpful observation by Hedlund: if a geodesic orbit of some point does not go off to infinity in the negative direction, then the horocycle corresponding to this point is everywhere dense. This result was subsequently generalized by Dani (see Sect. 4). Thus, for example if $\Gamma = SL(2,\mathbb{Z})$, then the horocycle of a point of $\Gamma \in SL(2,\mathbb{R})/\Gamma$ is periodic, and a geodesic from this point goes off to infinity in the negative direction.

With regard to a periodic flow, it is constructed in a very trivial manner: all its orbits are circles and, in the terminology of (Dani 1980), it is strictly non-ergodic. The partition of the surface into closed orbits of the flow forms an ergodic decomposition of it and may not be a smooth bundle (if $\Gamma/\mathbb{Z}_2$ contains elements of finite order).

The difference in the behaviour of these three flows from the algebraic point of view is occasioned by a very simple circumstance: a is a semisimple element with the real eigenvalues ± 1; the element n is nilpotent, while c is a semisimple element with the purely imaginary eigenvalues $\pm i$. The role of the eigenvalues of an element $x \in \mathfrak{g}$ in the behaviour of the homogeneous flow $(G/D, \exp(\mathbb{R}x))$ will be revealed in what follows.

Representation Theory and Ergodicity of Homogeneous Flows. The existence on G/D of a finite G-invariant measure μ makes it possible to use the algebraic methods of representation theory in the study of G-induced flows. For this we consider the Hilbert space $L^2(G/D,\mu)$ of complex-valued functions f on G/D that are square-integrable with respect to μ and with the scalar product

$$(f, f') = \int f(g)\overline{f'(g)}d\mu_g.$$

The left action of G on G/D induces the unitary representation $\rho : G \to U(L^2(G/D,\mu))$ in accordance with the formula $\rho(g)f(hD) = f(g^{-1}hD)$. In these terms, the ergodicity of the flow $(G/D, \exp(\mathbb{R}x))$ is equivalent to the property that any invariant function in $L^2(G/D,\mu)$ is a constant. On the other hand, weak mixing means that any eigenfunction of the flow is a constant. We recall the following definition.

Definition 1.5. A function $f \in L^2(G/D,\mu)$ is called an *eigenfunction of the flow* with eigenvalue $\lambda \in \mathbb{C}$ if $\rho(\exp(tx)f = e^{\lambda t} f$ for all $t \in \mathbb{R}$. The eigenvalue $\lambda = 0$ corresponds to an invariant function.

It is not difficult to prove the following.

Proposition 1.6. *Any eigenfunction for* $\exp(\mathbb{R}x)$ *is always invariant with respect to the horospherical subgroups* $G^+, G^- \subset G$.

In fact, if $h \in G^+$, then

$$\begin{aligned}(\rho(h)f,f) &= (\rho(\exp(tx))\rho(\exp(-tx)h\exp(tx))\rho(\exp(-tx))f,f)\\ &= (\rho(\exp(-tx)h\exp(tx))\rho(\exp(-tx))f,\rho(\exp(tx))f)\\ &= e^{\lambda t}e^{-\lambda t}(\rho(\exp(-tx)h\exp(tx))f,f)\to(f,f),\end{aligned}$$

as follows from the unitarity and continuity of ρ. Using Schwarz's inequality and the unitarity of ρ we deduce that $\rho(h)f=f$.

The invariance of f with respect to G^- is derived similarly. Note that, in fact, we have proved the weak mixing property of a geodesic flow, since its horospherical subgroups G^+ and G^- generate the entire group $G=SL(2,\mathbb{R})$.

The fact that the eigenfunctions of the subgroup $\exp(\mathbb{R}x)\subset G$ are always invariant with respect to certain other subgroups of G is called the *Mautner phenomenon.* We note that instead of $L^2(G/D,\mu)$ we could have considered any Hilbert space $\mathcal{H}$ with the unitary representation $\rho: G\to U(\mathcal{H})$. It is natural to try to find the largest subgroup $F_x\subset G$ with respect to which all the vectors of $\mathcal{H}$ that are eigenvectors for the subgroup $\exp(\mathbb{R}x)\subset G$ are invariant (independently of ρ and $\mathcal{H}$). It would seem that the definitive result was obtained by Moore (Moore 1980). We describe the construction of such a subgroup $F_x\subset G$ along with two other subgroups.

Definition 1.7. Consider the root decomposition $\mathfrak{g}=\sum_\alpha\mathfrak{g}^\alpha$ of the Lie algebra $\mathfrak{g}$ with respect to the operator $\operatorname{ad}x$. Let

$$\mathfrak{g}^+=\sum_{\operatorname{Re}\alpha>0}\mathfrak{g}^\alpha,\quad \mathfrak{g}^-=\sum_{\operatorname{Re}\alpha<0}\mathfrak{g}^\alpha.$$

Since $[\mathfrak{g}^\alpha,\mathfrak{g}^\beta]\subset\mathfrak{g}^{\alpha+\beta}$, it follows that $\mathfrak{g}^+$ and $\mathfrak{g}^-$ are subalgebras of $\mathfrak{g}$, while the subalgebra $\mathfrak{u}_x$ generated by $\mathfrak{g}^+$ and $\mathfrak{g}^-$ is an ideal in $\mathfrak{g}$, called the *unstable ideal* of $\mathfrak{g}$ with respect to $x\in\mathfrak{g}$.

An equivalent definition is that $\mathfrak{u}_x$ is the smallest ideal in $\mathfrak{g}$ for which the operator $\operatorname{ad}x$ on the quotient algebra $\mathfrak{g}/\mathfrak{u}_x$ has purely imaginary eigenvalues. It was first introduced by Auslander (Auslander, Green, and Hahn 1963). Note that the subalgebras $\mathfrak{g}^+$ and $\mathfrak{g}^-$ correspond to the horospherical subgroups G^+ and G^-.

Definition 1.8. Consider the Levi decomposition $\mathfrak{g}=\mathfrak{s}+\mathfrak{r}$ of the Lie algebra $\mathfrak{g}$ into a semisimple subalgebra $\mathfrak{s}$ and the radical $\mathfrak{r}$. We define $\mathfrak{i}_x$ as the smallest ideal in $\mathfrak{g}$ for which the operator $\operatorname{ad}x$ on $\mathfrak{g}/\mathfrak{i}_x$ has purely imaginary eigenvalues and, furthermore, is semisimple (completely reducible) on $\mathfrak{g}/(\mathfrak{r}+\mathfrak{i}_x)$.

The ideal $\mathfrak{i}_x$ is called the *Dani ideal* since it was first constructed in (Dani 1977a).

Definition 1.9. By the ad-*compact ideal* $\mathfrak{f}_x$ for x we mean the smallest ideal $\mathfrak{f}_x$ for which the operator $\operatorname{ad}x$ on $\mathfrak{g}/\mathfrak{f}_x$ is a semisimple operator with purely imaginary eigenvalues.

The term "ad-compact" for $\mathfrak{f}_x$ is explained by the property that, as is easily seen, the subgroup $\mathrm{Ad}(\exp(\mathbb{R}x))$ has a compact closure in the group of automorphisms $\mathrm{Aut}(\mathfrak{g}/\mathfrak{f}_x)$.

The ideal $\mathfrak{f}_x$ was introduced by Moore (Moore 1980). Clearly, $\mathfrak{u}_x \subset \mathfrak{i}_x \subset \mathfrak{f}_x$.

In the case of a nilpotent Lie algebra $\mathfrak{g}$ we always have $\mathfrak{u}_x = \mathfrak{i}_x = 0$; in the solvable case we have $\mathfrak{u}_x = \mathfrak{i}_x$, while in the semisimple case we have $\mathfrak{i}_x = \mathfrak{f}_x$. The Mautner phenomenon has been investigated in these three model cases by Green (Green 1961), Auslander (Auslander 1973) and Moore (Moore 1966), respectively.

We consider the corresponding analytic normal subgroups $U_x \subset I_x \subset F_x$. The following result is useful in the study of the ergodic and spectral properties of homogeneous flows.

Theorem 1.10 (Moore 1980). *Let $\rho : G \to U(\mathcal{H})$ be a unitary representation of the Lie group G in Hilbert space $\mathcal{H}$, f an eigenvector of the subgroup $\exp(\mathbb{R}x) \subset G$, and F_x the* ad-*compact normal subgroup for $x \in \mathfrak{g}$. Then f is invariant with respect to $F_x \subset G$. Furthermore, if $d\rho(x)$ is the infinitesimal operator of $\rho(\exp(\mathbb{R}x))$, then the restriction of $d\rho(x)$ to the space of $\rho(F_x)$-fixed vectors has an absolutely continuous spectrum.*

Corollary 1.11. *The flows $(G/D, \exp(\mathbb{R}x))$ and $(G/\overline{F_xD}, \exp(\mathbb{R}x))$ are either both ergodic or both non-ergodic.*

Proof. Let $f \in L^2(G/D, \mu)$ be an invariant function for $\exp(\mathbb{R}x) \subset G$. Then $\rho(F_x)f = f$ and it follows from the normality of F_x that f is constant on almost all fibres of the form $gF_xD \subset G/D$. We deduce in standard fashion from the continuity of the representation ρ on $L^2(G/D, \mu)$ that f is constant on almost all the closed fibres $g\overline{F_xD} \subset G/D$ and therefore defines a function $\tilde{f} \in L^2(G/\overline{F_xD}, \tilde{\mu})$, where $\tilde{\mu}$ is a G-invariant measure on $G/\overline{F_xD}$. But if the factor-flow on $G/\overline{F_xD}$ is ergodic, then $\tilde{f} \equiv \text{const}$, therefore $f \equiv \text{const}$ and the original flow is also ergodic. □

It is also easy to prove the following.

Corollary 1.12. *The closed subsets $\tilde{E}_g(x) = \overline{\exp(\mathbb{R}x)F_xgD} \subset G/D$ form a partition of G/D. This partition $\tilde{E}$ is the* ergodic decomposition *of the flow $(G/D, \exp(\mathbb{R}x))$.*

This means that on almost all the $\tilde{E}_g(x)$ there exists a finite $\exp(\mathbb{R}x)$-invariant ergodic measure $\tilde{\mu}_g$ and the measure μ is "pasted together" from the measures $\tilde{\mu}_g$. In fact, all the $\tilde{E}_g(x)$ are submanifolds of G/D with $\overline{\exp(\mathbb{R}x)F_x}$-invariant smooth measures $\tilde{\mu}_g$ with respect to which almost every smooth flow $(\tilde{E}_g(x), \exp(\mathbb{R}x))$ is ergodic (Starkov 1983).

We also consider the closed subsets $E_g(x) = \overline{\exp(\mathbb{R}x)I_xgD} \subset G/D$. It turns out that they too are submanifolds and form a partition of G/D. There exists a finite $\exp(\mathbb{R}x)I_x$-invariant smooth measure μ_g on each of them. Using the fact that the normal Dani subgroup I'_x of the closed Lie group $\overline{\exp(\mathbb{R}x)I_x}$ is the same as I_x, we conclude that all the flows $(E_g(x),\ \exp(\mathbb{R}x))$ are ergodic, therefore the partitions E and $\tilde{E}$ are the same outside some set of measure 0 in G/D, the partition E being a subpartition of $\tilde{E}$.

Corollary 1.13. *The homogeneous flow* $(G/D,\ \exp(\mathbb{R}x))$ *is ergodic if and only if it is topologically transitive.*

In fact, if a smooth flow with a finite invariant measure is ergodic, then almost all its orbits are dense (see Sect. 4). Conversely, if at least one orbit $\exp(\mathbb{R}x)gD$ is dense in G/D, then $E_g(x) \supset \overline{\exp(\mathbb{R}x)gD} = G/D$, which means that the flow $(G/D,\ \exp(\mathbb{R}x))$ is ergodic.

Corollary 1.14. *If* $G = \overline{I_x D}$, *then the flow* $(G/D,\ \exp(\mathbb{R}x))$ *is weakly mixing.*

This corollary is proved in the same way as Corollary 1.11 and, as will be shown in Sect. 2, it is a criterion for mixing.

§2. A Criterion for Ergodicity and the Ergodic Decomposition

Solvable Case. We consider in succession the problem of the ergodicity of G-induced flows for abelian, nilpotent and solvable Lie groups G.

In the abelian case we have the linear flow $(\mathbb{R}^n/\mathbb{Z}^n, \mathbb{R}v)$ on the torus with direction element $v \in \mathbb{R}^n$. Kronecker's ergodicity criterion is well known for such a flow: it is ergodic if and only if the numbers $v_1, \dots, v_n$ are independent over the field $\mathbb{Q}$. If a linear flow is ergodic, then it is minimal and uniquely ergodic. In the opposite case, the ergodic decomposition is a bundle of lower dimensionial subtori.

Consider a *nil-flow*, that is, an arbitrary homogeneous flow on a compact nil-space. In view of the Zariski density of uniform subgroups in a nilpotent Lie group (Appendix A, Theorem 9) and Remark 1.1, we can restrict our discussion to the nil-space N/Γ, where Γ is a lattice in a simply connected group N.

Theorem 2.1 (Green's Ergodicity Criterion (Auslander, Green, and Hahn 1963)). *The nil-flow* $(N/\Gamma,\ \exp(\mathbb{R}n))$ *is ergodic if and only if the linear flow* $(N/\Gamma[N,N],\ \exp(\mathbb{R}n))$ *is ergodic on the torus* $N/\Gamma[N,N]$.

We note that in the statement of this theorem the product $\Gamma[N,N]$ is always closed, the group $N/[N,N]$ is abelian and $N/\Gamma[N,N]$ is in fact a torus.

Theorem 2.2. *An ergodic nil-flow is minimal* (Auslander, see (Auslander, Green, and Hahn 1963)) *and uniquely ergodic* (Furstenberg 1963).

It is now a straightforward matter to construct ergodic partitions E and $\tilde{E}$ for nil-flows. By definition, $E_g(n) = \overline{\exp(\mathbb{R}n)g\Gamma}$, $\tilde{E}_g(n) = \overline{\exp(\mathbb{R}n)Fg\Gamma}$, where F is the smallest connected normal subgroup of N for which the subgroup $\exp(\mathbb{R}n)F$ is normal in N. Consequently, the submanifolds $\tilde{E}_g(n)$ are automatically homogeneous subspaces of the form $\tilde{E}_g(n) = Hg\Gamma = gH\Gamma$, where $H = (\overline{\exp(\mathbb{R}n)F\Gamma})_0$ is a connected normal subgroup of N. On the other hand, $E_g(n) = gH_g\Gamma$, where $H_g = (\overline{g^{-1}\exp(\mathbb{R}n)g\Gamma})_0 \subset N$. Here all the H_g are also subgroups of N. In fact, if H'_g is the smallest subgroup of N such that $g^{-1}\exp(\mathbb{R}n)g \subset H'_g$ and the product $H'_g\Gamma$ is closed, then in view of Green's ergodicity criterion, the flow $(H'_g/\Gamma \cap H'_g,\ g^{-1}\exp(\mathbb{R}n)g)$ must be ergodic and therefore minimal. Consequently, $H_g = H'_g$. Thus, we have proved the following result.

Proposition 2.3. *In the nilpotent case $\tilde{E}$ is a smooth fibering of G/D into the orbits of the action of the normal subgroup H, and all the submanifolds $E_g(n)$ are homogeneous.*

We note that E may not be a bundle in view of the fall of the dimension of some of the $E_g(n)$.

Example 2.4. Let

$$n = \begin{pmatrix} 0 & 1 & 0 \\ 0 & 0 & 0 \\ 0 & 0 & 0 \end{pmatrix}, \quad g = \begin{pmatrix} 1 & 0 & 0 \\ 0 & 1 & r \\ 0 & 0 & 1 \end{pmatrix}, \quad r \in \mathbb{R},$$

and let Γ be the lattice of integer-valued matrices in the 3-dimensional nilpotent group N of upper-triangular matrices. Then

$$g^{-1}\exp(tn)g = \begin{pmatrix} 1 & t & tr \\ 0 & 1 & 0 \\ 0 & 0 & 1 \end{pmatrix}$$

and therefore the dimension of $E_g(n) = gE_1(\mathrm{Ad}_{g^{-1}}n)$ is equal to either 2 or 1 depending on the irrationality of the number $r \in \mathbb{R}$.

The G-induced flow in a solvable Lie group G is called a *solvable* flow. First we consider the case when G is of class (R). In such Lie groups the lattice is also Zariski dense in any (for example, the adjoint) representation, therefore the subgroup $H = (\overline{F_x\Gamma})_0$ is always normal in G. We set $G' = G/H$, $\Gamma' = \Gamma H/H$, $\exp(\mathbb{R}x') = \exp(\mathbb{R}x)H/H$. According to Theorem 1.11, the flows $(G/\Gamma, \exp(\mathbb{R}x))$ and $(G'/\Gamma',\ \exp(\mathbb{R}x'))$ are either both ergodic or both non-ergodic. Since G' is also a solvable group of type (R), it follows that $\exp(\mathbb{R}x')$ is in the centre and in the nil-radical of G'. Therefore $\tilde{E}_1(x') = \overline{\exp(Rx')\Gamma'} = H'\Gamma'$ is a homogeneous subspace of G'/Γ' and in order that the original flow be ergodic it is necessary that G' be nilpotent. As regards an ergodicity criterion for nil-flows, it is already known. It follows from the homogeneity of $\tilde{E}_1(x')$ that $\tilde{E}_1(x)$ and all the $\tilde{E}_g(x)$ are homogeneous, in view of Remark 1.2. Similarly, by considering the normal subgroup $H = (\overline{U_x\Gamma})_0$ and the factor flow $(G/\overline{U_x\Gamma},\ \exp(\mathbb{R}x))$, we deduce that all the $E_g(x)$ are homogeneous. Thus the ergodic decomposition E for G-induced flows of type (R) remains homogeneous and, in addition, the decomposition $\tilde{E}$ is a smooth bundle. However, with regard to the behaviour of the orbits there are principal differences compared to the nilpotent case in which the closure of any orbit coincides with the submanifold $E_g(x)$. In the solvable case there occur exotic orbits whose closures are, in general, not manifolds.

Example 2.5. Consider the semidirect product $G = \mathbb{R}^1 \cdot \mathbb{R}^2$, where $\mathbb{R}^1$ acts on $\mathbb{R}^2$ as the group of hyperbolic rotations

$$\begin{pmatrix} e^t & 0 \\ 0 & e^{-t} \end{pmatrix}, \quad t \in \mathbb{R}.$$

The action of some $t_0 \in \mathbb{R}$ in some basis of $\mathbb{R}^2$ can be described by the integer-valued matrix $a = \left(\begin{smallmatrix} 2 & 1 \\ 1 & 1 \end{smallmatrix}\right)$. We set $\Gamma = Z\{a\} \cdot \mathbb{Z}^2$ (where $Z\{a\}$ is the cyclic group generated by a) and $\exp(x) = a$. As is easily verified, the flow $(G/\Gamma, \exp(\mathbb{R}x))$ is an Anosov flow and a suspension over the Anosov cascade $Z\{a\}$ of the torus $\mathbb{R}^2/\mathbb{Z}^2$. Among the orbits of this cascade one encounters orbits whose closures are homeomorphic to a Cantor set. Here the arguments are the same as for the geodesic flows above.

A useful instrument for investigating homogeneous spaces of Lie groups and G-induced flows is the construction of a semisimple splitting invented by Mal'tsev (see Appendix B).

An application of the construction of a semisimple splitting is well illustrated in Auslander's formulation of the ergodicity criterion for G-induced flows with a solvable Lie group G of type (I). If such a Lie group contains a uniform subgroup $D \subset G$ without connected normal subgroups of G then its semisimple splitting has the form $G_S = T_G \cdot N_G = T_G \cdot G$, where T_G is a torus in the algebraic group $\operatorname{Aut} N_G$ of all automorphisms of the nil-radical N_G. The projection $p : G \to T_G$ is a group epimorphism, while the projection $\pi : G \to N_G$ is a diffeomorphism of the manifolds. Since the image $p(D)$ of the subgroup $D \subset G$ is discrete in T_G, it follows that $D' = D \cap N_G$ is a subgroup of finite index in D. For a fixed element $x \in \mathfrak{g}$ inducing a flow on G/D we choose the torus T_G such that we have the decomposition $\exp(tx) = \exp(ta) \times \exp(tn)$, where the one-parameter subgroups $\exp(\mathbb{R}a) \subset T_G$ and $\exp(\mathbb{R}n) \subset N_G$ commute. The ergodicity of the flow $(T_G/p(D),\ \exp(\mathbb{R}a))$ on the torus is clearly a necessary condition for the ergodicity of the flow $(G/D,\ \exp(\mathbb{R}x))$. If this condition holds, then $T_G = \overline{\exp(\mathbb{R}a)}$ and the projection $\pi : G \to N_G$ induces the equivalence of flows

$$\pi^* : (G/D',\ \exp(\mathbb{R}x)) \to (N_G/D',\ \exp(\mathbb{R}n)).$$

In fact,

$$\begin{aligned} \pi(\exp(tx)gD') &= \pi(\exp(ta)\exp(tn)p(g)\pi(g)D') \\ &= \exp(tn)\pi(g)D', \end{aligned}$$

where we have used the commutability of the subgroups $\exp(\mathbb{R}n) \subset N_G$ and $T_G = \overline{\exp(\mathbb{R}a)}$. As a consequence, we have proved the following.

Proposition 2.6. *If the group G is of type (I), then the ergodic flow $(G/D,\ \exp(\mathbb{R}x))$ is finite-sheetedly covered by the nil-flow $(N_G/D', \exp(\mathbb{R}n))$.*

A similar argument shows that in the non-ergodic case a flow on the ergodic component $E_1(x) = \overline{\exp(\mathbb{R}x)D} \subset G/D$ is finite-sheetedly covered by the ergodic nil-flow $(H_1/D \cap H_1, \exp(\mathbb{R}n))$, where $H_1 = (\overline{\exp(\mathbb{R}n)D'})_0$ is a subgroup of N_G. Note that $E_1(x)$ is a submanifold of G/D (since $E_1(x)$ is the orbit space of the free action of a finite group on the manifold $H_1/D' \cap H_1$; for details, see (Starkov 1984)), but is not necessarily a homogeneous subspace in G/D.

Example 2.7. Let $G = \mathbb{R}^1 \cdot \mathbb{R}^2$, where $\mathbb{R}^1$ acts on $\mathbb{R}^2$ as a group of orthogonal rotations with kernel $2\pi\mathbb{Z} \subset \mathbb{R}^1$. We set $\Gamma = Z\{a\} \times \mathbb{Z}^2$, where $\mathbb{Z}^2 \subset \mathbb{R}^2$ and $a = 2\pi \times b$, $2\pi \in \mathbb{R}^1$, $b = (\pi, 0) \in \mathbb{R}^2$. Then Γ is a lattice in G and $G_s = SO(2) \cdot N_G = SO(2) \cdot G$, where $SO(2) = \mathbb{R}^1/2\pi\mathbb{Z}$, $N_G = \Delta \times \mathbb{R}^2$ and Δ is the antidiagonal in $SO(2) \times \mathbb{R}^1$ (where $\Delta \cap \mathbb{R}^1 = 2\pi\mathbb{Z}$). If $\exp(\mathbb{R}x) = \mathbb{R}^1$, then the diffeomorphism $\pi : G \to N_G$ induces an equivalence of the original flow with the linear flow $(\Delta \times \mathbb{R}^2/\Gamma,\ \exp(\mathbb{R}n))$, where $\exp(\mathbb{R}n) = \Delta$. The lattice Γ is generated by the elements $(2\pi, \pi, 0)$, $(0, \pi, 0)$, $(0, 0, 1)$ and $\exp(tn) = (t, 0, 0)$. Therefore $H = (\overline{\exp(\mathbb{R}n)\Gamma})_0 = \Delta \times \mathbb{R}'$ (where $\mathbb{R}'$ is the linear span of $b = (0, \pi, 0)$) and $E_1(x) = \mathbb{R}^1\mathbb{R}'\Gamma$. The manifold $E_1(x)$ is not a homogeneous subspace of G/Γ, since the set $\mathbb{R}^1\mathbb{R}'$ is not a subgroup of G. We note that in general there are no G-induced ergodic flows on G/Γ.

Finally, we consider an arbitrary solvable flow $(G/D, \exp(\mathbb{R}x))$ on a compact space G/D. According to the Mautner phenomenon (Corollary 1.11), a necessary and sufficient condition for its ergodicity is the ergodicity of the factor-flow $(G/\overline{U_x D}, \exp(\mathbb{R}x))$, where U_x is the unstable normal subgroup of G for the element x. Let H be a maximal connected subgroup of $\overline{U_x D}$ that is normal in G and let $G' = G/H$, $D' = \overline{U_x D}/H$, $\exp(\mathbb{R}x') = \exp\,(\mathbb{R}x)H/H$. Then the flows $(G/\overline{U_x D},\ \exp(\mathbb{R}x))$ and $(G'/D',\ \exp(\mathbb{R}x'))$ are equivalent, where G'/D' is a non-cancellable compact solvmanifold and the element x' has purely imaginary eigenvalues on the Lie algebra $\mathfrak{g}'$ of the Lie group G'. Consider the semisimple splitting $G'_s = T'_G \cdot N'_G = T'_G \cdot G'$ of G', where T'_G is a closed abelian group with lattice $p(D')$. As always, we choose T'_G so that we obtain a Jordan decomposition for x', that is,

$$\exp(tx') = \exp(ta') \times \exp(tn'),$$

where the one-parameter subgroups $\exp(\mathbb{R}a') \subset T'_G$ and $\exp(\mathbb{R}n') \subset N'_G$ commute. The eigenvalues of x' and of the semisimple element a' are the same, therefore the closure $\overline{\exp(\mathbb{R}a')}$ is a torus in T'_G. In order that $(G'/D', \exp(\mathbb{R}x'))$ be ergodic it is necessary that the factor-flow $(T'_G/p(D'),\ \exp(\mathbb{R}a'))$ be ergodic, which in the present case is equivalent to the equality $T'_G = \overline{\exp(\mathbb{R}a')}$. Consequently, a necessary condition for the ergodicity of the original flow is that the quotient group G' be of type (I). This type of solvable group has just been considered by us.

Auslander's ergodicity criterion for solvable flows essentially consists in this reduction.

The ergodic manifold $E_1(x)$ is constructed in completely similar fashion; here $E_1(x) = \alpha^{-1}(E_1(x'))$ and $\alpha : G/D \to G'/D'$, where the ergodic flow $(E_1(x'),\ \exp(tx'))$ is finite-sheetedly covered by the nil-flow $(H_1/D' \cap H_1,$ $\exp(\mathbb{R}n'))$, where $H_1 = \overline{(\exp(\mathbb{R}n')D')}_0$ is a subgroup of N'_G. However, this still does not guarantee that a flow on $E_1(x)$ is covered by some homogeneous flow and, generally speaking, we cannot reduce the study of solvable flows to the ergodic case. This reduction is completely carried out in (Starkov 1987b).

Semisimple and General Cases. We now prove Moore's ergodicity criterion for the homogeneous flow $(G/D, \exp(\mathbb{R}x))$ on a quotient space G/D of finite volume of a semisimple Lie group G. For this we decompose G into a product $G = S \times K$ of non-compact and compact parts and denote by $p : G \to K$ the projection onto the compact part K. We set $H = \overline{p(D)}$, $T = \overline{p(\exp \mathbb{R}x)}$. If the original flow is ergodic, then the action of the torus T on K/H is also ergodic and topologically transitive. Consequently for some $k \in K$ we have $K = TkH$ and $K/H = k^{-1}Tk/k^{-1}Tk \cap H$. However, K/H has a finite fundamental group, therefore $K = H$, that is, the subgroup D is Zariski dense in any representation of G. In particular, D can be regarded as a lattice in G.

According to the Mautner phenomenon, it suffices for us to find an ergodicity criterion for the factor-flow $(G/\overline{I_xD},\ \exp(\mathbb{R}x))$. In view of A.21, the subgroup $(\overline{I_xD})_0$ is normal in G and we can consider the projection $q : G \to G' = G/(\overline{I_xD})_0$. We set $D' = q(\overline{I_xD})$, $\exp(\mathbb{R}x') = q(\exp(\mathbb{R}x))$ and consider the flow $(G'/D',\ \exp(\mathbb{R}x'))$, which is equivalent to our factor-flow. By construction, the subgroup $\exp(\mathbb{R}x')$ has the property that the closure $\overline{\mathrm{Ad}(\exp(\mathbb{R}x'))}$ in the adjoint group $\mathrm{Ad}\, G' = G'/Z'$ is a torus. But the intersection $Z' \cap D'$ has finite index in Z', so that the closure of any orbit of this flow is also a torus in G'/D'.

If this flow is ergodic, then G'/D' is also a torus. Consequently, the lattice $D' \subset G'$ is the fundamental group of the torus, which is only possible if $G' = 1$ and $G = \overline{I_xD}$. Thus we have proved the following assertion.

Proposition 2.8. *In the semisimple case the ergodicity criterion is the mixing condition* $G = \overline{I_xD}$.

However, this criterion cannot be regarded as effective, since there is a closure operation in it. To circumvent this problem we decompose D into irreducible components: $D = \prod_i D_i$, $G = \prod_i G_i$, where the D_i are irreducible in G_i. Let $p_i : G \to G_i$ be the corresponding projection and let $\exp(\mathbb{R}x_i) = p_i(\exp \mathbb{R}x)$. If each flow $(G_i/D_i,\ \exp(\mathbb{R}x_i))$ is ergodic, then it is mixing, which means that the original flow $(G/D,\ \exp \mathbb{R}x)$ is also ergodic. In view of the irreducibility of $D_i \subset G_i$, the flow on G_i/D_i is always ergodic if $I_{x_i} \neq 1$. This is equivalent to the requirement that the subgroup $\mathrm{Ad}(\exp(\mathbb{R}x_i)) \subset \mathrm{Ad}\, G$ is closed and non-compact. Thus we have deduced Moore's ergodicity criterion for the semisimple case.

Finally, we find the ergodicity criterion for the homogeneous flow $(G/D, \exp(\mathbb{R}x))$ on an arbitrary space G/D of finite volume by reducing it to the semisimple and solvable cases which have already been discussed. Such a reduction was first carried out by Dani (Dani 1977a) for so-called admissible spaces G/D of finite volume. It was then modified by Brezin and Moore (Brezin and Moore 1981). Here the admissibility condition was weakened somewhat, which made it possible to prove the admissibility of any space of finite volume (Starkov 1986), (Witte 1987). In fact, this reduction can be carried out by a simpler method without using the admissibility condition (Starkov 1987a). We now give an outline of this method.

Let $G = P \cdot R$ be the Levi decomposition of a connected simply connected Lie group G into a semisimple Levi subgroup P and the radical R. We decompose P into non-compact and compact parts $P = S \times K$ and denote by G_∞ the smallest normal subgroup of G containing P.

We note that the semisimple group P acts transitively on the space $G/\overline{RD} = P/P \cap \overline{RD}$, and the solvable group R acts on the space $G/\overline{G_\infty D} = R/R \cap \overline{G_\infty D}$. Our aim is to establish the following.

Theorem 2.9. *The flow on the space G/D of finite volume is ergodic if and only if the factor-flows on $G/\overline{RD}$ and $G/\overline{G_\infty D}$ are ergodic.*

Proof. We denote by $p : G \to P$ the projection of G onto P and suppose that the factor-flow $(P/P \cap \overline{RD}, p(\exp(\mathbb{R}x))$ is ergodic.

Since the Dani normal subgroup for the projection $p(\exp(\mathbb{R}x))$ is the projection of the Dani normal subgroup I_x for $\exp(\mathbb{R}x)$, it follows that $P = \overline{p(I_x)(P \cap \overline{RD})}$, therefore $G = \overline{I_x RD}$. Since $I_x \subset SR$, it follows that $G = \overline{SRD}$, which by Theorem 27 of Appendix A implies the compactness of $R/D \cap R$ and $N/D \cap N$. *A fortiori* the space $R/\overline{F_x D} \cap R$ is compact, which implies that the product $R \cdot \overline{F_x D}$ is closed. Consequently, $G = R\overline{F_x D}$ and for some choice of the Levi subgroup $P \subset G$ we have $P \subset \overline{F_x D}$. The group $D \cap N$ normalizes $(\overline{F_x D})_0$ and in view of A.9, N also normalizes $(\overline{F_x D})_0$. By Mal'tsev's theorem there exists for any $g \in G$ an element $n \in N$ such that $gPg^{-1} = nPn^{-1}$. But $nPn^{-1} \subset (\overline{F_x D})_0$ and, since the subgroup G_∞ is generated by all subgroups of the form gPg^{-1}, $g \in G$, it follows that $G_\infty \subset (\overline{F_x D})_0$ and $(\overline{G_\infty D})_0 \subset (\overline{F_x D})_0$. It is now clear that the ergodicity of the flow on $G/\overline{G_\infty D}$ implies the ergodicity on $G/\overline{F_x D}$, which, by the Mautner phenomenon suffices for ergodicity on G/D. □

Theorem 2.10. *The condition $G = \overline{I_x D}$ is a criterion for weak mixing of a flow on a space of finite volume.*

Proof. In view of Corollary 1.14, it suffices to assume that the flow is mixing. Let H be a maximal connected subgroup of $(\overline{I_x D})_0$ that is normal in G, and let $G' = G/H$ and $D' = \overline{I_x D}/H$. Since the flow $(G'/D', \exp(\mathbb{R}x'))$ is ergodic, it follows that the projection $p(D')$ of the subgroup $D' \subset G'$ onto the semisimple Levi subgroup is a lattice, where the Dani normal subgroup for x' is trivial. Consequently, the ergodicity of the flow $(P'/p(D'), p(\exp(\mathbb{R}x')))$ is possible only if $P' = 1$, that is, if G' is solvable. But a solvable flow cannot be mixing. Hence $G' = 1$. □

Thus we have deduced the ergodicity criterion for a flow on a homogeneous space of finite volume. We can now consider the question of the existence on G/D of ergodic G-induced flows.

It is easy to see that in the abelian and nilpotent cases almost all the elements of the Lie algebra $\mathfrak{g}$ induce ergodic flows on G/D for any lattice $D \subset G$. The situation changes in the solvable case. It is not difficult to prove that if D is an arbitrary lattice in a solvable Lie group of type (R), then again almost all $x \in \mathfrak{g}$ induce ergodic flows. By contrast, for the lattice D in the solvable group G of type (I) of Example 2.7, no G-induced flow on G/D is ergodic (although by taking another lattice $D' \subset G$ we could guarantee the ergodicity of almost all G-induced flows on G/D').

If D is an irreducible lattice in a non-compact semisimple group G, then non-ergodic flows on G/D are induced only by semisimple elements $x \in \mathfrak{g}$ with purely imaginary eigenvalues of the operator $\operatorname{ad} x$. However, these elements form a set of positive measure with non-empty interior and, roughly speaking, only half the elements in $\mathfrak{g}$ induce ergodic flows (where the ergodicity does not depend on the choice of the irreducible lattice). If, on the other hand, the Levi semisimple group P of a Lie group G is compact and non-trivial, then, independently of the choice of the lattice $D \subset G$, all the G-induced flows are non-ergodic.

These observations show that non-ergodicity is a very widespread phenomenon and it is of interest to reduce the study of a non-ergodic homogeneous flow to a family of ergodic homogeneous flows. We have already constructed the ergodic decompositions E and $\tilde{E}$ in the nilpotent case and have shown that all the manifolds $E_g(x) = \overline{\exp(\mathbb{R}x)I_x gD}$ and $\tilde{E}_g(x) = \overline{\exp(\mathbb{R}x)F_x gD}$ are homogeneous subspaces of G/D (closed orbits of the action of subgroups of G). In particular, any flow $(E_g(x),\ \exp(\mathbb{R}x))$ is equivalent to the flow $(H_g/D \cap H_g,\ \exp(\mathbb{R}\operatorname{Ad}_{g^{-1}}x))$. Furthermore, in the nilpotent case the ergodic decomposition $\tilde{E}$ turns out to be a smooth bundle of G/D. Even in the solvable case, the submanifolds $E_g(x)$ and $\tilde{E}_g(x)$ lose their homogeneous structure in general. However, we have the following result.

Theorem 2.11 (on the ergodic decomposition (Starkov 1989)). *Let $(G/D, \exp(\mathbb{R}x))$ be a G-induced flow on a homogeneous space of finite volume of the connected simply connected Lie group G. Then there exist:*

a) *a Lie group $G^* \supset G$, where G is a normal subgroup of G^* and G^*/G is a torus;*

b) *a connected subgroup $H \subset G^*$;*

c) *a one-parameter subgroup $\exp(\mathbb{R}y) \subset H$ such that:*

1) *for all $g \in G$ we have the cover of the flows*

$$(H/D \cap H,\ \exp(\mathbb{R}\operatorname{Ad}_{g^{-1}}y)) \to (\tilde{E}_g(x),\ \exp(\mathbb{R}x)),$$

which is finite-sheeted for almost all $g \in G$;

2) *for any* $g \in G$ *there exists a connected subgroup* $H_g \subset H$ *such that we have the finite-sheeted cover of flows*

$$(H_g/D \cap H_g,\ \exp(\mathbb{R}\,\mathrm{Ad}_{g^{-1}}y)) \to (E_g(x),\ \exp(\mathbb{R}x)),$$

where $H_g = H$ *for almost all* $g \in G$;

3) *the subgroup* D_0 *is normal in* H *and in all* $H_g \subset H$.

In particular, to within a finite-sheeted cover, the study of the flow reduces to the study of the family of ergodic homogeneous flows on the quotient space of the Lie group H_g/D_0 with respect to the lattice $D \cap H_g/D_0$.

Furthermore, almost all the submanifolds $E_g(x)$ and $\tilde{E}_g(x)$ have the same dimension as $H/D \cap H$. However, although even in the nilpotent case the dimension of the submanifolds $E_g(x)$ falls on an everywhere dense subset of G/D, the partition $\tilde{E}$ in a neighbourhood of a typical point is always a smooth bundle (Starkov 1989). In particular, the dimension of $\tilde{E}_g(x)$ can fall only on a closed set of measure 0 in G/D.

§3. Spectrum of Ergodic Flows on Homogeneous Spaces

We keep our previous notation and, in addition, we denote by $T(g)$ the transformation

$$G/D \to G/D, \quad hD \mapsto ghD.$$

The *spectrum of the flow* $\{T_t\} = \{T(\exp tx) : t \in \mathbb{R}\}$ on G/D is the spectrum of the one-parameter group of unitary operators $\{\mathcal{V}_t\} = \{\rho(\exp tx)\}$ on the space of functions in $L^2(G/D, \mu)$ with zero mean value.

The spectrum of an ergodic flow on a nil-manifold was computed by Green (Green 1961), (Auslander 1966) and Stepin (Stepin 1969), on a solvmanifold by Stepin (Stepin 1969) and Safonov (Safonov 1982), and on a semisimple manifold by Stepin (Stepin 1973). The spectrum of an ergodic flow on an arbitrary homogeneous space under the assumption that certain conditions with respect to the space hold (admissibility) was calculated by Dani (Dani 1977a) and Brezin and Moore (Brezin and Moore 1981). Below we calculate the spectrum of an ergodic flow on an arbitrary homogeneous space.

Theorem 3.1. *If the homogeneous space* G/D *has finite volume, then the spectrum of an ergodic flow* $\{T_t\}$ *on* G/D *is the sum of the point and countably multiple Lebesgue spectra. If* G *is semisimple, then there is no point component; if the quotient space* G/D *is non-cancellable, then the countably multiple Lebesgue component is absent if and only if* G *is solvable of type* (I) *with abelian nil-radical.*

Proof. We suppose without loss of generality that G/D is non-cancellable.

a) G is nilpotent.

We decompose the representation $g \mapsto \rho(g)$ into irreducible representations ρ_α. Let $\int \oplus H_\alpha d\sigma(\alpha)$ be the corresponding decomposition of $L^2(G/D)$. We denote by σ_1 the part of the measure σ concentrated on the set of those α for which H_α is one-dimensional and we set $\sigma_2 = \sigma - \sigma_1$. If $\dim H_\alpha = \infty$, then ρ_α can be realized in $L^2(\mathbb{R}^d)$ with some $d < \infty$ as a transformation of the form

$$(\rho_\alpha(g)f)(r) = \theta(g, r)f(r + S(g)),$$

where $S : G \to \mathbb{R}^d$ is a homomorphism (see Auslander 1966, Chap. V, Corollary 3.2). The infinitesimal operator of the group $\{\rho_\alpha(\exp tx)\}$ is equal to

$$S(\exp x) \cdot \operatorname{grad} + \frac{d}{dt}(\theta(\exp tx, r))\Big|_{t=0}. \tag{1}$$

In view of the infinite dimensionality of ρ_α, $S(\exp x) \neq 0$ and therefore the spectrum of the operator (1) is Lebesgue. If the support of the measure σ_2 is infinite, then the multiplicity of the Lebesgue component in the spectrum of the flow $\{T_t\}$ is infinite. The case when ρ contains only finitely many infinite-dimensional irreducible representations is not possible. In fact, the infinitesimal operator of the group $\{\rho(\exp tz) : t \in \mathbb{R}\}$ is bounded whenever z is a central element. On the other hand, there exists a central one-parameter subgroup $\{g_t\}$ with non-trivial vector field $dT(g_t)/dt|_{t=0}$ and, consequently, the operator $d\rho(g_t)/dt|_{t=0}$ is unbounded. Finally, it remains to observe that the spectrum of an ergodic flow in $\int \oplus H_\alpha d\sigma_1(\alpha)$ is a point spectrum. Note that the point component in the spectrum of a nil-flow is non-trivial, since it is fibred over the flow by the torus $G/D[G, G]$.

b) G is solvable.

The investigation of the spectrum of an ergodic flow on a solvmanifold can be reduced to the case of a flow on a solvmanifold of type (I) by means of the following device proposed by Stepin (Stepin 1969).

Let π be a maximal measurable invariant partition with respect to the flow such that the flow on the quotient space with respect to it has zero entropy. The presence of expanding and contracting foliations implies that $\pi \leq \xi_1$ and $\pi \leq \xi_2$, where the ξ_i are the partitions of G/D into trajectories of the groups $G^{\pm}$. Consequently, $\pi \leq \xi$, where ξ is the partition into fibres of the bundle $G/D \to G/\overline{U_x D}$ and U_x is the unstable normal subgroup. As was noted in Sect. 2, in view of the ergodicity of the flow, $G/\overline{U_x D}$ is a solvmanifold of type (I). It is well known (Sinai 1966) that in the orthocomplement of the space of functions that are constant on elements of the π-partition the spectrum of the flow is a Lebesgue spectrum of countable multiplicity.

We turn to the case of a solvable flow on a manifold of type (I). As was pointed out in Sect. 1, in this case the flow $(G/D, \exp(\mathbb{R}x))$ admits a finite-sheeted cover by the nil-flow $(N_G/D', \exp(\mathbb{R}n))$, where $G_S = T_G \cdot N_G = T_G \cdot G$ is a semisimple splitting of G, T_G is a torus, $T_G \subset \operatorname{Aut}(N_G)$, N_G is a maximal connected nilpotent normal subgroup of G_S, $D' = D \cap N_G$ is a normal subgroup of D of finite index, $\exp(tx) = \exp(ta)\exp(tn)$, and the subgroups $\exp(\mathbb{R}a)$ and $\exp(\mathbb{R}n)$ commute.

The natural projection $p : G \to N_G$ is a homeomorphism and takes the (two-sided) invariant Haar measure on G to an invariant Haar measure on N_G, since T_G is compact and the right shift by the element $g = t_0 n_0$ is taken under the projection p to the affine action of G_S on $N_G : A_g(n) = t_0(n) \cdot n_0$. In fact,

$$p(tn \cdot t_0 n_0) = p(tt_0 \cdot t_0(n)n_0) = t_0(n)n_0. \tag{2}$$

Thus the finite-sheeted cover of the nil-flow is also realized with preservation of the measure. Consequently, the infinitesimal operator of a solvable flow is the restriction of the infinitesimal operator of the nil-flow acting in the space $L^2(N_G \cdot D')$ on the subspace of functions that are constant on the fibres of the cover. It is known from a) that the spectrum of an ergodic nil-flow is the sum of a discrete component and a Lebesgue component of countable multiplicity. We claim that the spectrum of the infinitesimal operator in the subspace is also of the same type. To prove this, it suffices to show that there are discrete and Lebesgue components of countable multiplicity in the subspace.

The non-triviality of the discrete component for a non-nilpotent flow follows from the fact that the flow $(G/D, \exp(\mathbb{R}x))$ is a fibre bundle over the flow on the torus $(G/MD,\ \exp(\mathbb{R}x'))$, where M is a maximal connected nilpotent normal subgroup of G (see Appendix A with regard to the closedness of MD in G and the discreteness of the image of D under the quotient map $G \to G/M$), and $\exp(\mathbb{R}x')$ is the image of the subgroup $\exp(\mathbb{R}x)$ under the factorization by M.

We prove that the spectrum has a Lebesgue component of countable multiplicity.

The finite group $H = D/D'$ can be regarded as a subgroup of T_G: if $t_0 n_0$, then $t_0 \in H$. The affine action A_d, $d \in D$, on N_G is taken under the factorization $N_G \to N_g/D'$ to the action $\{\tilde{A}_h : h \in H\}$ of H on N_G/D'. Thus, if $t_0 n_0 \in D$, $h = t_0 \in H$, then $\tilde{A}_h(nD') = t_0(nD')n_0 = t_0(n)t_0(D')n_0 = t_0(n)D'n_0 = t_0(n)n_0 D'$. (The equality $t_0(D') = D'$ follows from the fact that D' is normal in D.)

The fibres of the cover G/D' over G/D consist of the orbits of the action of H on G/D' obtained under the factorization of the action of D by the right shifts on G. It follows from this remark and from (2) that the fibres of the cover N_G/D' over G/D consist of the orbits of the action $\{\tilde{A}_h : h \in H\}$. The action $\tilde{A}$ is measure-preserving; consequently, it corresponds to a finite group of unitary operators $\{U(h) : h \in H\}$ in $L^2(N_G/D')$. The subspace $L^2(G/D)$ consists of functions that are fixed with respect to $\rho(h)$. Let l_n be the action of N_G onto itself by left shifts, $\tilde{l}_n$ the factor-action on N_G/D', and $R(n)$ the corresponding unitary representation of N_G in $L^2(N_G/D')$. If $g = t_0 n_0 \in D$, then we have the following commutation relation for the actions on N_G:

$$A_g l_n A_g^{-1} = l_{t_0(n)}.$$

Therefore a similar relation holds for the factor-action on N_G/D':

$$\tilde{A}_h \tilde{l}_n \tilde{A}_h^{-1} = \tilde{l}_{h(n)},$$

where $h = t_0$. Consequently, we have the following equality for the unitary operators $\rho(h)$ and $R(n)$:

$$\rho(h)R(n)\rho(h)^{-1} = R(h(n)). \tag{3}$$

It follows from a) that the representation R decomposes into a direct sum of countably many irreducibles, each of which has finite multiplicity. In this sum there are countably many infinite-dimensional representations, where the spectrum of $\{R(\exp(tn) : t \in \mathbb{R}\}$ in each space V_i of the irreducible representation is Lebesgue. It follows from (3) that $\rho(h)V_i = V_j$. The subspace $\Lambda_1 = \sum_{h\in H} \rho(h)V_1$ is invariant with respect to ρ and $R(N_G)$. If $x \in V_1$, then $x_1 = \sum_{h\in H} \rho(h)x$ is a fixed vector with respect to ρ, that is, an element of $L^2(G/D)$.

We denote the subspace of vectors of such a form in Λ_1 by Λ_1'. The element x_1 is a sum of vectors of Lebesgue spectral type; consequently, the spectral type of x_1 is Lebesgue and the group $\{R(\exp(tn)) : t \in \mathbb{R}\}$ has a Lebesgue spectrum in the subspace $\Lambda_1' \subset L^2(G/D)$. In view of the fact that there is a countable set of subspaces V_i and H is finite, the above construction enables us to obtain infinitely many subspaces Λ_i', in each of which the spectrum of $\{R(\exp(tn)) : t \in \mathbb{R}\}$ is Lebesgue. Thus we have established the existence of a Lebesgue component of countable multiplicity in the spectrum of the original flow $\{T_t\}$ and, consequently, the spectrum of the solvable flow has been completely analysed.

We note that in the case when the Lebesgue component of the spectrum of the flow $\{T_t\}$ is missing, G is of necessity of type (I) and N_G/D' is a torus (see item a)). We denote by D_0' the component of the identity in D'. We then have the chain of inclusions

$$[M, M] \subset [N_G, N_G] \subset D_0' \subset D.$$

In the case when the quotient space is non-cancellable, it follows that the commutator $[M, M]$ is trivial, that is, M is abelian.

c) G is semisimple.

Step 1. First we make a reduction to the case when the operator $\operatorname{ad} x : y \mapsto [x, y]$ in $\mathfrak{g}$ has only purely imaginary eigenvalues. For this we use the contracting and expanding transversal foliation of the flow $\{T_t\}$. The leaves of these foliations turn out to be orbits of the horospherical subgroups for $\{T_t\}$ (cf. Sect. 1).

We denote by $\mathfrak{g}^{\mathbb{C}}$ the complexification of the real Lie algebra $\mathfrak{g}$. We decompose $\mathfrak{g}^{\mathbb{C}}$ into a sum of root subspaces V_λ of $\operatorname{ad} x$. Since $\operatorname{ad} x$ is a real endomorphism of complex space $\mathfrak{g}^{\mathbb{C}}$, it follows that the spectrum of $\operatorname{ad} x$ contains along with λ its complex conjugate $\bar{\lambda}$. The subspace $V_\lambda + V_{\bar{\lambda}}$ is invariant with respect to complex conjugation in $\mathfrak{g}^{\mathbb{C}}$, therefore it is representable in the form $\mathfrak{g}_\lambda + i\mathfrak{g}_\lambda$, $\mathfrak{g}_\lambda \subset \mathfrak{g}$. Clearly, $\mathfrak{g} = \sum_{\lambda\in\mathbb{C}} \mathfrak{g}_\lambda$ (the summation being taken over the spectrum of $\operatorname{ad} x$). The subspaces $\mathfrak{g}_\lambda$ satisfy the following commutation relations:

$$[\mathfrak{g}_{\lambda_1}, \mathfrak{g}_{\lambda_2}] \subset \mathfrak{g}_{\lambda_1+\lambda_2} + \mathfrak{g}_{\lambda_1+\bar{\lambda}_2}. \tag{4}$$

(Here it is helpful to take into account the fact that $\mathfrak{g}_\lambda = \mathfrak{g}_{\bar{\lambda}}$.)

Let $\mathfrak{h}$ be the smallest subalgebra of $\mathfrak{g}$ containing $\sum_{\operatorname{Re}\lambda\neq 0} \mathfrak{g}_\lambda$. It follows from the commutation relations (4) that $\mathfrak{h}$ is an ideal. It is generated by the subalgebras

$$\mathfrak{h}^+ = \sum_{\operatorname{Re}\lambda>0} \mathfrak{g}_\lambda, \quad \mathfrak{h}^- = \sum_{\operatorname{Re}\lambda<0} \mathfrak{g}_\lambda.$$

We denote by H, H^+ and H^- the connected subgroups of G with Lie algebras $\mathfrak{h}, \mathfrak{h}^+$ and $\mathfrak{h}^-$ respectively. The subgroup H is normal in G.

The partition ζ of the space with measure $(G/D, \mu)$ into the ergodic components of the action of H is the same as the partition into the inverse images of the points under the natural map

$$G/D \to G/D',$$

where $D' = \overline{DH}$ is the closure in G of the subgroup DH generated by the subgroups D and H.

Here the space $L^2(G/D')$ is identified with the subspace $F_H \subset L^2(G/D)$ of H-fixed elements.

Since the partition ζ is invariant with respect to the flow $\{T_t\}$, it follows that the subspace F_H and, consequently, its orthocomplement $F_H^\perp$ in $L^2(G/D)$ are invariant with respect to the one-parameter group $\{\mathcal{V}_t\}$.

To calculate the spectrum of $\{\mathcal{V}_t\}$ in $F_H^\perp$ we consider the subspace F_+ of $L^2(G/D)$ consisting of the H^+-fixed elements. It is invariant with respect to $\{\mathcal{V}_t\}$, as is the subspace $F_+^\perp = L^2(G/D) \ominus F_+$.

Since the group H^+ is horospherical for the one-parameter subgroup $\{\exp tx\}_{t\in\mathbb{R}}$, it follows that the spectrum of $\{\mathcal{V}_t\}$ in $F_+^\perp$ is absolutely continuous (provided, of course, that $F_+^\perp \neq \{0\}$).

If $\operatorname{ad} x$ has an eigenvector, say y, in $\mathfrak{h}^+$, then the flow $\{T(\exp sy) : s \in \mathbb{R}\}$ satisfies the commutation relation

$$T(\exp tx) \cdot T(\exp sy) = T(\exp s \cdot e^{\lambda t} y) \cdot T(\exp tx), \quad \lambda > 0,$$

therefore the spectrum of the flow $\{T_t\} = \{T(\exp tx)\}$ in the subspace $F_+^\perp$ has a Lebesgue component of countable multiplicity (Sinai 1966).

If, on the other hand, $\operatorname{ad} x$ has no eigenvector in $\mathfrak{h}^+$, then there exist commuting elements $y_1, y_2 \in \mathfrak{h}^+$ such that the two-parameter flow $\{T(\exp(s_1 y_1 + s_2 y_2) : (s_1, s_2 \in \mathbb{R}^2\}$ satisfies the commutation relation

$$\begin{aligned} &T(\exp tx) \cdot T(\exp(s_1 y_1 + s_2 y_2)) \\ &= T(\exp\ e^{\lambda t}[(s_1 \cos\phi t + s_2 \sin\phi t) y_1 + (-s_1 \sin\phi t \\ &\quad + s_2 \cos\phi t) y_2]) \cdot T(\exp tx), \end{aligned}$$

where λ is positive. As in the previous case, it follows that the spectrum of $\{T_t\}$ in $F_+^\perp$ contains a Lebesgue component of countable multiplicity. It remains to note that an absolutely continuous spectrum containing a Lebesgue component of countable multiplicity is Lebesgue of countable multiplicity.

The subspace F_- of $L^2(G/D)$ consisting of the H^--fixed elements changes roles with F_+ under time reversal in the flow $\{T_t\}$. Therefore in $F_-^\perp = L^2(G/D) \ominus F_-$ the group $\{\mathcal{V}_t\}$ has a homogeneous Lebesgue spectrum of infinite multiplicity.

Furthermore, in view of the equality $F_H = F_+ \cap F_-$, the linear manifold $F_+^\perp + F_-^\perp$ is everywhere dense in $F_H^\perp$. Hence it follows that the group $\{\mathcal{V}_t\}$ has a Lebesgue spectrum of countable multiplicity in $F_H^\perp$.

We now consider the action of the one-parameter group $\{\mathcal{V}_t\}$ in the invariant subspace $F_H = L^2(G/\overline{DH})$. The quotient group G/H acts transitively on the homogeneous manifold $G/\overline{DH}$. Since G is semisimple, it follows that G/H is as well. The flow $\{T_t\}$ is a fibre bundle on G/D (extension, see (Kornfel'd, Sinai and Fomin 1980)) in the base of which the G-induced flow on the quotient space of G/H is generated by the image $\tilde{x} \in \mathfrak{g}/\mathfrak{h}$ of the element $x \in \mathfrak{g}$ under the natural factoring. This base G-induced flow is clearly ergodic and the operator $\operatorname{ad}\tilde{x}$ in $\mathfrak{g}/\mathfrak{h}$ has purely imaginary eigenvalues (by the construction of $\mathfrak{h}$).

Step 2. Thus the matter reduces to the case of an ergodic flow $\{T_t\}$ on the quotient space G/D of a simply connected semisimple Lie group G when the operator $\operatorname{ad} x$ in $\mathfrak{g}$ has purely imaginary eigenvalues.

We represent x in the form x_1+x_2, where x_1 is semisimple, x_2 is a nilpotent element of $\mathfrak{g}$ and $[x_1, x_2] = 0$. We note the following properties of x_1 and x_2:

(i) the restriction of an irreducible unitary representation of G to the subgroup $\{\exp tx_1 : t \in \mathbb{R}\}$ has a point spectrum (since the image of $\{\exp tx_1 : t \in \mathbb{R}\}$ in the adjoint group $\operatorname{Ad} G$ has a compact closure);

(ii) the flow $\{T(\exp tx_2) : t \in \mathbb{R}\}$ is ergodic (according to the ergodicity criterion, see Sect. 2).

We choose an element $y \in \mathfrak{g}$ such that $[y, x_2] = x_2$ (by the Morosov-Jacobson lemma).[3] We denote by L the two-dimensional solvable subgroup of G corresponding to the Lie subalgebra generated by y and x_2. For almost all the irreducible unitary representations W occurring in ρ, the restriction of W to L contains no one-dimensional representations (otherwise there would exist non-constant $\{\exp tx : t \in \mathbb{R}\}$-invariant elements in $L^2(G/D)$, which would contradict the ergodicity of $\{T_t\}$). Thus, W/L contains only infinite-dimensional irreducible representations $\pi^\pm$ of L. Hence it follows that the spectrum of $W\big|_{\{\exp tx_2 : t \in \mathbb{R}\}}$ is absolutely continuous (more precisely, either Lebesgue or empty on the semi-axes $(-\infty, 0]$ and $[0, +\infty)$). Furthermore, we see that the spectrum of the flow $\{T(\exp tx_2)\}$ is Lebesgue of countable multiplicity (since the multiplicity function of the spectrum is symmetric and $\{T(\exp sy) : s \in \mathbb{R}\}$ is a K-flow).

[3] Here we prefer the chronological to the lexicographical ordering. For the statement and proof of this lemma see (Jacobson 1962, Chap. III, No. 11, Lemma 7) or (Bourbaki 1971, Chap. 1, Sect. 6, Exercise 15b); hints are provided.

We now restrict W to the subgroup

$$C = \{\exp(t_1 x_1 + t_2 x_2) : (t_1, t_2 \in \mathbb{R}^2\}.$$

Since the spectrum of $W|_{\{\exp t_1 x_1 : t_1 \in \mathbb{R}\}}$ is a point spectrum (property (i)), it follows that the spectrum of $W|_C$ is concentrated on a countable family of "vertical" lines and is absolutely continuous on them. Hence it follows that the spectrum of the restriction of W to the diagonal subgroup $\{\exp t(x_1+x_2) : t \in \mathbb{R}\}$ is absolutely continuous. Thus we have shown that the spectrum of the flow $\{T_t\}$ is absolutely continuous.

The fact that it is Lebesgue of countable multiplicity is established in this way. Note that the element y can be chosen to commute with x_1 (see Jacobson 1962). Then the action of L can be "lowered" to $(G/D)/\eta$, where η is the partition of $\{T_t\}$ into ergodic components. Since $\{T_t\}$ is taken to $\{T(\exp t x_2)\}$ under the factoring with respect to η, the assertion of the theorem is completely proved for semisimple groups.[4]

d) G is an arbitrary Lie group.

Let H_x be an ad-compact subgroup of G. The manifold G/D is fibred over the manifold $G/\overline{DH_x}$. Here, in accordance with the Mautner phenomenon (see Sect. 1), the flow on G/D has an absolutely continuous spectrum in the orthocomplement of the space of functions that are constant on the fibres of this fibering.

Applying the ergodicity criterion to the factor-flow on the semisimple manifold $G/\overline{DR}$ (see Sect. 2), we have

$$G/\overline{DH_x R} = \{1\}. \tag{5}$$

Let H_1 be a subgroup of $\overline{DH_x}$ that is maximal connected and normal in G. The group $G' = G/H_1$ is solvable. In fact, let $D' = \overline{DH_x}/H_1$, $R' = RH_1/H_1$. From (5) we have $G' = \overline{D'R'}$. Since the quotient space G'/D' is non-cancellable and the factor-flow is ergodic on it, we infer from Corollary 29 of Theorem 27 in Appendix A that G' is solvable.

Thus the investigation of the spectrum has been reduced to the case of the solvmanifold $G/\overline{DH_x}$ of G/H_1, which was considered in item b). If the spectrum of a solvable flow on $G/\overline{DH_x}$ has a Lebesgue component of countable multiplicity, then the theorem is proved. We consider the case when the spectrum of the factor-flow on $G/\overline{DH_x}$ is a pure point spectrum. There are two possible cases: either 1) G/D has a non-trivial semisimple quotient $G/\overline{RD}$, or 2) $G/\overline{RD} = \{1\}$. In the first case the spectrum of the flow on G/D has a Lebesgue component of countable multiplicity in the subspace of functions that are constant on the fibres of the bundle $G/D \to G/\overline{RD}$ (item c)), and the theorem is proved. The second case has in fact been considered in item b) since G/D is a solvmanifold. □

---

[4] Here the non-triviality of $(G/D)/\eta$ is vital. In fact, the ergodic flow $\{T(\exp t x_1)\}$ may not commute with the K-flow $\{T(\exp sy) : s \in \mathbb{R}\}$.

Remark 1. Following (Brezin and Moore 1981), we call a solvable group G of type (I) with abelian nil-radical a Euclidean group. A homogeneous space of a Euclidean group is called a Euclidean homogeneous space. Every solvable group G has a maximal Euclidean factor. It follows from the proof that the spectrum of the factor-flow on it is a pure point spectrum, while on the orthocomplement it is a Lebesgue spectrum of countable multiplicity.

An arbitrary homogeneous space G/D has a maximal solvable factor and consequently, it also has a maximal Euclidean factor G/L, where the point spectrum is concentrated on the subspace of functions that are constant on the fibres of the bundle $G/D \to G/L$ (Brezin and Moore 1981).

Remark 2. The properties of the spectrum of a flow on a solvmanifold of type (I) can be established by another method using a result in (Auslander, Green, and Hahn 1963) which asserts that a flow on a homogeneous manifold of type (I) is minimal if and only if it is isomorphic to a nil-flow. Bearing in mind that minimality and ergodicity are equivalent for nil-flows, we see that the cover $G/D' \to G/D$ in item b) is one-sheeted.

Remark 3. If the solvable space $G/\overline{DH_x}$ is non-trivial, then in view of b), the factor-flow on $G/\overline{DH_x}$, along with the original flow on G/D has a non-trivial point-spectrum component. Consequently a criterion for the spectrum to be Lebesgue of countable multiplicity is the triviality of the solvable space $G/\overline{DH_x}$, that is, the mixing condition $G = \overline{DH_x}$. Here, G is not necessarily a semisimple Lie group, as the following example shows:

$$G = SL(2,\mathbb{R}) \cdot \mathbb{R}^2, \quad D = SL(2,\mathbb{Z}) \cdot \mathbb{Z}^2,$$
$$\exp(tx) = \begin{pmatrix} e^t & 0 \\ 0 & e^{-t} \end{pmatrix} \subset SL(2,\mathbb{R}).$$

§4. Orbits of Homogeneous Flows

The question of the structure of the orbits is one of the most complex problems in the theory of homogeneous flows. If out of measure theoretic considerations it is not difficult to give an explicit form for the closure of an orbit in general position, then, as a rule, the study of the structures of the individual orbits relies on subtle questions of the algebraic and arithmetic theory of groups. First we study the structure of typical orbits of homogeneous flows. In the ergodic case, the answer is provided by the following well known result.

Theorem 4.1. *The set A of non-dense orbits of a smooth ergodic flow $(M, \{\phi_t\})$ on a manifold M with smooth invariant measure μ has measure 0 and is a set of the first category in M.*

Proof. Let V be an open subset of M and $A(V) = \{m \in M;\ \phi_{\mathbb{R}} m \cap V = \emptyset\}$. Then $A(V)$ is a closed invariant subset of M and, in view of the ergodicity,

$\mu(A(V)) = 0$. If now $\{V_i\}$ is a countable basis of open subsets of M and $A = \{x \in M : \overline{\phi_{\mathbb{R}}x} \neq M\}$, then clearly $A = \bigcup_i A(V_i)$ is a set of measure 0 and of the first category in M. □

We have the following result for homogeneous flows in the non-ergodic case.

Theorem 4.2 (Starkov 1983). *For the flow $(G/D, \exp(\mathbb{R}x))$ on a space G/D of finite volume there exists a set A of measure zero and of the first category in G/D such that for any point $gD \notin A$ we have $\overline{\exp(\mathbb{R}x)gD} = E_g(x)$, where $E_g(x)$ is the ergodic submanifold containing $gD \in G/D$.*

In essence, the measure theoretic assertion on the "smallness" of the set follows from the previous theorem and the properties of the ergodic decomposition. The topological part requires a more detailed proof, since it is not true that by "pasting together" the first-category sets in the various $E_g(x)$ we necessarily obtain a set of the first category.

Thus, the closure of a typical orbit of a homogeneous flow coincides with the corresponding ergodic submanifold. If, however, we take an arbitrary point $gD \in G/D$, then we cannot guarantee that it is generic and we cannot describe in advance the arrangement of its orbits. We consider in more detail such properties of orbits as smoothness and homogeneity of their closure, recurrence, boundedness, divergence (that is, going off to infinity), and so on.

The examples given in Sect. 1, Sect. 2 show that the closure of the orbit of a homogeneous flow can be locally arranged as the product of a Cantor set and an interval, so that it is not a smooth manifold. Common to these examples is the presence in the flow of non-trivial expanding and contracting foliations which, in the homogeneous case, is equivalent to the non-triviality of its entropy. One can establish the following general law of behaviour:

Lemma 4.3 (Starkov 1990). *Let $\{\phi_t\}$ be a smooth conservative*[5] *flow on a Riemannian manifold M that has non-trivial expanding and contracting foliations. Then there exists $x \in M$ such that the closure $\overline{\phi_{\mathbb{R}}x}$ of its orbit is not a submanifold of M.*

The proof depends on the proof of Lemma 4.2 in (Starkov 1990) proposed by Margulis, where a (not necessarily conservative) flow is considered on a compact manifold. Let R be an invariant k-dimensional expanding foliation. We denote by $R_r m$ the closed k-dimensional disc with centre $m \in M$ of radius r lying entirely in a leaf of R.

We also consider the three closed $(n-1)$-dimensional discs $V_r \subset V_{2r} \subset V_{4r}$ with centre at $m \in M$ lying transversally to the flow $\{\phi_t\}$ and locally invariant with respect to R. We may suppose without loss of generality that:

a) over the interval of time $\Delta t = 1$ the distance between any two points in $R_r m$ increases at least twofold;

b) the number $r > 0$ is such that we have the homeomorphism

[5] In the sense as understood in ergodic theory (decomposition into conservative and dissipative parts).

$$\phi_{(-1,1)}V_{4r} \simeq (-1,1) \times V_{4r};$$

c) almost all the points of $R_r m$ return to V_{2r} infinitely often as $t \to +\infty$ (the flow is conservative!).

We define on V_{4r} the function $\tau(x) = \min\{t > 0 | \phi_t x \in V_{2r}\}$, which is the first-return time to V_{2r}. Let $\tau_0 = \inf\{\tau(x) : x \in R_r m\}$. Then $1 < \tau_0 < \infty$. Since the diameter of the disc $\phi_{\tau_0}(R_r m)$ is at least greater than $4r$ and $\phi_{t_0}(R_r m) \cap V_{2r} \neq \emptyset$, it follows from geometric considerations that there exists a disc $R_r m_1 \subset \phi_{\tau_0}(R_r m) \cap (V_{4r} \backslash V_{2r})$. We set $C_1 = \phi_{-\tau_0}(R_r m_1) \subset R_r m$ and let $\tau_1 = \inf\{\tau(x) : x \in C_1\}$. Then $2 < \tau_1 < \infty$ and there exists a disc $R_r m_2 \subset \phi_{\tau_1}(C_1) \cap (V_{4r} \backslash V_{2r})$. We construct by induction a sequence of closed sets $R_r m \supset C_1 \supset \ldots \supset C_k \supset \ldots$ such that $\tau|_{C_k} > k+1$. Let $\rho_0 = \bigcap_k C_k \in V_r$. Then the positive semi-orbit $\phi_{\mathbb{R}_+} p_0$ returns to V_{4r} infinitely often but not once to V_{2r}. Thus it is locally closed but is not closed.

In fact, this argument proves that at least one point of the expanding foliation is not a limit point on its positive semi-orbit. The behaviour of the negative semi-orbits of points of the contracting leaf at p_0 is considered in similar fashion. This time there exists a point p_1, the negative semi-orbit of which is locally closed. (Since we cannot guarantee that almost all the points of the given leaf return to a neighbour of it infinitely often as $t \to -\infty$, we cannot rule out the possibility that the negative semi-orbit of p_1 goes off to infinity; in this case it is a closed set.) Since the positive semi-orbit $\phi_{\mathbb{R}_+} p_1$ approaches $\phi_{\mathbb{R}_+} p_0$ asymptotically, it follows that the orbit $\phi_{\mathbb{R}} p_1$ is locally closed but not closed. Consequently, its closure is not a submanifold of M.

Remark. It would be interesting in the conditions of the lemma to show the existence of recurrent[6] orbits whose closures are not manifolds (as in the examples in Sect. 1 and Sect. 2.5).

Conjecture 4.4 (Starkov 1987b). *If $(G/D, \exp \mathbb{R}x)$ is a flow on a space of finite volume, then the closure of each of its orbits is a manifold if and only if it has zero entropy.*

The proof of the "only if" half of this conjecture follows from Lemma 4.3. There is a proof of the "if" half for the solvable case (Starkov 1987b). The semisimple case rests on a proof of the following conjecture:

Conjecture 4.5 (Raghunathan) (see Dani 1981). *If U is a unipotent subgroup of a Lie group G, then the closure of each orbit of the flow $(G/D, U)$ on a space G/D of finite volume is a homogeneous subspace of finite volume, that is, for any $g \in G$ there exists a subgroup $L \subset G$ such that $\overline{UgD} = LgD \subset G/D$ and there exists on LgD a finite L-invariant measure.*

[6] Here and in what follows, by recurrence we mean the property that the orbit returns to an arbitrary neighbourhood of the original point (and not the stronger Birkhoff recurrence property in connection with minimal sets).

Remark. This conjecture was stated in (Dani 1981) for a reductive Lie group G and a lattice $D \subset G$, although there is no reason to restrict oneself to this case.

The following "measure theoretic" version is closely related to the "algebraic" conjecture.

Conjecture 4.6 (Dani 1981). *Any locally finite*[7] *ergodic measure for a unipotent flow on a space G/D of finite volume is supported on some closed homogeneous subspace $FgD \subset G/D$ and is a finite F-invariant volume on it.*

Note that since each compact minimal set has an invariant ergodic measure, Conjecture 4.5 follows from Conjecture 4.6 for such invariant sets. The greatest progress in the proof of Raghunathan's conjecture has been achieved for horospherical unipotent subgroups $G^+ \subset G$. The first result of this kind was the following.

Theorem 4.7 (Hedlund 1936). *If $G = SL(2,\mathbb{R})$, Γ is a lattice in G, $\exp(\mathbb{R}n)$ is a unipotent subgroup of G and $g \in G$, then the orbit $\exp(\mathbb{R}n)g\Gamma$ is either dense in G/Γ or periodic. If Γ is uniform, then all the orbits of the flow are dense in G/Γ.*

In the proof of this theorem Hedlund makes essential use of the horospherical property of unipotent subgroups of $SL(2,\mathbb{R})$. This theorem was strengthened for uniform lattices by Furstenberg who showed that a horocycle flow on a compact surface of constant negative curvature is uniquely ergodic. Using the papers of Veech (Veech 1977) and Bowen (Bowen 1976) the following general result of such a kind has been obtained:

Theorem 4.8 (Ellis and Perrizo 1978). *If G^+ is a horospherical subgroup for an element $a \in \mathfrak{g}$ and Γ is a uniform lattice in G, then the following conditions are equivalent:*

1) *the flow $(G/\Gamma,\ \exp(\mathbb{R}a))$ is mixing;*
2) *the flow $(G/\Gamma, G^+)$ is ergodic;*
3) *the flow $(G/\Gamma, G^+)$ is uniquely ergodic;*
4) *the flow $(G/\Gamma, G^+)$ is minimal.*

In the general case, a horospherical flow is not, of course, necessarily ergodic but its ergodic decomposition is very easily constructed.

Theorem 4.9 (Starkov 1991). *If $(G/D, G^+)$ is a horospherical flow for an element $a \in \mathfrak{g}$ on a space of finite volume and U_a is an unstable normal subgroup for $a \in \mathfrak{g}$, then each flow $(\overline{U_a gD}, G^+)$ is ergodic.*

In particular, the ergodic submanifolds $E_g(G^+) \subset \tilde{E}_g(G^+)$ in Theorem 2.11 (on the ergodic expansion) are the same and are isomorphic to $H/D \cap H$, where $H = (\overline{U_a D})_0$.

[7] That is, finite on any compact set.

The explicit form of the ergodic expansion of a horospherical flow suggests the idea of the following generalization of Theorem 4.9.

Theorem 4.10 (Starkov 1991). *If $(G/D, G^+)$ is a horospherical flow on a compact space of finite volume, then for each $g \in G$ we have $\overline{G^+gD} = \overline{U_a gD}$.*

This theorem is not, however, a consequence of Theorems 4.8 and 4.9. In fact, the flow $(\overline{U_a gD}, G^+)$ is not necessarily horospherical since, in general, the homogeneous subspace $\overline{U_a gD}$ is not $\exp(\mathbb{R}a)$-invariant.

Thus Raghunathan's conjecture has been proved for the action of horospherical subgroups on compact spaces of finite volume. We have the following result for the non-compact case.

Theorem 4.11 (Dani 1986a). *Raghunathan's conjecture is true for a horospherical flow $(G/D, G^+)$ on a quotient space of finite volume of a reductive Lie group G.*

The proof of this theorem, in contrast to the previous ones, makes essential use of results of the theory of algebraic groups and Margulis's arithmeticity theorem (see Appendix A). The behaviour of a horospherical flow in the non-compact case is not as simple as in the compact case. The theorem does not establish the explicit form of the closure of an arbitrary orbit, but merely asserts that it is homogeneous.

The next result plays an essential role in the proof of this theorem.

Theorem 4.12 (Dani 1986a). *Let G^- be a contracting horospherical subgroup for the semisimple element $a \in \mathfrak{g}$. Let Γ be a lattice in G and suppose that the flow $(G/\Gamma, G^-)$ is ergodic. If the orbit $\exp(\mathbb{R}a)g\Gamma \subset G/\Gamma$ does not go off to infinity as $t \to +\infty$, then the horospherical orbit of $G^-g\Gamma$ is dense in G/Γ.*

It is not difficult to see that in the solvable case Theorem 4.10 establishes the truth of Raghunathan's conjecture for horospherical flows, while in the semisimple case, Theorem 4.11 establishes the truth of the conjecture. In the general situation it appears that Raghunathan's conjecture has not yet been proved even for horospherical subgroups.

On the other hand, progress has recently been made in the proof of Raghunathan's conjecture for certain non-horospherical unipotent subgroups. We have in mind the series of papers by Margulis (some jointly with Dani) on the study of unipotent flows on the homogeneous space $SL(3,\mathbb{R})/SL(3,\mathbb{Z})$. Not every unipotent subgroup of $SL(3,\mathbb{R})$ is horospherical. An example is the one-parameter subgroup

$$\exp(\mathbb{R}m) = \left\{ \begin{pmatrix} 1 & t & t^2/2 \\ 0 & 1 & t \\ 0 & 0 & 1 \end{pmatrix} : t \in \mathbb{R} \right\} \subset SL(3,\mathbb{R}).$$

One's interest in the study of the flow $(SL(3,\mathbb{R})/SL(3,\mathbb{Z}),\ \exp(\mathbb{R}m))$ is explained by the fact that the proof of Raghunathan's conjecture for it gives an affirmative answer to the classical Oppenheim-Davenport conjecture for quadratic forms:

Conjecture 4.13 (Oppenheim). *Let B be a non-degenerate indefinite quadratic form in $n > 2$ variables that is not proportional to a form with rational coefficients. Then for any $\epsilon > 0$ there exist integers $x_1, \ldots, x_n$, not all equal to 0 such that $B(x_1, \ldots, x_n) < \epsilon$.*

To date, a proof of this conjecture has only been available for $n > 20$ using analytic number theory. It turns out that it can be deduced in its entirety from the following assertion.

Theorem 4.14 (Margulis 1987). *Let $G = SL(3, \mathbb{R})$, $\Gamma = SL(3, \mathbb{Z})$ and let H be the connected component of the identity of the subgroup of G that preserves the form $2x_1x_3 - x_2^2$. Then any relatively compact H-orbit in G/Γ is compact.*

A detailed derivation of Oppenheim's conjecture from this theorem is set out in (Margulis 1987). We merely note briefly that it suffices to consider the case $n = 3$. If $B(x_1, x_2, x_3)$ is a non-degenerate indefinite quadratic form and H_B is a subgroup of G preserving B, then there exists $g \in G$ for which $H_B = gHg^{-1}$, since the form B can be reduced to the form $\pm(2x_1x_3 - x_2^2)$ by a change of variables. It turns out that if the set of values $B(\mathbb{Z}^3 \backslash 0)$ is separated from zero, then the orbit $H_B\Gamma = gHg^{-1}\Gamma \subset G/\Gamma$ is relatively compact. Consequently, it is compact, which in turn implies that B is proportional to a form with rational coefficients.

It is not difficult to see that $\exp(\mathbb{R}m) \subset H$, where H is locally isomorphic to $SL(2, \mathbb{R})$. The proof of the theorem on H-orbits basically depends on the study of the orbits of the unipotent flow $(G/\Gamma, \exp(\mathbb{R}m))$. Here it can be proved that any H-orbit in G/Γ is either dense or compact, which enables one to prove (Dani and Margulis 1989) the strengthened version of Oppenheim's conjecture asserting that the set of values of $B(P(\mathbb{Z}^n))$ is dense in $\mathbb{R}$ for all the above-described quadratic forms B (here

$$P(\mathbb{Z}^n) = \{x \in \mathbb{Z}^n | x \neq ky,\ k \in \mathbb{Z},\ |k| > 1,\ y \in \mathbb{Z}^n\}$$

is the set of primitive elements in $\mathbb{Z}^n$).

Remark. Margulis has suggested that in Raghunathan's conjecture the requirement that the subgroup U be unipotent may be replaced by the weaker requirement: the subgroup U is generated by unipotent elements. An example of such a group is the subgroup H, and the study of its orbits supports the conjecture that the closure of the orbits of the subgroups generated by the unipotent elements is homogeneous.

We note that Raghunathan's conjecture implicitly asserts the recurrence of all orbits of one-parameter unipotent flows, and even this conjecture has not been proved in the general case. If it is true, then it is not difficult to prove the recurrence of all orbits of a homogeneous flow with zero entropy. By contrast, if the flow has non-zero entropy (non-trivial horospherical foliations), then there always exist non-recurrent orbits (see the proof of Lemma 4.3).

We now turn to the (measure theoretic) Conjecture 4.6 describing all the ergodic invariant measures of unipotent flows. For ergodic horospherical flows on compact spaces of finite volume it is trivial, since such flows are uniquely ergodic. In the non-compact case this conjecture has been proved by Dani (Dani 1986a) for reductive Lie groups G and maximal horospherical subgroups U. As communicated by Margulis, Ratner has recently announced a proof of the conjecture in the general case, although the details on this remarkable result are still unknown.

The following result is a necessary step in the proof of the measure theoretic conjecture.

Theorem 4.15 (Dani 1986b). *Any locally finite ergodic measure of a unipotent flow is finite.*

For one-parameter unipotent flows this theorem depends on the following result on the frequency of return of unipotent orbits to some compact set.

Theorem 4.16 (Dani 1986b). *If Γ is a lattice in a semisimple Lie group G and* $\exp(\mathbb{R}n)$ *is a unipotent subgroup, then for any $g \in G$ there exist a compact set $C \subset G/\Gamma$ and $\epsilon > 0$ such that*

$$\lim_{T\to\infty} \frac{1}{T}\int_0^T \chi_C(\exp(tn)g\Gamma)dt > \epsilon,$$

where χ_C is the characteristic function of C in G/Γ.

We deduce from this result the assertion on the finiteness of ergodic measures. Let π be a locally finite ergodic measure for the flow $(G/\Gamma, \exp(\mathbb{R}n))$ and let f be a continuous positive function on G/Γ such that $\int_{G/\Gamma} f d\pi = 1$.

We set

$$f^*(g\Gamma) = \lim_{T\to\infty} \frac{1}{T}\int_0^T f(\exp(tn)g\Gamma)dt.$$

Then by the individual ergodic theorem $f^* \in L^1(G/\Gamma, \pi)$ and f^* is $\exp(\mathbb{R}n)$-invariant almost everywhere with respect to π. It follows from Theorem 4.16 that $f^* > 0$ everywhere on G/Γ. But π is ergodic, therefore $f^* \equiv \text{const} > 0$ π-almost everywhere. On the other hand, $f^* \in L^1(G/\Gamma, \pi)$; consequently π is finite.

Remark. There is a strengthening (Dani 1986b) of Theorem 4.16 in which the compact set C can be chosen to be the same for all orbits of the unipotent flow that are not contained in homogeneous subspaces of lower dimension.

Theorems 4.15 and 4.16 have applications to simultaneous diophantine approximations of linear forms (Dani 1986b). They arose as generalizations of the well known "Margulis lemma" (Margulis 1971) which asserts that no semi-orbit of a one-parameter unipotent flow on the quotient space of a Lie group with respect to an arithmetic lattice can go off to infinity. Another strengthening of this fact is the assertion (Margulis 1989) that any closed minimal set of a unipotent flow is compact.

It is not difficult to deduce from these results that the orbit of a homogeneous flow of zero entropy is not divergent, that is, it cannot go off to infinity in any direction. By contrast, a flow with non-zero entropy can have such orbits and infinite ergodic measures. An example is the geodesic flow on the non-compact surface of constant negative curvature $(G/\Gamma,\ \exp(\mathbb{R}a))$, where

$$G = SL(2,\mathbb{R}),\ \Gamma = SL(2,\mathbb{Z}),\ \exp(ta) = \begin{pmatrix} e^t & 0 \\ 0 & e^{-t} \end{pmatrix}.$$

The orbit $\exp(\mathbb{R}a)\Gamma$ is divergent because the horocycle orbit $\exp(\mathbb{R}n)\Gamma$ is periodic (otherwise, in accordance with Theorem 4.12, the given horocycle would be everywhere dense). Dani proved the following sufficient condition for the existence of divergent orbits:

Theorem 4.17 (Dani 1985). *If Γ is an irreducible non-uniform lattice in a connected semisimple Lie group G, the element $a \in \mathfrak{g}$ is semisimple and its horospherical subgroups are non-trivial (that is, $U_a \neq 1$) then the flow $(G/\Gamma, \exp(\mathbb{R}a))$ has divergent orbits.*

The condition that the element a be semisimple seems to be inessential (at least, this is so for $G = SL(n,\mathbb{R}),\ \Gamma = SL(n,\mathbb{Z})$, see (Dani 1985)). However, the proof of this theorem rests on Theorem 4.12 (in which the semisimplicity of $a \in \mathfrak{g}$ is required) and the following result:

Theorem 4.18 (Dani 1985). *A horospherical flow on a non-compact space of finite volume has orbits that are contained in closed homogeneous subspaces of lower dimension. In particular, a horospherical flow on a non-compact space of finite volume is not minimal.*

It follows from the last theorem that a necessary condition for a one-parameter homogeneous flow to be minimal is that the space be compact. According to Lemma 4.3, another necessary condition is that the flow have zero entropy and, of course, be ergodic. The criterion itself that the homogeneous flow be minimal and uniquely ergodic has so far not been stated even in the form of a conjecture.

We turn to the geodesic flow $(SL(2,\mathbb{R})/SL(2,\mathbb{Z}),\ \exp(\mathbb{R}a))$. For this one can describe in explicit form the set of bounded (relatively compact) divergent orbits. Namely, for each element $g \in SL(2,\mathbb{R})$ there is a decomposition $g = p\binom{1\ 0}{\alpha\ 1}\gamma$, where p is an upper-triangular matrix, $\gamma \in SL(2,\mathbb{Z})$ and $\alpha \in \mathbb{R}$. It turns out that the criterion for the divergence of the orbit $\exp(\mathbb{R}a)gSL(2,\mathbb{Z})$ is the rationality of α, while the criterion for boundedness is the bad approximability of α by rational numbers. We recall that a number α is badly approximable if there exists $\delta > 0$ such that for all $m, n \in \mathbb{Z}$ we have $|a - m/n| > \delta/n^2$.

Bounded divergent orbits of a flow form a set of measure zero (since almost all orbits are dense), which is in agreement with the fact that the sets of rationals and badly approximated numbers in $\mathbb{R}$ are of zero measure. However, it is known (W.M. Schmidt 1966) that the set of badly approximable numbers has Hausdorff dimension 1. Consequently, the bounded orbits of a geodesic flow form a set of Hausdorff dimension 3. The following theorem is a generalization of this fact.

Theorem 4.19 (Dani 1986c). *If G is a connected semisimple Lie group of $\mathbb{R}$-rank 1, Γ is a lattice in G, $a \in \mathfrak{g}$ and $U_a \neq 1$, then the set of bounded orbits of $(G/\Gamma, \exp(\mathbb{R}a))$ has the Hausdorff dimension of the space G/Γ.*

Apparently, the restriction on the Lie group G in this theorem is superfluous, but has so far not been proved in general form.

§5. Statistical Properties of G-Induced Flows (and Actions)

The first in the hierarchy of statistical properties of dynamical systems is ergodicity. The ergodicity criterion for G-induced flows was discussed in Sect. 2. Naturally, they were stated in algebraic and geometric terms.

Generally speaking, in the spirit of this book we discuss measure theoretic properties of homogeneous flows in connection with their topological (geometric) properties.

The classical result on the minimality of a horocycle flow (Hedlund) placed the question of the unique ergodicity of this flow on the agenda.

An affirmative answer using the technique of harmonic functions was given by Furstenberg in (Furstenberg 1973). We give the author's sketch of the proof.

Let $G = SL(2, \mathbb{R})$, Γ a uniform lattice in G, and H a "horocycle" one-parameter subgroup of G. It is required to prove the one-dimensionality of the space of Γ-invariant (Borel σ-finite) measures on $\mathbb{R}^2\backslash\{0\} = G/H$, since the dimensions of the spaces of H-invariant measures on G/Γ and Γ-invariant measures on G/H are the same. Furthermore, the ergodicity of a horocycle flow implies the ergodicity of the action of Γ on G/H; therefore it suffices to establish that the Γ-invariant measures π on $\mathbb{R}^2\backslash\{0\}$ are absolutely continuous (with respect to Lebesgue measure). By averaging the measures obtained from π by similarities of $\mathbb{R}^2\backslash\{0\}$, the problem reduces to a consideration of Γ-invariant measures with the property of radial absolute continuity. Thus we are required to show that any Γ-invariant measure π has an absolutely continuous projection of π onto the circle P^1 with respect to the natural bundle $\mathbb{R}^2\backslash\{0\} \to P^1$. Next, the class π is defined by the projection onto P^1 of the restriction π_M of π to a sufficiently large compact set $M \subset \mathbb{R}^2\backslash\{0\}$; for this purpose any set with the property $T(\Gamma)M = \mathbb{R}^2\backslash\{0\}$ is suitable and the existence of such a set follows from the compactness of Γ.

We consider on the unit disc $D \cong G/K$, where K is the subgroup of rotations in G, the harmonic function

$$h(gK) = \int \|g^{-1}x\|^{-2} d\tilde{\pi}_M(x).$$

The measure $\tilde{\pi}_M$ is equivalent to the measure μ on ∂D in the representation of h as a Poisson integral. It is well known that μ is absolutely continuous if for some $p \in (1, \infty]$ the integrals $\int_0^{2\pi} |h(re^{i\phi})|^p d\phi$ are bounded as $r \nearrow 1$ (so-called L^p-boundedness of h). To estimate these integrals we consider the integral

$$\int_{D_R} h^2 dm, \tag{6}$$

where m is an invariant measure on the Lobachevskij plane Λ^2 and $D_R \subset \Lambda^2$ is a (non-Euclidean) disc of radius R with centre at the origin. On the one hand, the integral (6) is estimated by the sum

$$\sum_R = \sum_{\gamma(0)\in D_R} |h(\gamma(0))|^2.$$

On the other hand, it turns out that $\Sigma_R = O(S_R)$, where S_R is the non-Euclidean area of D_R; here the Γ-invariance of π is used. Hence and from the fact that the main contribution to the integral (6) is given by a neighbourhood of the boundary of D_R, it follows that h is L^2-bounded, which means that the horocycle flow is strictly ergodic.

This result, as well as the strict ergodicity of minimal nil-flows has led to the conjecture that all minimal G-induced flows (actions) are strictly ergodic.

Marcus (Marcus 1975) considered the property of strict ergodicity of a horocycle flow in the case of non-constant negative curvature. More precisely, let M be a smooth compact connected manifold, and $\{f_t\} : M \to M$ an Anosov flow on M of class C^2 for which the unstable subbundle $E^{\mathrm{u}} \subset TM$ is one-dimensional and orientable. We suppose that each unstable manifold $W^{\mathrm{u}}(x)$ is dense in M and $\{f_t\}$ preserves the smooth measure. Under these assumptions, we can construct a continuous vector field X such that $X(x) \in E^{\mathrm{u}}_x$ for $x \in M$. It turns out that such a vector field induces a strictly ergodic flow. In particular, for a Riemannian metric of class C^3 on a compact connected surface of negative curvature the horocycle flow is strictly ergodic.

The action of horocycle subgroups in connection with the strict ergodicity property has been studied by Bowen (Bowen 1976) and Veech (Veech 1977). Ellis and Perrizo (Ellis and Perrizo 1978) have investigated the question of the connection between the strict ergodicity of the action of a horospherical subgroup of a semisimple element a and other topological and metric properties of this action and the properties of the shift transformation T_a on a. The structure of compact minimal sets for non-strict ergodic actions of horospherical subgroups on non-compact manifolds has been studied by Dani (Dani 1981). A precise statement of the results of the above papers was given in Sect. 4.

A generalization of the results of Bowen and Veech going beyond the framework of G-induced flows was obtained by Bowen and Marcus (Bowen and Marcus 1977). They introduced the notion of an invariant measure for foliations and proved the strict ergodicity of the stable and unstable foliations of a basic set M of a dynamical system satisfying Smale's Axiom A under the assumption that this system is topologically mixing on M.

The question of the number of invariant measures of a dynamical system naturally arises in the case when strict ergodicity is absent. The answer to this question for the action of $SL(2, \mathbb{Z})$ on $SL(2, \mathbb{R})/\Gamma$, where Γ is a uniform lattice, was obtained in (K. Schmidt 1980). It is proved that on each such homogeneous space there is an infinite (even uncountably infinite) set of pairwise singular continuous infinite σ-finite invariant measures.

Further Stochastic Properties of G-Induced Flows (Actions). The mixing property for geodesic and horocycle flows on surfaces of constant negative curvature was established by Hopf (Hopf 1936) and Hedlund (Hedlund 1936); the K-property of a geodesic flow was established by Sinai (Sinai 1960), and the Bernoulli property of a geodesic flow by Ornstein and Weiss (Ornstein 1974), (Ornstein and Weiss 1973).

Marcus (Marcus 1978) proved that a horocycle flow on a surface of finite volume of constant negative curvature has the property of mixing of any multiplicity. A more general assertion is as follows (Marcus 1978): let G be a semisimple Lie group without compact factors, D a closed subgroup, let G/D have finite volume, and let $\{T_t\} = \{\exp tx\}$ be an ergodic flow on G/D. Then $\{T_t\}$ is mixing of all degrees.

As is well known, the mixing property of a flow $\{g_t\}$ implies that for any functions $\phi, \psi \in L^2(G/D)$ such that $(\psi, 1) = 0$, the quantity $(\phi, \psi \circ g_t)$ converges to zero as $t \to \infty$. There is the question of the rate of decrease of the correlations. Ratner (Ratner 1987) proved that for a geodesic flow $\{g_t\}$ on a surface of constant negative curvature and when ϕ, ψ are Hölder continuous in the direction of the subgroup of rotations, the rate of decrease of the correlation is exponential.

Dani (Dani 1977a) obtained sufficient conditions for mixing for an arbitrary G-induced automorphism. Namely, let D be a lattice in the Lie group G, T_g the shift acting on G/D. If there exists at least one weakly mixing shift on G/D, then every ergodic shift is strongly mixing.

In connection with the statistical properties of G-induced transformations (flows) it is natural to consider the analogous properties of a DS on a G/D of more general algebraic origin, namely, affine transformations. Let A be an automorphism of G preserving D, and $\overline{A}$ the corresponding automorphism of G/D; T_g is the shift by the element g. The transformation $T = T_g \circ \overline{A}$ is called affine. When is T mixing? A sufficient condition for this was obtained in (Dani 1977a). He proved that every weakly mixing affine transformation is mixing. Let $G_\infty \subset G$ be the smallest connected normal subgroup such that G/G_∞ is solvable and let R be the radical of G. If $G_\infty D$ is dense in G

and G/R has no compact factors, then each ergodic affine transformation is mixing. Furthermore, if G is a connected solvable Lie group, D is a closed uniform subgroup of G and T, which is an affine transformation on the non-cancellable homogeneous subspace G/D, is weakly mixing, then G is nilpotent, D is a lattice and T is a K-automorphism.

Finally, in connection with the mixing property the question of mixing of a transformation group is of interest. Sufficient conditions for a multiparameter abelian group to be mixing were obtained in (Dani 1977c). More precisely, let G be a semisimple Lie group without compact factors, T a closed subgroup of G isomorphic to $\mathbb{R}^r \times \mathbb{Z}^s$ $(r, s \geq 0)$, K a compact subgroup of G contained in the centralizer of T, Γ a torsionless lattice in G, where T contains no non-trivial elements all of whose eigenvalues are equal to 1 in modulus, while if L is a connected normal subgroup of G with Lie algebra $\mathfrak{l}$ and $L\Gamma$ is closed, then for all $g \in T\backslash\{1\}$ the element $\operatorname{ad} g|\mathfrak{l}$ has an eigenvalue with modulus not equal to 1. Then the action of T on $K\backslash G/\Gamma$ is mixing.

The paper (Bowen 1971) is devoted to the calculation of the entropy of shifts and affine transformations of Lie groups and homogeneous spaces. Let G be a Lie group, μ the right-invariant Haar measure on G, d a right-invariant metric, and A an endomorphism of G. Although G may well be non-compact, one can nevertheless define the topological entropy $h_d(A)$ (which in the compact case is the same as the usual topological entropy, while in the non-compact case it may depend on the metric being used; this is reflected in the notation. It is, so to speak, the entropy "with respect to d", in fact, with respect to the corresponding uniform structure). It turns out that

$$h_d(A) = \sum_{|\lambda_i|>1} \log|\lambda_i|,$$

where $\lambda_1, \ldots, \lambda_n$ are the eigenvalues of the differential $dA|T_eG$. Furthermore, if R_g is the right shift on G, $R_g x = xg$ and $T = R_g \circ A$, then $h_d(T) = h_d(A) = h_\mu(T) = h_\mu(A)$, where h_μ is the metric entropy (also defined in a special way even for an infinite measure). In particular, for the inner automorphism $A_g : x \mapsto gxg^{-1}$ and the left shift L_g the equality $h_d(A_g) = h_d(L_d)$ always holds. Next, let Γ be a uniform lattice in $G, A\Gamma \subset \Gamma$, and suppose that the map $T : x \mapsto gA(x)$ induces the affine transformation $S : G/\Gamma \to G/\Gamma'$, $S(x\Gamma) = gA(x)\Gamma$. Then the ordinary topological entropy $h(S)$ is the same as $h_d(T)$. In particular, we have the following equality for the shift $T_g : x\Gamma \mapsto gx\Gamma$:

$$h(T_g) = \sum_{|\lambda_i|>1} \log|\lambda_i|,$$

where $\lambda_1, \ldots, \lambda_n$ are the eigenvalues of $\operatorname{ad} g$.

Following mixing in the hierarchy of stochastic properties is the K-property. Earlier we recalled that a geodesic flow enjoys this property. Under what conditions does a G-induced shift (flow) or an affine transformation possess the K-property? A G-induced flow on a solvable homogeneous space fibres over a linear flow on a torus and therefore does not possess a K-property. By contrast, an affine transformation can possess the K-property in this case. Conditions for the K-property to hold for automorphisms of the torus were obtained by Sinai and generalized to the case of an affine transformation of a nil-manifold by Parry (Parry 1969). Let $T = T_g \circ \overline{A}$ be an affine transformation of the nil-manifold G/D. Then T possesses the K-property if the operator $dA\big|_{J|[J,J]}$ has no eigenvalues equal to unity in modulus.

For homogeneous spaces of semisimple Lie groups conditions under which an affine transformation enjoys the K-property were obtained by Dani in (Dani 1976b). We give the main result in (Dani 1976b). Let G be a nilpotent Lie group or a unipotent p-additive algebraic group, and let σ be an expanding endomorphism of G. Suppose that G acts in $(X, \mathfrak{M}, \mu)$, preserving the normalized measure μ and let T be an automorphism of $(X, \mathfrak{M}, \mu)$ satisfying the commutation relation $T_g x = \sigma(g) T x$, $x \in X$. Then the Pinsker partition $\pi(T) \leq \xi(G)$, where $\xi(G)$ is the measurable hull of the partition of X into orbits of G. As a corollary of this main assertion, sufficient conditions are obtained for the K-property for homogeneous semisimple flows (shifts). Let G be a semisimple k-algebraic Lie group over a locally compact field k of characteristic 0, $\tilde{G}$ the closed subgroup generated by all the unipotent elements (if $k = \mathbb{R}$ or $\mathbb{C}$, then $\tilde{G} = G$), let $\Gamma \subset \tilde{G}$ be an irreducible lattice in G, and suppose that the element $g \in G$ is such that the operator $\operatorname{ad} g$ has an eigenvalue with modulus not equal to 1. Then the shift of $\tilde{G}/\Gamma$ by g is a K-automorphism. We note that as a consequence of the main theorem in (Dani 1976b) one can obtain a criterion for a homogeneous flow to possess the K-property in terms of the unstable normal subgroup U_x (see Sect. 1). Namely, the flow $\{T_t\}$ possesses the K-property on the non-cancellable homogeneous space G/D if and only if $G = \overline{U_x D}$. In fact, it follows from the main result of (Dani 1976b) that $\pi(T) \leq \xi(G^+)$, $\pi(T) \leq \xi(G^-)$, where G^+, G^- are the horospherical subgroups of G corresponding to the one-parameter subgroup $\{\exp tx\}$. Therefore $\pi(T) \leq \xi$, where ξ is the decomposition into fibres of the bundle $G/D \to G/\overline{U_x D}$. But if $\overline{U_x D} = G$, then ξ is the trivial partition ν, therefore $\pi(T) = \nu$ and $\{T_t\}$ is a K-flow. Conversely, if $G \neq \overline{U_x D}$, then the factor-flow $\{T_t\}$ on $G/\overline{U_x D}$ has zero entropy and $\pi(T) \neq \nu$.

It can be proved by similar arguments that $\{T_t\}$ has zero entropy if and only if $U_x = \{1\}$.

The Bernoulli property of a geodesic flow (Ornstein 1974), (Ornstein and Weiss 1973) leads in natural fashion to the question of conditions under which a shift or an affine transformation is Bernoullian. Sufficient conditions for the Bernoulli property are obtained in (Dani 1976a). Let $T = T_g \circ A$ be an affine transformation of G/Γ (Γ is a lattice in G), and σ the automorphism of the Lie algebra $\mathfrak{g}$ given by $\operatorname{ad} g \circ dA$. Suppose that the following condition holds:

(i) The restriction of σ to a maximal σ-invariant subspace on which all the eigenvalues of σ are equal to 1 in modulus is a semisimple automorphism, and let T be a K-automorphism.

Then T is a Bernoulli shift. In this same paper it is proved that all the orbits of the horospherical group associated with T are dense in the case when G/Γ is compact and T satisfies condition (i) and is weakly mixing.

The ergodic properties of geodesic flows in the non-homogeneous case have been investigated by Ballmann and Brin (Ballmann and Brin 1982). They proved that a geodesic flow on a compact Riemannian manifold of non-positive curvature is ergodic and Bernoullian if there exists a geodesic γ for which there exists no parallel Jacobian field orthogonal to γ.

We note also the following stochastic properties of horocycle flows:

• The dense trajectories of a horocycle flow are uniformly distributed (Dani 1982). As a corollary one obtains the following formula for every irrational number θ:

$$\lim_{t\to\infty} \frac{1}{T} \sum_{\substack{0<m<T\{m\theta\} \\ (m,[m\theta])=1}} \{m\theta\}^{-1} = \frac{6}{\pi^2},$$

where $\{\alpha\}$, $[\alpha]$ denote the fractional and integer parts of α respectively. At the time of publication of (Dani 1982) this number-theoretic formula was not known.

• A horocycle flow is monotonically loosely Bernoullian (Ratner 1978).

• The Cartesian square of a horocycle flow on the Lobachevskij plane is not monotonically loosely Bernoullian (Ratner 1979).

Of interest is the investigation of the properties of G-induced DSs $\{T_t\}$ (with discrete or continuous time) for the case when $\exp \mathbb{Z}x$ or $\exp \mathbb{R}x$ is contained in some horospherical subgroup. This action of the one-parameter subgroup is called a generalized horocycle flow (shift) in (Dani 1977b). For every unipotent element u the shift $T(u)$ has zero entropy (Dani 1977a). If $\{T(\exp tx)\}$ is a generalized horocycle flow, then the transformation $T(\exp tx)$ is isomorphic to $T(\exp x)$ for any $t \neq 0$. Nevertheless there exist infinitely many non-isomorphic ergodic generalized horocycle shifts on homogeneous spaces of simple Lie groups (Dani 1977b).

A survey of the results on ergodic properties of discrete conformal transformation groups is contained in (Sullivan 1982).

To conclude this section we note the paper (K. Schmidt 1980) in which an example of the G-induced action T of $SL(2,\mathbb{Z})$ is considered, namely, the restriction to this group of the natural action of $SL(2,\mathbb{R})$ on the Riemann sphere. It is proved in (K. Schmidt 1980) that T has no asymptotically invariant sequences. Let G be a countable group of transformations of the space (X,μ) with quasi-invariant measure μ. A sequence $\{B_n\}$ of subsets of X is said to be asymptotically invariant if for all $g \in G$, $g \neq 1$,

$$\mu(B_n \Delta g B_n)/\mu(B_n) \to 0 \text{ as } n \to \infty.$$

This notion proves useful in the theory of invariant means.

§6. Rigidity of Homogeneous Flows

Any two geodesic flows on surfaces of equal constant negative curvature and finite volume are measure theoretically isomorphic, since they are Bernoullian with the same entropy.

It is of interest to classify horocycle flows to within measure theoretic isomorphism. It turns out (Ratner 1982a) that every such isomorphism of horocycle flows is of an algebraic nature. More precisely: let Γ_1, Γ_2 be lattices in $G = SL(2,\mathbb{R})$. The spaces $(G/\Gamma_1, \mu_1)$ and $(G_2/\Gamma_2, \mu_2)$ are isomorphic if there exists a measurable invertible map $\psi : G/\Gamma_1 \to G/\Gamma_2$ taking the measure μ_1 to the measure μ_2. Let $\{h_t^{(i)}\}$ be a horocycle flow on G/Γ_i. The flows $\{h_t^{(1)}\}$ and $\{h_t^{(2)}\}$ are called (measure theoretically) isomorphic if there exists an isomorphism $\psi : G/\Gamma_1 \to G/\Gamma_2$ such that $\psi h_t^{(1)}(x) = h_t^{(2)}\psi(x)$ for μ_1-almost all $x \in \Gamma_1$ and all $t \in \mathbb{R}$. If $\{h_t^{(1)}\}$ and $\{h_t^{(2)}\}$ are isomorphic, then there exist $c \in G$ and $\sigma \in \mathbb{R}$ such that $c\Gamma_1 c^{-1} = \Gamma_2$ and

$$\psi(x) = \psi(\Gamma_1 g) = h_\sigma^{(2)}(\Gamma_2 cg) \tag{7}$$

for μ_1-almost all $x = \Gamma_1 g \in G/\Gamma_1$.

Furthermore, the transformations over unit time $h_1^{(1)}$ and $h_1^{(2)}$ are isomorphic if and only if $\Gamma_2 = c\Gamma_1 c^{-1}$ for some $c \in G$ and each isomorphism of $h_1^{(1)}$ and $h_1^{(2)}$ has the form (7).

Flows obtained from horocycle flows by change of time also possess the property of rigidity (Ratner 1986). More precisely, let $\{h_t\}$ be a flow on G/Γ, and $\tau : G/\Gamma \to \mathbb{R}$ a positive integrable function. The flow $\{h_t^\tau\}$, $h_t^\tau : x \mapsto h_{w(h,t)}(x)$, where $\int_0^{w(x,t)} \tau(h_u(x))du = t$, is called a flow obtained from $\{h_t\}$ by change of time. Let $h_t^{(1)}, h_t^{(2)}$ be horocycle flows corresponding to the lattices Γ_1 and Γ_2, and let $\tau_i : G/\Gamma_i \to \mathbb{R}$ be functions with the same means. It is proved in (Ratner 1986) that if certain regularity conditions hold for τ_1 and τ_2, then for every isomorphism ψ of the flows $\{(h_t^{(1)})^{\tau_1}\}$ and $\{(h_t^{(2)})^{\tau_2}\}$ there exist $c \in G$ and a measurable function $\sigma : G/\Gamma_2 \to \mathbb{R}$ such that $c\Gamma_1 c^{-1} = \Gamma_2$ and $\psi(x) = h^{(2)}_{\sigma(\psi_c(x))}(\psi_c(x))$ for μ_1-almost all $x \in G/\Gamma_1$, where $\psi_c(\Gamma_1 g) = \Gamma_2 cg$. Furthermore, if for some $p \neq 0$ the automorphisms $(h_p^{(1)})^{\tau_1}$ and $(h_p^{(2)})^{\tau_2}$ are ergodic and ψ realizes an isomorphism of them, then ψ realizes an isomorphism of the flows $\{(h_t^{(1)})^{\tau_1}\}$ and $\{(h_t^{(2)})^{\tau_2}\}$.

The series of papers (Ratner 1982b), (Ratner 1983) are devoted to an investigation of Cartesian powers of horocycle flows and their factors. In particular, Ratner gives a classification to within isomorphism of the factors of Cartesian powers of horocycle flows.

In the non-homogeneous case, horocycle flows on compact surfaces of variable negative curvature also possess a certain weakened rigidity property (Feldman and Ornstein 1987) (semi-rigidity in their terminology). Namely, let M_1 and M_2 be compact surfaces of negative curvature, $\{g_t^{(1)}\}$ and $\{g_t^{(2)}\}$

geodesic flows on M_1 and M_2 respectively, let μ_i be $\{g_t^{(i)}\}$-invariant measures of maximal entropy, and $\{h_t^{(i)}\}$ uniformly parametrized flows along the horocycle foliation. Then every isomorphism of $(\{h_t^{(1)}\}, \mu_1)$ and $(\{h_t^{(2)}\}, \mu_2)$ has the form $h_\sigma^{(2)}\theta$, where θ is a homeomorphism (in fact, a C^1-diffeomorphism) conjugating $\{g_t^{(1)}\}$ and $\{g_t^{(2)}\}$.

Ratner's results on rigidity can be stated in terms of affine maps. A map $\psi : G/\Gamma \to H/\Lambda$ is said to be affine for an element $g \in G$ if there exists $h \in H$ such that $\psi(\Gamma xg) = \psi(\Gamma x)h$ for almost all $\Gamma x \in G/\Gamma$. A map ψ is said to be affine for G if it is affine for any $g \in G$. Ratner's rigidity theorem stated above implies that every measure theoretic isomorphism of horocycle flows on a surface of constant negative curvature is an affine map almost everywhere. How are the properties of affineness with respect to an element $g \in G$ and with respect to G in the large related? An answer was obtained by Witte (Witte 1985). Let Γ and Λ be lattices in the connected semisimple Lie groups G and H respectively. Let $\psi : G/\Gamma \to H/\Lambda$ be a Borel map that preserves the invariant measure and is affine for a shift of G/Γ of zero entropy. Then ψ is affine for G. The following assertion (Witte 1985) is obtained as a corollary. Let G, H_1, H_2 be connected non-compact simple Lie groups with trivial centre, Λ_i a lattice in H_i, with G embedded in H_1 and H_2. Then every measure theoretic isomorphism of the G-actions on H_1/Λ_1 and H_2/Λ_2 is an affine map almost everywhere.

The paper (Witte 1985) contains a generalization of Ratner's rigidity theorem. Let T_1, T_2 be measure-preserving invertible ergodic unipotent affine actions on the connected non-cancellable homogeneous spaces G_1/Γ_1 and G_2/Γ_2, respectively. If $\psi : G_1/\Gamma_1 \to G_2/\Gamma_2$ is a measure-preserving Borel map conjugating T_1 and T_2 (that is, $T_1\psi = \psi T_2$), then ψ is affine almost everywhere.

The unipotency condition can be weakened by replacing it by the requirement that T_1 and T_2 have zero entropy. More precisely, Let T_1, T_2 be measure-preserving invertible ergodic affine maps of zero entropy of the connected non-cancellable homogeneous spaces G_1/Γ_1 and G_2/Γ_2, respectively, and let $(T_1, G_1/\Gamma_1)$ be weakly mixing. Then any measure-preserving Borel map $\psi : (T_1, G_1/\Gamma_1) \to (T_2, G_2/\Gamma_2)$ is affine almost everywhere.

Topological Rigidity of Homogeneous Flows. For the rest of this section we shall be discussing the two recent papers by Benardete (Benardete 1988) and Witte (Witte 1990) in which the connection between topological and affine equivalence of homogeneous flows is considered.

Definition 6.1. By a *topological equivalence* $\phi : (G/D, F) \to (G'/D', F')$ of homogeneous flows we mean a homeomorphism $\phi : G/D \to G'/D'$ taking orbits to orbits, that is, $\phi(FgD) = F'\phi(gD)$ for all $g \in G$.

Definition 6.2. An affine equivalence of homogeneous spaces is a composite of an isomorphism and a translation. By an affine equivalence of homogeneous flows we mean an affine equivalence of the homogeneous spaces taking orbits to orbits.

Remark. In Benardete's paper only one-parameter flows are considered and in Definitions 6.1, 6.2 it is further required that the maps preserve the orientation of the orbits, that is, positive semi-orbits are taken to positive ones and negative semi-orbits to negative ones.

Both papers are based on a remarkable simple proof of the classical result for linear flows on a torus.

Theorem 6.3. *If the flows* $(\mathbb{R}^n/\mathbb{Z}^n, \mathbb{R}x)$ *and* $(\mathbb{R}^n, \mathbb{Z}^n, \mathbb{R}x')$ *are topologically equivalent, then there exists an element* $\sigma \in GL(n, \mathbb{Z})$ *such that* $\sigma(\mathbb{R}x) = \mathbb{R}x'$, *that is, the flows are affinely equivalent.*

Proof. Let $\phi : \mathbb{R}^n/\mathbb{Z}^n \to \mathbb{R}^n/\mathbb{Z}^n$ be a homeomorphism realizing the topological equivalence. By taking if necessary the composite of ϕ and a translation, we may suppose that $\phi(\mathbb{Z}^n) = \mathbb{Z}^n$ and extend ϕ to a homeomorphism $\tilde{\phi} : \mathbb{R}^n \to \mathbb{R}^n$. The restriction $\tilde{\phi}|\mathbb{Z}^n$ is a lattice automorphism of $\mathbb{Z}^n$, that is, $\sigma = \tilde{\phi}|\mathbb{Z}^n \in GL(n, \mathbb{Z})$. The composite $\sigma^{-1} \circ \tilde{\phi}$ is a homeomorphism of $\mathbb{R}^n$ whose restriction to $\mathbb{Z}^n$ is the identity. In view of the compactness of the torus $\mathbb{R}^n/\mathbb{Z}^n$, there exists $c \in \mathbb{R}$ such that $\rho(v, \sigma^{-1}\phi(v)) < c$ for all $v \in \mathbb{R}^n$. On the other hand, $\sigma^{-1} \circ \tilde{\phi}$ takes $\mathbb{R}x$-orbits to $\sigma^{-1}(\mathbb{R}x')$-orbits. Since two distinct one-parameter subgroups of $\mathbb{R}^n$ diverge at an arbitrarily large distance, it is necessary that $\mathbb{R}x = \sigma^{-1}(\mathbb{R}x')$. □

Definition 6.4. Two one-parameter subgroups $\exp(\mathbb{R}x)$, $\exp(\mathbb{R}y)$ of a Lie group G are said to be non-divergent from each other if there exists $c \in \mathbb{R}$ such that $\rho(\exp(tx), \exp(\mathbb{R}y)) < c$ for all $t \in \mathbb{R}$.

Clearly this theorem carries over to the general case if the following conditions hold:

a) the discrete subgroups $\Gamma \subset G, \Gamma' \subset G'$ are uniform lattices in G and G', respectively;

b) the lattice isomorphism $\psi : \Gamma \to \Gamma'$ extends to an isomorphism $\tilde{\psi} : G \to G'$ of the enveloping Lie groups;

c) any two distinct one-parameter subgroups in G diverge from each other.

In fact, condition a) can be weakened by doing away with the uniformity of Γ, Γ' and Benardete shows how this can be achieved. Condition b) always holds when the theorem on strong rigidity of lattices holds. Finally, condition c) holds for nilpotent and solvable groups of type (R) (see (Benardete 1988), and we have proved the following result:

Theorem 6.5. *If* Γ, Γ' *are lattices in simply connected solvable Lie groups* G, G' *of type* (R) *and the homogeneous flows* $(G/\Gamma, \exp(\mathbb{R}x))$ *and* $(G'/\Gamma',$ $\exp(\mathbb{R}x'))$ *are topologically equivalent, then they are affinely equivalent.*

In the semisimple case, condition c) may not hold. For example, if the torus T commutes with $\exp(\mathbb{R}x)$ and $\exp(\mathbb{R}y) \subset T \times \exp(\mathbb{R}x)$, then the subgroups $\exp(\mathbb{R}x)$ and $\exp(\mathbb{R}y)$ do not diverge from each other. However, as is shown in (Benardete 1988), condition c) can be replaced by the condition that the flow $(G/\Gamma, \exp(\mathbb{R}x))$ be ergodic (note that in Theorem 6.5 it is not required that the flows be ergodic). Thus we have the following result.

Theorem 6.6. *Let Γ, Γ' be lattices in connected semisimple Lie groups G, G' having no centre or compact factors. Suppose further that G contains no normal subgroup $N \simeq PSL(2, \mathbb{R})$ for which $N\Gamma$ is closed. Then the topological equivalence of the ergodic flows $(G/\Gamma, \exp(\mathbb{R}x))$ and $(G'/\Gamma', \exp(\mathbb{R}x'))$ implies their affine equivalence.*

In the paper (Witte 1990) the theme started in (Benardete 1988) is continued and the following result is proved.

Theorem 6.7. *Let Γ, Γ' be lattices in the connected Lie groups G, G', and let F, F' be connected unimodular subgroups of G, G' inducing ergodic flows on G/Γ and G'/Γ', respectively. Then if*

a) *G, G' are simply connected solvable Lie groups of type (R), or*

b) *G, G' are semisimple Lie groups without a centre or compact factors, where G contains no normal subgroup $N \simeq PSL(2, \mathbb{R})$ for which $N\Gamma$ is closed, then the topological equivalence $\phi : (G/\Gamma, F) \to (G'/\Gamma', F')$ is a composite of an affine equivalence and a homeomorphism of G'/Γ' that preserves F'-orbits.*

For the semisimple case this theorem clearly strengthens Theorem 6.6, since it operates with possibly multiparameter flows and establishes a connection between topological and affine equivalence. In the solvable case, however, this connection must be paid for by the extra condition of ergodicity. We show how this is established for flows on a torus.

Let F, F' be two connected subgroups of G, G', and let $\phi : \mathbb{R}^n/\mathbb{Z}^n \to \mathbb{R}^n/\mathbb{Z}^n$ be a homeomorphism taking F-orbits to F'-orbits. Using the divergence of any distinct connected subgroups of $\mathbb{R}^n$, we can, as in the one-parameter case, obtain an element $\sigma \in GL(n, \mathbb{Z})$ such that $\sigma^{-1} \circ \phi$ is a homeomorphism of $\mathbb{R}^n$ that is the identity on $\mathbb{Z}^n$, stabilizes the subgroup $F \subset \mathbb{R}^n$ and takes F-orbits to F'-orbits. But in view of the density of the orbit $F\mathbb{Z}^n \subset \mathbb{R}^n$, the map $\sigma^{-1} \circ \phi$ must preserve F-orbits. □

The proof in the solvable case of type (R) is carried out in completely similar fashion using the property of the rigidity of lattices and the divergence of connected subgroups. The discussion of the semisimple case is considerably more complicated. We briefly describe the basic idea. In view of Theorem 23 (on rigidity) in Appendix A, we may suppose that $\Gamma = \Gamma', G = G'$ and we have a homeomorphism $\sigma^{-1} \circ \phi : G \to G$ that is the identity on Γ and establishes a topological equivalence of $(G/\Gamma, F)$ and $(G/\Gamma, \sigma^{-1}(F'))$. The subgroups F and $\sigma^{-1}(F')$ cannot diverge, which proves that their non-elliptic parts are the same (the precise definitions of elliptic and non-elliptic parts of unimodular groups are given in (Witte 1990)). In the semisimple case ergodicity implies mixing, and this is used in the proof that the elliptic parts of F and F' are the same. This in turn enables one to establish the equality $F = \sigma^{-1}(F')$ which completes the proof of the theorem. □

In conclusion we consider two counterexamples to topological rigidity of homogeneous flows.

Example 6.8. Let $G = \mathbb{R}^1 \cdot \mathbb{R}^2$ be a solvable Lie group of type (I), where $\mathbb{R}^1$ acts on $\mathbb{R}^2$ as the group of orthogonal rotations with centre $2\pi\mathbb{Z} \subset \mathbb{R}^1$, and let $\Gamma = 2\pi\mathbb{Z} \times \mathbb{Z}^2$. Then the semisimple splitting of G has the form $G_S = T_G \cdot G = T_G \cdot N_G$, where $T_G = SO(2)$, $N_G = \Delta \times \mathbb{R}^2$ and Δ is the "antidiagonal" in $SO(2) \times \mathbb{R}^1$ (see Appendix B). The projection $\pi : G_s \to N_G$ along the *torus* T_G induces a homeomorphism $p : G \to N_G$ together with a topological equivalence of flows $(G/\Gamma, \mathbb{R}^1) \to (N_G/\Gamma, \Delta)$. However, the given flows are not affinely equivalent, since the solvable subgroup of G is not isomorphic to the abelian group N_G. In the present instance the topologically equivalent flows are not ergodic, although it is not difficult to achieve ergodicity as well by setting $\Gamma = Z\{b\} \times \mathbb{Z}^2$, where $b \in 2\pi\mathbb{Z} \times \mathbb{R}^2$ is a suitably chosen element.

Example 6.9. Consider flows on homogeneous spaces of the group $G = PSL(2,\mathbb{R}) = SL(2,\mathbb{R})/Z_2$. It is well known that there exist isomorphic non-conjugate uniform lattices $\Gamma, \Gamma' \subset G$ (it is for such lattices that rigidity breaks down). This is in accordance with the fact that two compact surfaces of constant negative curvature can be diffeomorphic but not isometric. As follows from (Anosov 1967), geodesic flows on such surfaces are topologically equivalent (but not affinely equivalent).

It is highly non-trivial, however, that for horocycle flows on compact surfaces of constant negative curvature affine equivalence follows from topological equivalence (Marcus 1983). Consequently, Theorems 6.6 and 6.7 do not cover all the cases of topological rigidity of homogeneous flows.

Appendix A. Structure of Spaces of Finite Volume

We give a brief résumé of the results on the structure of spaces of finite volume. A detailed account of these results (apart from Theorem 27) can be found in the monograph (Raghunathan 1972) or the survey (Vinberg, Gorbatsevich and Shvartsman 1988).

Definition 1. The quotient space G/D of a Lie group G with respect to a closed subgroup $D \subset G$ is called a space of finite volume if there exists on G/D a smooth G-invariant measure μ normalized by the condition $\mu(G/D) = 1$.

Proposition 2. *If D_1 and D_2 are closed subgroups of G and $D_1 \subset D_2$, then G/D_1 is of finite volume if and only if the spaces G/D_2 and D_2/D_1 have finite volume.*

Definition 3. A discrete subgroup Γ is called a lattice in G if G/Γ has finite volume.

Definition 4. A subgroup $D \subset G$ is called uniform if G/D is compact (and non-uniform otherwise).

Proposition 5. *A discrete uniform subgroup is a lattice.*

Definition 6. We say that a homogeneous space G/D is non-cancellable if D contains no non-trivial connected normal subgroups of G.

In some works such spaces are called locally faithful or presentations, and when G/D is of finite volume, the subgroup $D \subset G$ is called a quasilattice.

Definition 7. A subgroup $D \subset G$ is said to be Zariski dense in any finite-dimensional representation of G if for any representation ρ of G in the matrix group $GL(V)$ of a finite-dimensional space V the algebraic span $A(\rho(D))$ contains $\rho(G)$.

Definition 8. Let M be a class of Lie groups. We say that a lattice Γ in a Lie group G of class M is rigid if for any Lie group G' of class M with lattice Γ' and isomorphism $\phi : \Gamma \to \Gamma'$ there exists an isomorphism $\tilde{\phi} : G \to G'$ such that $\tilde{\phi}|_{\Gamma} = \phi$.

Theorem 9 (Mal'tsev). *Let D be a closed subgroup of a connected nilpotent Lie group N. Then*

a) *the space N/D is of finite volume if and only if it is compact;*

b) *if N/D is of finite volume, then D is Zariski dense in any representation of the group N.*

Corollary 10. *If N/D is a non-cancellable nil-space of finite volume, then D is a lattice in N.*

Proof. Consider the adjoint representation $\mathrm{Ad} : N \to \mathrm{Aut}\,\mathfrak{n}$ of N in the algebraic group of all automorphisms of its Lie algebra $\mathfrak{n}$. Then $\mathrm{Ad}(D)$ normalizes the Lie subalgebra $\mathfrak{d} \subset \mathfrak{n}$ of the connected component D_0, therefore $\mathrm{Ad}(N)$ also normalizes $\mathfrak{d}$, so that the subgroup D_0 is normal in N. Consequently, $D_0 = 1$. □

Theorem 11 (Mal'tsev). *If Γ is a lattice in a connected nilpotent Lie group N, then the intersection $\Gamma \cap [N, N]$ is a lattice in the commutator-group $[N, N]$. In particular, the compact nil-space N/Γ fibres over the torus $N/\Gamma[N, N]$.*

Theorem 12 (Mal'tsev). *Lattices in the class of connected simply connected nilpotent Lie groups are rigid.*

Definition 13. A solvable Lie algebra $\mathfrak{g}$ is called an algebra of type (I) or of type (R) if all the eigenvalues of any operator $\mathrm{ad}\, x$, $x \in \mathfrak{g}$, are purely imaginary, or purely real, respectively. A solvable Lie group predetermines the type of its Lie algebra. A simply connected solvable Lie group G is called a group of type (E) if the exponential map $\exp : \mathfrak{g} \to G$ is a homeomorphism.

It is known that any solvable Lie group of type (R) is also of type (E). A solvable Lie group G is of types (I) and (E) simultaneously only if it is nilpotent.

Theorem 14 (Mostow). *Let D be a closed subgroup of a solvable Lie group G. Then:*

a) *the space G/D is of finite volume if and only if it is compact;*

b) *if G/D is a non-cancellable compact solvable space and N is the nilradical in G, then $N/D \cap N$ is compact. In particular, a compact solvable space fibres over the torus G/ND.*

Remark 15. Corollary 10 does not hold for solvable groups that are not of type (R). For example, if $G = SO(2) \cdot \mathbb{R}^2$, where $SO(2)$ acts on $\mathbb{R}^2$ as a group of orthogonal rotations and $D = \mathbb{Z} \times \mathbb{R} \subset \mathbb{R}^2 \subset G$, then G is of type (I), the space G/D is compact and non-cancellable, but the subgroup D is non-discrete.

It is known that not every nilpotent or solvable Lie group contains a lattice (see (Vinberg, Gorbatsevich and Shvartsman 1988) for details).

Theorem 16 (Saito). *Lattices in connected simply connected solvable Lie groups of type (R) are rigid.*

Remark 17. As is easily verified, the lattice $\Gamma = \mathbb{Z}^2$ in the Lie group $G = SO(2) \cdot \mathbb{R}^2$ of type (I) in Remark 15 is non-rigid (since not every automorphism of Γ extends to an automorphism of G). Examples are also known of non-rigid lattices in solvable groups of type (E) (Milovanov).

Definition 18. A lattice Γ in a connected semisimple Lie group G is said to be irreducible if for any connected normal subgroup $H \subset G$ the product $H\Gamma$ is dense in G.

Theorem 19. *If G is a connected simply connected semisimple Lie group without compact factors and Γ is a lattice in G, then there exist connected normal subgroups $G_1, \ldots, G_n \subset G$ such that:*

1) $G = G_1 \times \ldots \times G_n$,

2) *$\Gamma \cap G_i$ is an irreducible lattice in G_i for all i.*

Definition 20. By a compact part K of a connected semisimple Lie group P we mean a maximal connected compact normal subgroup of it; by a non-compact part S we mean a connected normal subgroup complementary to K.

Theorem 21 (Borel). *If P is a semisimple Lie group with non-compact part S, the space P/D is of finite volume and $P = \overline{SD}$, then the subgroup D is Zariski dense in any representation of P.*

Corollary 22. *If under the conditions of Theorem* 21 *the space P/D is non-cancellable, then*

a) *the subgroup D is a lattice in P;*

b) *if Z is the centre of P, then the product ZD is closed and $Z \cap D$ is a subgroup of Z of finite index.*

The proof consists in the verification of the triviality of the connected components D_0 and $(\overline{ZD})_0$ and then proceeding along the lines of the proof of Corollary 10.

It is known (Raghunathan 1972) that in any semisimple Lie group there exist both uniform and non-uniform lattices.

Theorem 23 (Mostow, Prasad, Margulis). *Irreducible lattices in the class of connected semisimple Lie groups without a centre or compact factors and which are not locally isomorphic to $SL(2, \mathbb{R})$ are rigid.*

All three restrictions on the class M of semisimple Lie groups in this theorem are essential, since on removing any of them one can construct an example of a non-rigid irreducible lattice Γ in a Lie group P of class M (see Vinberg, Gorbatsevich, and Shvartsman 1988).

Definition 24. Let P be a connected semisimple Lie group without compact factors, and Γ a discrete subgroup of P. Then Γ is called an arithmetic subgroup of P if there exist a $\mathbb{Q}$-defined algebraic group G and an epimorphism $\phi : G_{\mathbb{R}} \to \operatorname{Ad} P$ defined over $\mathbb{R}$ with compact kernel such that the group $\phi(G_{\mathbb{Z}})$ is commensurable with $\operatorname{Ad} \Gamma$.

Theorem 25 (Margulis). *If a semisimple group P of R-rank greater than 1 has no centre or compact factors, then any irreducible lattice $\Gamma \subset P$ is arithmetic.*

The first examples of non-arithmetic lattices in semisimple Lie groups of $\mathbb{R}$-rank 1 were constructed by Makarov and Vinberg.

Theorem 26 (Auslander). *If R is the radical of a Lie group G and D is the closed subgroup of G with solvable connected components of the identity D_0, then the subgroup $(\overline{RD})_0$ is solvable.*

Theorem 27 (Starkov 1987a), (Witte 1987). *Let $G = P \cdot R$ be the Levi decomposition of the connected Lie group G and let $P = S \times K$ be a decomposition of the semisimple Levi subgroup P into non-compact and compact parts. Suppose that the space G/D has finite volume and $G = \overline{SRD}$. Then:*

a) *the space $R/D \cap R$ is compact;*

b) *if G/D is non-cancellable, then $R/D \cap R$ is also non-cancellable and the image $p(D)$ of the subgroup $D \subset G$ under the projection $p : G \to P$ is a lattice in P.*

Corollary 28 (Wang's theorem). *If Γ is a lattice in the Lie group $G = S \cdot R$, where the semisimple Levy subgroup S has no compact factors, then $\Gamma \cap R$ is a lattice in R and $p(\Gamma)$ a lattice in S.*

Corollary 29. *If G/D is a non-cancellable space of finite volume, G is connected and $G = \overline{RD}$, then G is solvable.*

Proof. As a consequence of Theorem 27, the product RD is closed and $D_0 \subset R$. Therefore $G = \overline{RD} = RD = (RD)_0 = R$. □

We give the proof of Theorem 27 in several stages, following (Starkov 1987a).

Proposition 30. *If under the conditions of Theorem 27 the subgroup D normalizes the connected subgroup $H \subset G$, then P also normalizes H for some choice of $P \subset G$.*

Proof. Consider the adjoint representation $\mathrm{Ad} : G \to \mathrm{Aut}(\mathfrak{g})$. Since $G = \overline{SRD}$, it follows that $P = S(P \cap \overline{RD})$. Since, in view of Proposition 2, the space $P/(P \cap \overline{RD})$ is of finite volume, it follows from Borel's Theorem 21 that the algebraic span $A(\mathrm{Ad}(P \cap \overline{RD}))$ contains $\mathrm{Ad}\, P$. But $A(\mathrm{Ad}\, D)A(\mathrm{Ad}\, R) = A(\mathrm{Ad}\, \overline{RD}) \supset A(\mathrm{Ad}(P \cap \overline{RD})) \supset \mathrm{Ad}\, P$, therefore for some choice of $P \subset G$ we have $\mathrm{Ad}\, P \subset A(\mathrm{Ad}\, G)$. The required result is now trivial. □

We now prove part a) of Theorem 27. Without loss of generality we suppose that the Lie group G is simply connected. Let $D_0 = P^* \cdot R^*$ be the Levi decomposition of the connected Lie group D_0. Since D normalizes R^*, it follows from Proposition 30 that P can be assumed to normalize R^* and R^*R. Consequently, R^*R is a connected solvable normal subgroup of G, therefore $R^* \subset R$. On the other hand, D and P normalize P^*R, therefore P^* is a connected normal subgroup of some Levi subgroup of P. Let $P = P^* \times P'$ where P' is a connected normal subgroup of P with decomposition $P' = S' \times K'$ into non-compact and compact parts, $G' = P'R$ and $D' = D \cap G'$. Then $P^* \cap G' = 1$, $\overline{S'RD'} \subset G'$ and the product $P^*\overline{S'RD'}$ is closed. Therefore $P^*\overline{S'RD'} = P^*\overline{SRD} = G$, since $P^*S' \supset S$ and $P^*D' = D$. Consequently, $G' = \overline{S'RD'}$ and, in view of Proposition 30, $(\overline{RD'})_0$ is a normal subgroup of G'. But $(D')_0 = R^*$ and by Auslander's Theorem 26, the group $(\overline{RD'})_0$ is solvable. Consequently, $(\overline{RD'})_0 = R$ and the space $R/D \cap R = R/D' \cap R$ is compact. □

We now prove part b). First we let H be a maximal connected subgroup of $(D \cap R)_0$ that is normal in R. Then the set $\mathrm{Int}_D(H) = \{dhd^{-1} : d \in D, h \in H\}$ lies in $(D \cap R)_0$, is normalized by D and R and generates the subgroup $H^* \subset (R \cap D)_0$. The subgroup H^* is normalized by D and R, where $(H^*)_0 \supset H$. But according to Proposition 30, the connected subgroup $(H^*)_0$ is normal in G and, since G/D is non-cancellable, it follows that $H \subset (H^*)_0 = 1$.

Thus $R/D \cap R$ is a non-cancellable compact solvable space. Consequently, by Mostow's Theorem 14, the space $N/D \cap N$ is also compact. Let H be the subgroup of D generated by $\mathrm{Int}_{D \cap N}(P^*)$. Then by Mal'tsev's Theorem 9, the nil-radical N normalizes H, therefore H is the subgroup generated by $\mathrm{Int}_N(P^*)$. But by Mal'tsev's second theorem (on the conjugacy of Levi subgroups by elements of the nil-radical) $\mathrm{Int}_N(P^*) = \mathrm{Int}_R(P^*)$, and since P^* is normal in P, it follows that $\mathrm{Int}_R(P^*) = \mathrm{Int}_G(P^*)$ and H is normal in G. Consequently $P^* \subset H = 1$ and Theorem 27 is proved. □

Corollary 31. *If G/D has finite volume, then $KR/D \cap KR$ is compact.*

Proof. Consider the group $G' = \overline{SRD}$. Then $G' = P'R'$, where $P' = S \times K'$, $K' \subset K$, $R \subset R' \subset RK$ and $G' = \overline{SR'D}$. Consequently, $R'/D \cap R'$ is compact and since KR/R' is compact, it follows that $KR/D \cap KR$ is compact. □

Remark 32. If Γ is a lattice in an arbitrary Lie group $G = P \cdot R$, then in general the compactness of $R/\Gamma \cap R$ cannot be guaranteed. For example, let $G = SO(3) \times \mathbb{R}$ and Γ a lattice in the cylinder $SO(2) \times \mathbb{R}$ such that $\overline{\Gamma R} = SO(2) \times \mathbb{R}$. Furthermore, even if $G = K \cdot R$, where the compact semisimple Levi subgroup K acts effectively on the radical R, the intersection $\Gamma \cap R$ of Γ with R is not necessarily a lattice in R, in spite of Corollary 8.28 in (Raghunathan 1972). At the same time, the intersection $\Gamma \cap N$ with the nil-radical N is a lattice in N (see the Appendix to (Starkov 1989) for details).

As was shown in Sect. 4, if at least one ergodic G-induced flow acts on the space G/D, then the condition $G = \overline{SRD}$ always holds, so that the space $R/D \cap R$ is compact. But in the general case, $R/D \cap R$ is not compact, which makes the study of non-ergodic homogeneous flows more difficult.

Definition 33. We say that the space G/D is *admissible* if there exists a connected solvable subgroup $A \subset G$ containing the radical R and normalized by D with compact quotient-space $A/D \cap A$.

This concept was introduced in (Brezin and Moore 1981), where the ergodicity criterion is proved for homogeneous flows on admissible spaces of finite volume. It subsequently turned out that any space of finite volume is admissible (Starkov 1986), (Witte 1987) and the ergodicity criterion can be proved without using the notion of admissibility (see Sect. 2).

Appendix B. Construction of Semisimple Splitting of Simply Connected Lie Groups

The idea of embedding simply connected Lie groups into splittable Lie groups is due to Mal'tsev. This idea was modified by Auslander (Auslander, Green and Hahn 1963), (Auslander 1973) for the study of homogeneous flows and we give the corresponding non-constructive definition:

Definition B.1. Suppose that R is a connected simply connected solvable Lie group, and T_R an abelian group of non-trivial semisimple (that is, diagonalizable over the field $\mathbb{C}$) automorphisms of R. Let $R_S = T_R \cdot R$ and let N_R be the nil-radical in R_S. If

1) $R_S = T_R \cdot N_R$ (that is, $T_R \cap N_R = 1$ and $R_S = T_R N_R$) and
2) $R_S = RN_R$,

then the group R_S is called a semisimple splitting of R.

Note that if $\pi : R_S = T_R \cdot N_R \to N_R$ is the projection along T_R and $p : R_S \to T_R$ is the projection along N_R, then conditions 1) and 2) imply that $\pi : R \to N_R$ is a diffeomorphism of manifolds and $p : R \to T_R$ is a group epimorphism.

It is well known (Auslander 1973), (Auslander and Brezin 1968) that a semisimple splitting R_S exists and is unique to within isomorphism. We note that the subgroup T_R is defined to within conjugacy by elements of N_R.

Since T_R is an abelian group of non-trivial automorphisms N_R, there exists an embedding of T_R into the algebraic group of automorphisms of the simply connected nilpotent group $\operatorname{Aut} N_R$ and we can consider the algebraic span $A(T_R) \subset \operatorname{Aut} N_R$. We set $A(R_S) = A(T_R) \cdot N_R$. Then we can introduce an algebraic structure on $A(R_S)$ in which N_R is the unipotent radical and $A(T_R)$ is a maximal abelian subgroup of semisimple elements. Starting from the definition of this structure, the group R is Zariski dense in $A(R_S)$ (even if R itself is algebraic).

For a splittable solvable group R a semisimple splitting can be given in explicit form. Let $R = A \cdot N$, where N is the nil-radical of R and let A be an abelian group of semisimple automorphisms of N with kernel $Z \subset A$. We set $A^* = A/Z$ and $R_S = A^* \cdot R = (A^* \cdot N) \times \Delta_A$, where $\Delta_A = \{(a^*, a^{-1}) : a \in A\}$ is the "antidiagonal" in $A^* \times A$. Then as is easily seen, $R_S = T_R \cdot N_R$, where $T_R = A^*$ and $N_R = \Delta_A \times N$.

Theorem B.2 (Mostow-Auslander (Auslander 1973)). *If D is a closed subgroup of a simply connected solvable group R with non-cancellable compact space R/D, then:*

1) *T_R is a closed subgroup of $A(T_R)$ in the Euclidean topology,*
2) *$p(D)$ is a lattice in T_R.*

Apart from the Chevalley decomposition $R_S = T_R \cdot N_R$, the group R_S has the further property that each of its elements admits a Jordan decomposition $x = a \times n$ into a semisimple element a and a unipotent element $n \in N_R$ which commute (here $a \in T_R$ for some choice of T_R).

We now consider an arbitrary simply connected Lie group G with Levi decomposition $G = P \cdot R$ into the radical R and a semisimple Levi subgroup P. If $R_S = T_R \cdot N_R$ is a semisimple splitting of R, then for some choice of T_R the actions of P and T_R on R commute (Auslander and Brezin 1968) and we can form the semidirect product $G_S = (P \times T_R) \cdot N_R$.

Definition B.3 The group G_S constructed above is called the semisimple splitting of the simply connected Lie group G.

Consider the Lie group $A(G_S) = (P \times A(T_R)) \cdot N_R = P \cdot A(R_S)$. By contrast with $A(R_S)$, the group $A(G_S)$ may not have an algebraic group structure in view of the possible presence of an infinite centre $Z \subset P$. However in a certain sense such a structure can be emulated on $A(G_S)$, which can be used in the construction of the ergodic decomposition of homogeneous flows. Furthermore, it is useful for this purpose to know the arrangement of the semisimple splitting of the connected normal subgroups $H \subset G$, the quotient groups $G' = G/H$ and the extensions $G' = A \cdot G$ by the abelian group of semisimple automorphisms A (Starkov 1984), (Starkov 1989).

Added to the English Translation

Great progress in the theory of homogeneous flows has recently been made due to fundamental results of M. Ratner. She has proved the measure and topological Raghunathan conjectures in a general form (see Sect. 4 or the lecture (Margulis 1991) given at the 1990 International Congress of Mathematicians). Let us state briefly the formulations of the main theorems.

Theorem A (Ratner 1991a). *Let G be a connected Lie group, Γ a discrete subgroup of G, and U a subgroup of G of the form $U = \bigcup_i u_i U_0$, where the u_i are unipotent in G, $i = 1, 2, \ldots \infty$, U/U_0 is finitely generated and the connected component U_0 of U is generated by unipotent elements. Then for each finite U-invariant ergodic finite measure μ on G/Γ there exists a closed subgroup $H \subset G$ and an element $g \in G$ such that μ is an H-invariant measure supported on the closed subspace $Hg\Gamma \subset G/\Gamma$.*

Theorem B (Ratner 1991b). *Let G and U be as above and Γ a lattice in G. Then for every $g \in G$ there exists a closed subgroup $H \subset G$ such that $\overline{Ug\Gamma} = Hg\Gamma$ and $Hg\Gamma$ has a finite H-invariant measure.*

This theorem is deduced from the following result on distribution rigidity of unipotent flows.

Theorem C (Ratner 1991b). *Let G be a connected Lie group, Γ a lattice in G and $\exp(\mathbb{R}x)$ a unipotent subgroup of G. Then for any point $g\Gamma \in G/\Gamma$ there exists a closed subgroup $H \subset G$ such that $\overline{\exp(\mathbb{R}x)g\Gamma} = Hg\Gamma$, $Hg\Gamma$ has an H-invariant probability measure ν_H and*

$$\lim_{t \to \infty} \frac{1}{t} \int_0^t f(\exp(sx)g\Gamma)ds = \int_{G/\Gamma} f d\nu_H$$

for all bounded continuous functions f on G/Γ.

For a semisimple Lie group G with $\mathrm{rank}_{\mathbb{R}}(G) = 1$ Theorem C was independently proved by Shah (Shah 1991). The proof of the main Theorem A relies on the so-called R-property (Ratner property) for unipotent flows (Ratner 1983); (Witte 1987). In (Dani and Margulis 1992) an alternative derivation of Theorem C from Theorem A and new important number theoretic applications were found.

Making use of Ratner's fundamental theorems enables one to obtain some interesting results. Conjecture 4.4 for closures of all orbits of homogeneous flow to be smooth easily follows from Theorem B (Starkov 1990). Witte generalized

Ratner's theorem on quotients of a horocycle flow (Ratner 1982b) by proving in (Witte 1992) that all measurable quotients of an ergodic unipotent flow have an algebraic origin.

As is known, Marcus (Marcus 1978) proved Sinai's conjecture that a horocycle flow is mixing of all degrees and conjectured that any mixing homogeneous flow is mixing of all degrees (see also (Margulis 1991)). Various generalizations of Marcus's result to non-homogeneous measurable actions were found in (Ryzhikov 1991) and (Mozes 1992). Theorem A has allowed Starkov (Starkov 1993) to prove Marcus's conjecture for any one-parameter homogeneous flow.

Borel and Prasad proved the S-arithmetic analogue of the Oppenheim conjecture on values of quadratic forms (see Sect. 4) using topological Theorem B in (Borel and Prasad 1992). Margulis and Tomanov extended measure Theorem A to products of algebraic groups over local fields of characteristic 0 and announced the S-arithmetic generalization of topological Theorem B (Margulis and Tomanov 1992). Ratner announced in (Ratner 1993) the extension of all results to p-adic Lie groups and to the S-arithmetic case.

It should be mentioned also that Ratner's theorem on measure rigidity of horocycle flows (see Sect. 6 or (Ratner 1982a)) was generalized in different directions to horospherical foliations on hyperbolic manifolds in (Flaminio 1987); (Witte 1989) and (Flaminio and Spatzier 1990).

Starkov announced in (Starkov 1992) the criterion for rigidity of lattices in solvable Lie groups and constructed new examples of rigid and nonrigid lattices (see Appendix A).

References*

Alekseev, V.M. (1976): Symbolic Dynamics (Eleventh Mathematical School). Inst. Mat. Akad. Nauk Ukr. SSR

Anosov, D.V. (1967): Geodesic flows on closed Riemannian manifolds of negative curvature. Tr. Mat. Inst. Steklova *90* (209 pp.). [English transl.: Proc. Steklov Inst. Math. *90* (1969)] Zbl. 163,436

Anosov, D.V., Sinai, Ya.G. (1967): Some smooth ergodic systems. Usp. Mat. Nauk *22*, No. 5, 107–172. [English transl.: Russ. Math. Surv. *22*, No. 5, 103–167 (1967)] Zbl. 177,420

Auslander, L. (1973): An exposition of the structure of solvmanifolds. Bull. Am. Math. Soc. *79*, No. 2, 227–285. Zbl. 265.22016–017

Auslander, L., Brezin, J. (1968): Almost algebraic Lie algebras. J. Algebra *8*, No. 1, 295–313. Zbl. 197,30

Auslander, L., Green, L., Hahn, F. (1963): Flows on homogeneous spaces. Ann. Math. Stud. *53*. Princeton Univ. Press, Princeton N.J. Zbl. 106,368

* For the convenience of the reader, references to reviews in Zentralblatt für Mathematik (Zbl.), compiled using the MATH database, and Jahrbuch über die Fortschritte der Mathematik (FdM.) have, as far as possible, been included in this bibliography.

Ballmann, W., Brin, M. (1982): On the ergodicity of geodesic flows. Ergodic Theory Dyn. Syst. *2*, No. 3–4, 311–315. Zbl. 519.58036

Benardete, D. (1988): Topological equivalence of flows on homogeneous spaces and divergence of one-parameter subgroups of Lie groups. Trans. Am. Math. Soc. *306*, 499–527. Zbl. 652.58036

Borel, A., Prasad, G. (1992): Values of isotropic quadratic forms at S-integral points. Compos. Math. *83*, No. 3, 347–372. Zbl. 777.11008

Bourbaki, N. (1971): Groupes et algèbres de Lie. Ch. 1,2. Eléments Math. No. 26, Hermann, Paris. Zbl. 213,41

Bowen, R. (1971): Entropy for group endomorphisms and homogeneous spaces. Trans. Am. Math. Soc. *153*, 401–414; *181* (1973), 509–510. Zbl. 212,292

Bowen, R. (1976): Weak mixing and unique ergodicity on homogeneous spaces. Isr. J. Math. *23*, No. 3–4, 267–273. Zbl. 338.43014

Bowen R. (1979): Methods of Symbolic Dynamics. Mir, Moscow (Russian). (Russian translation of several papers by Bowen not listed in this bibliography)

Bowen, R., Marcus, B. (1977): Unique ergodicity for horocycle foliations. Isr. J. Math. *26*, No. 1, 43–67. Zbl. 346.58009

Brezin, J., Moore, C.C. (1981): Flows on homogeneous spaces: a new look. Am. J. Math. *103*, 571–613. Zbl. 506.22008

Bunimovich, L.A., Pesin, Ya.B., Sinai, Ya.G., Yakobson, M.V. (1985): Dynamical Systems II: Ergodic Theory of Smooth Dynamical Systems. Itogi Nauki Tekh., Ser. Sovrem. Probl. Mat., Fundam. Napravleniya *2*, 113–232. [English transl. in: Encycl. Math. Sci. *2*, 99–206. Springer-Verlag, Berlin Heidelberg New York 1989] Zbl. 781.58018

Chumak, M.L., Stepin, A.M. (1989): Partial hyperbolicity of G-induced flows. Dynamical Systems and Ergodic Theory. Banach Cent. Publ. *23*, 475–479. Zbl. 705.58040

Dani, S.G. (1976a): Bernoullian translations and minimal horospheres on homogeneous spaces. J. Indian Math. Soc. *40*, No. 1, 245–284. Zbl. 435.28015

Dani, S.G. (1976b): Kolmogorov automorphisms on homogeneous spaces 5. Am. J. Math. *98*, 119–163. Zbl. 325.22011

Dani, S.G. (1977a): Spectrum of an affine transformation. Duke Math. J. *44*, No. 1, 129–155. Zbl. 351.22005

Dani, S.G. (1977b): Sequential entropy of generalized horocycles. J. Indian Math. Soc. *41*, No. 1–2, 17–25. Zbl. 445.28017

Dani, S.G. (1977c): Some mixing multi-parameter dynamical systems. J. Indian Math. Soc. *41*, No. 1–2, 1–15. Zbl. 447.28018

Dani, S.G. (1980): Strictly non-ergodic actions on homogeneous spaces. Duke Math. J. *47*, No. 3, 633–639. Zbl. 447.28019

Dani, S.G. (1981): Invariant measures and minimal sets of horospherical flows. Invent. Math. *64*, 357–385. Zbl. 498.58013

Dani, S.G. (1982): On uniformly distributed orbits of certain horocycle flows. Ergodic Theory Dyn. Syst. *2*, No. 2, 139–158. Zbl. 504.22006

Dani, S.G. (1985): Divergent trajectories of flows on homogeneous spaces and diophantine approximation. J. Reine Angew. Math. *359*, 55–89. Zbl. 578.22012

Dani, S.G. (1986a): Orbits of horospherical flows. Duke Math. J. *53*, No. 1, 177–188. Zbl. 609.58038

Dani, S.G. (1986b): On orbits of unipotent flows on homogeneous spaces II. Ergodic Theory Dyn. Syst. *6*, No. 2, 167–182. Zbl. 601.22003

Dani, S.G. (1986c): Bounded orbits of flows on homogeneous spaces. Comment. Math. Helv. *61*, 636–660. Zbl. 627.22013

Dani, S.G., Keane, M. (1979): Ergodic invariant measures for actions of $SL(2,\mathbb{Z})$. Ann. Inst. Henri Poincaré, New Ser., Sect. B *15*, No. 1, 79–84. Zbl. 392.22018

Dani, S.G., Margulis, G.A. (1989): Values of quadratic forms at primitive integral points. Invent. Math. *98*, 405–424. Zbl. 682.22008

Dani, S.G., Margulis, G.A. (1992): On the limit distributions of orbits of unipotent flows and integral solutions of quadratic inequalities. C. R. Acad. Sci. Paris, Ser. I *314*, 699–704. Zbl. 780.22002

Ellis, R., Perrizo, W. (1978): Unique ergodicity of flows on homogeneous spaces. Isr. J. Math. *29*, No. 2–3, 276–284. Zbl. 383.22004

Feldman, J., Ornstein, D.E. (1987): Semi-rigidity of horocycle flows over compact surfaces of variable negative curvature. Ergodic Theory Dyn. Syst. *7*, No. 1, 49–72. Zbl. 633.58024

Flaminio, L. (1987): An extension of Ratner's rigidity theorem to n-dimensional hyperbolic space. Ergodic Theory Dyn. Syst. *7*, 73–92. Zbl. 623.57030

Flaminio, L., Spatzier, R.J. (1990): Geometrically finite groups, Patterson-Sullivan measures and Ratner's rigidity theorem. Invent. Math. *99*, No. 3, 601–626. Zbl. 667.57014

Furstenberg, H. (1963): The structure of distal flows. Am. J. Math. *85*, 477–515. Zbl. 199,272

Furstenberg, H. (1973): The unique ergodicity of the horocycle flow. In A. Beck (ed.): Recent Advances in Topological Dynamics, Proc. Conf. 1972. Lect. Notes Math. *318*, 95–115. Zbl. 256.58009

Gel'fand, I.M., Fomin, S.V. (1952): Geodesic flows on manifolds of constant negative curvature. Usp. Mat. Nauk *7*, No. 1, 118–137. Zbl. 48,92

Green, L. (1961): Spectra of nilflows. Bull. Am. Math. Soc. *67*, 414–415. Zbl. 99,391

Gurevich, B.M. (1961): Entropy of a horocycle flow. Dokl. Akad. Nauk SSSR *136*, 768–770. [English transl.: Sov. Math., Dokl. *2*, 124–126 (1961)] Zbl. 134,338

Hedlund, G.A. (1936): Fuchsian groups and transitive horocycles. Duke Math. J. *2*, No. 1, 530–542. Zbl. 15,102

Hedlund, G.A. (1939): The dynamics of geodesic flows. Bull. Am. Math. Soc. *45*, No. 4, 241–260. Zbl. 20,403

Hopf, E. (1936): Fuchsian groups and ergodic theory. Trans. Am. Math. Soc. *39*, 299–314. Zbl. 14,83

Hopf, E. (1939): Statistik der geodätischen Linien in Mannigfaltigkeiten negativer Krümmung. Ber. Verh. Sachs. Akad. Wiss., Leipzig, Math.-Nat. Kl. *91*, 261–304. Zbl. 24,80

Jacobson, N. (1962): Lie algebras. Interscience, New York London. Zbl. 121,275

Kornfel'd, I.P., Sinai, Ya.G., Fomin, S.V. (1980): Ergodic Theory. Nauka, Moscow. [English transl.: Springer-Verlag, Berlin Heidelberg New York 1982] Zbl. 493.28007

Kushnirenko, A.G. (1965): An estimate from above of the entropy of a classical dynamical system. Dokl. Akad. Nauk SSSR *161*, 37–38. [English transl.: Sov. Math., Dokl. *6*, 360–362 (1965)] Zbl. 136,429

Mañé, R. (1981): A proof of Pesin's formula. Ergodic Theory Dyn. Syst. *1*, No. 1, 95–102; (1983): *3*, No. 1, 159–160. Zbl. 489.58018; Zbl. 522.58031

Mañé, R. (1987): Ergodic Theory and Differentiable Dynamics. Springer-Verlag, Berlin Heidelberg New York. Zbl. 616.28007

Marcus, B. (1975): Unique ergodicity of the horocycle flow: the variable negative curvature case. Isr. J. Math. *21*, No. 2–3, 133–144. Zbl. 314.58013

Marcus, B. (1978): The horocycle flow is mixing of all degrees. Invent. Math. *46*, No. 3, 201–209. Zbl. 395.28012

Marcus, B. (1983): Topological conjugacy of horocycle flows. Am. J. Math. *105*, 623–632. Zbl. 533.58029

Margulis, G.A. (1971): On the action of unipotent groups in a space of lattices. Mat. Sb., Nov. Ser. *86*, 552–556. [English transl.: Math. USSR, Sb. *15*, 549–554 (1972)] Zbl. 257.54037

Margulis, G.A. (1987): Formes quadratiques indéfinies et flots unipotents sur les espaces homogènes. C. R. Acad. Sci. Paris, Ser. I *304*, No. 10, 249–253. Zbl. 624.10011

Margulis, G.A. (1989): Compactness of minimal closed invariant sets of actions of unipotent groups. [Preprint Math. No. 78, Inst. Hautes Etud. Sci., Bures sur Yvette;] see also Geom. Dedicata *37*, No, 1, 1–7 (1991). 733.22005

Margulis, G.A. (1991): Dynamical and ergodic properties of subgroup actions on homogeneous spaces with applications to number theory. Proc. ICM, Kyoto Japan 1990, Vol. I, 193–215. Zbl. 747.58017

Margulis, G.A., Tomanov, G.M. (1992): Measure rigidity for algebraic groups over local fields. C. R. Acad. Sci., Paris, Ser. I *315*, No. 12, 1221–1226

Mautner, F.J. (1957): Geodesic flows on symmetric Riemann spaces. Ann. Math., II. Ser. *65*, 416–431. Zbl. 84,375

Millionshchikov, V.M. (1976): A formula for the entropy of a smooth dynamical system. Differ. Uravn. *12*, 2188–2192. [English transl.: Differ. Equations *12*, 1527–1530 (1977)] Zbl. 357.34043

Moore, C.C. (1966): Ergodicity of flows on homogeneous spaces. Am. J. Math. *88*, 154–178. Zbl. 148,379

Moore, C.C. (1980): The Mautner pheonomenon for general unitary representations. Pac. J. Math. *86*, No. 1, 155–169. Zbl. 446.22014

Morse, H.M. (1921): Recurrent geodesics on a surface of negative curvature. Trans. Am. Math. Soc. *22*, 84-100. FdM 48,786

Mozes, S. (1992): Mixing of all orders of Lie group actions. Invent. Math. *107*, No. 2, 235–241. Zbl. 783.28011

Nicholls, P.J. (1989): The Ergodic Theory of Discrete Groups. Lond. Math. Soc. Lect. Note Ser. *143*, Cambridge. Zbl. 674.58001

Ornstein, D.S. (1974): Ergodic Theory, Randomness, and Dynamical Systems. Yale Univ. Press, New Haven London. Zbl. 296.28016

Ornstein, D.S., Weiss, B. (1973): Geodesic flows are Bernoullian. Isr. J. Math. *14*, No. 1, 184–198. Zbl. 256.58006

Parasyuk, O.S. (1953): Horocycle flows on surfaces of constant negative curvature. Usp. Mat. Nauk *8*, No. 3, 125–126. Zbl. 52,342

Parry, W. (1969): Ergodic properties of affine transformations and flows on nilmanifolds. Am. J. Math. *91*, 751–777. Zbl. 183,515

Parry, W. (1971): Metric classification of ergodic nilflows and unipotent affines. Am. J. Math. *93*, 819–828. Zbl. 222.22010

Pesin, Ya.B. (1977): Characteristic Lyapunov exponents and smooth ergodic theory. Usp. Mat. Nauk *32*, No. 4, 55–111. [English transl.: Russ. Math. Surv. *32*, No. 4, 55–114 (1977)] Zbl. 359.58010

Pesin, Ya.B. (1981): Geodesic flows with hyperbolic behaviour of the trajectories and objects associated with them. Usp. Mat. Nauk *36*, No. 4, 3–51. [English transl.: Russ. Math. Surv. *36*, No. 4, 1–59 (1981)] Zbl. 482.58002

Raghunathan, M.S. (1972): Discrete Subgroups of Lie Groups. Springer-Verlag, Berlin Heidelberg New York. Zbl. 254.22005

Ratner, M. (1978): Horocycle flows are loosely Bernoullian. Isr. J. Math. *31*, No. 2, 122–132. Zbl. 417.58013

Ratner, M. (1979): The Cartesian square of the horocycle flow is not loosely Bernoullian. Isr. J. Math. *34*, No. 1–2, 72–96. Zbl. 437.28010

Ratner, M. (1982a): Rigidity of horocycle flows. Ann. Math.. II. Ser. *115*, 597–614. Zbl. 506.58030

Ratner, M. (1982b): Factors of horocycle flows. Ergodic Theory Dyn. Syst. *2*, No. 3–4, 465–485. Zbl. 536.58029

Ratner, M. (1983): Horocycle flows, joinings and rigidity of products. Ann. Math., II. Ser. *118*, 277–313. Zbl. 556.28020

Ratner, M. (1987): The rate of mixing for geodesic and horocycle flows. Ergodic Theory Dyn. Syst. *7*, No. 2, 267–288. Zbl. 623.22008

Ratner, M. (1986): Rigidity of time changes for horocycle flows. Acta Math. *156*, No. 1–2, 1–32. Zbl. 694.58036

Ratner, M. (1990): Invariant measures for unipotent translations on homogeneous spaces. Proc. Natl. Acad. Sci. USA, 4309–4311

Ratner, M.(1991a): On Raghunathan's measure conjecture. Ann. Math., II. Ser. *134*, 545–607. Zbl. 763.28012

Ratner, M.(1991b): Raghunathan's topological conjecture and distributions of unipotent flows. Duke Math. J. *63*, No. 1, 235–280. Zbl. 733.22007

Ratner, M. (1993): to appear in Bull. Am. Math. Soc

Ryzhikov, V. (1991): Mixing property of a flow and its connection with the isomorphism between the transformation in the flow. Mat. Zametki *49*, No. 6, 98–106. [English transl.: Math. Notes *49*, 621–627 (1991)] Zbl. 777.28008

Safonov, A.V. (1982): On the spectral type of ergodic G-induced flows. Izv. Vyssh. Uchebn. Zaved., Mat. 1982, No. 6, (241), 42–47. [English transl.: Sov. Math. *26*, No. 6, 47–54 (1982)] Zbl. 498.22008

Schmidt, K. (1980): Asymptotically invariant sequences and an action of $SL(2,\mathbb{Z})$ on the 2-sphere. Isr. J. Math. *37*, No. 3, 193–208. Zbl. 485.28018

Schmidt, W.M. (1966): On badly approximable numbers and certain games. Trans. Am. Math. Soc. *123*, 178–199. Zbl. 232.10029

Shah, N. (1991): Uniformly distributed orbits of certain flows on homogeneous spaces. Math. Ann. *289*, 315–334. Zbl. 702.22014

Sinai, Ya.G. (1960): Geodesic flows on manifolds of constant negative curvature. Dokl. Akad. Nauk SSSR *131*, 752–755. [English transl.: Sov. Math., Dokl. *1*, 335–339 (1960)] Zbl. 129,311

Sinai, Ya.G. (1961): Geodesic flows on compact surfaces of negative curvature. Dokl. Akad. Nauk SSSR *136*, 549–552. [English transl.: Sov. Math., Dokl. *2*, 106–109 (1961)] Zbl. 133,110

Sinai, Ya.G. (1966): Classical dynamical systems with Lebesgue spectrum of countable multiplicity II. Izv. Akad. Nauk SSSR, Ser. Mat. *30*, No. 1, 15–68. [English transl.: Am. Math. Soc., Transl., II. Ser. *68*, 34–88 (1968)] Zbl. 213,340

Starkov, A.N. (1983): Ergodic behaviour of flows on homogeneous spaces. Dokl. Akad. Nauk SSSR *273*, 538–540. [English transl.: Sov. Math., Dokl. *28*, 675-676 (1983)] Zbl. 563.43010

Starkov, A.N. (1984): Flows on compact solvmanifolds. Mat. Sb., Nov. Ser. *123*, 549–556. [English transl.: Math USSR, Sb. *51*, 549–556 (1985)] Zbl. 545.28013

Starkov, A.N. (1986): On spaces of finite volume. Vestn. Mosk. Univ., Ser. I 1986, No. 5, 64–66. [English transl.: Mosc. Univ. Math. Bull. *41*, No. 5, 56–58 (1986)] Zbl. 638.22007

Starkov, A.N. (1987a): On an ergodicity criterion for G-induced flows. Usp. Mat. Nauk *42*, No. 3, 197–198. [English transl.: Russ. Math. Surv. *42*, No. 3, 233–234 (1987)] Zbl. 629.22009

Starkov, A.N. (1987b): Solvable homogeneous flows. Mat. Sb., Nov. Ser. *134*, 242–259. [English transl.: Math. USSR, Sb. *62*, No. 1, 243–260 (1989)] Zbl. 653.58034

Starkov, A.N. (1989): Ergodic decomposition of flows on homogeneous spaces of finite volume. Mat. Sb. *180*, 1614–1633. [English transl.: Math. USSR, Sb. *68*, No. 2, 438–502 (1989)]

Starkov, A.N. (1990): Structure of orbits of homogeneous flows and Raghunathan's conjecture. Usp. Mat. Nauk *45*, No. 2, 219–220. [English transl.: Russ. Math. Surv. *45*, No. 2, 227–228 (1990)] Zbl. 722.58034

Starkov, A.N. (1991): Horospherical flows on homogeneous spaces of finite volume. Mat. Sb. *182*, No. 5, 774–784. [English transl.: Math. USSR, Sb. *73*, No. 1, 161–170 (1992)] Zbl. 742.58042

Starkov, A.N. (1992): Rigidity criterion for lattices in solvable Lie groups. Workshop in Ergodic Theory and Lie groups, MSRI, April 1992, 59–60

Starkov, A.N. (1993): On mixing of higher orders of homogeneous flows. To appear in Dokl. Ross. Akad. Nauk

Stepin, A.M. (1969): Flows on solvmanifolds. Usp. Mat. Nauk *14*, No. 5, 241–242. Zbl. 216,345

Stepin, A.M. (1973): Dynamical systems on homogeneous spaces of semisimple Lie groups. Izv. Akad. Nauk SSSR, Ser. Mat. *37*, No. 5, 1091–1107. [English transl.: Math. USSR, Izv. *7*, 1089–1104 (1975)] Zbl. 314.28014

Sullivan, D. (1981): On the ergodic theory at infinity of an arbitrary discrete group of hyperbolic motions. Riemann surfaces and related topics. Proc. 1978 Stony Brook Conf., Ann. Math. Stud. *97*, 465–496. Zbl. 567.58015

Sullivan, D. (1982): Discrete conformal groups and measurable dynamics. Bull. Am. Math. Soc., New Ser. *6*, No. 1, 57–73. Zbl. 489.58027

Tomter, P. (1975): On the classification of Anosov flows. Topology *14*, No. 2, 179–189. Zbl. 365.58013

Veech, W.A. (1977): Unique ergodicity of horospherical flows. Am. J. Math. *99*, 827–859. Zbl. 365.28012

Vinberg, Eh.B., Gorbatsevich, V.V., Shvartsman, O.V. (1988): Lie Groups and Lie Algebras II: Discrete Subgroups of Lie Groups. Itogi Nauki Tekh., Ser. Sovrem. Probl. Mat., Fundam. Napravleniya *21*, 5–120. [English transl. in: Encycl. Math. Sci. *21*. Springer-Verlag, Berlin Heidelberg New York, in preparation] Zbl. 656.22004

Witte, D. (1985): Rigidity of some translations on homogeneous spaces. Bull. Am. Math. Soc., New Ser. *12*, No. 1, 117–119 [see also Invent. Math. *81*, 1–27 (1985)] Zbl. 558.58025

Witte, D. (1987): Zero-entropy affine maps on homogeneous spaces. Am. J. Math. *109*, 927–961. Zbl. 653.22005

Witte, D. (1989): Rigidity of horospherical foliations. Ergodic Theory Dyn. Syst. *9*, 191–205. Zbl. 667.58053

Witte, D. (1990): Topological equivalence of foliations of homogeneous spaces. Trans. Am. Math. Soc. *317*, 143–166. Zbl. 689.22006

Witte, D. (1992): Measurable quotients of unipotent translations on homogeneous spaces. To appear in Trans. Am. Math. Soc

Zimmer, R.J. (1984): Ergodic Theory and Semisimple Groups. Birkhäuser, Boston. Zbl. 571.58015

Author Index

Subject Index

Encyclopaedia of Mathematical Sciences
Editor-in-Chief: R. V. Gamkrelidze

Dynamical Systems

Volume 1: **D.V. Anosov, V.I. Arnol'd** (Eds.)
Dynamical Systems I
Ordinary Differential Equations and Smooth Dynamical Systems
2nd printing 1994. IX, 233 pp. 25 figs. ISBN 3-540-17000-6

Volume 2: **Ya.G. Sinai** (Ed.)
Dynamical Systems II
Ergodic Theory with Applications to Dynamical Systems and Statistical Mechanics
1989. IX, 281 pp. 25 figs. ISBN 3-540-17001-4

Volume 3: **V.I. Arnold, V.V. Kozlov, A.I. Neishtadt**
Dynamical Systems III
Mathematical Aspects of Classical and Celestial Mechanics
2nd ed. 1993. XIV, 291 pp. 81 figs. ISBN 3-540-57241-4

Volume 4: **V.I. Arnol'd, S.P. Novikov** (Eds.)
Dynamical Systems IV
Symplectic Geometry and its Applications
1990. VII, 283 pp. 62 figs. ISBN 3-540-17003-0

Volume 5: **V.I. Arnol'd** (Ed.)
Dynamical Systems V
Bifurcation Theory and Catastrophe Theory
1994. IX, 271 pp. 130 figs. ISBN 3-540-18173-3

Volume 6: **V.I. Arnol'd** (Ed.)
Dynamical Systems VI
Singularity Theory I
1993. V, 245 pp. 55 figs. ISBN 3-540-50583-0

Volume 16: **V.I. Arnol'd, S.P. Novikov** (Eds.)
Dynamical Systems VII
Nonholonomic Dynamical Systems. Integrable Hamiltonian Systems.
1994. VII, 341 pp. 9 figs. ISBN 3-540-18176-8

Vol. 39: **V.I. Arnol'd** (Ed.)
Dynamical Systems VIII
Singularity Theory II. Applications
1993. V, 235 pp. 134 figs. ISBN 3-540-53376-1

Vol. 66: **D.V. Anosov** (Ed.)
Dynamical Systems IX
Dynamical Systems with Hyperbolic Behavior
1995. VII, 235 pp. 39 figs. ISBN 3-540-57043-8

Partial Differential Equations

Volume 30: **Yu.V. Egorov, M.A. Shubin** (Eds.)
Partial Differential Equations I
Foundations of the Classical Theory
1991. V, 259 pp. 4 figs. ISBN 3-540-52002-3

Volume 31: **Yu.V. Egorov, M.A. Shubin** (Eds.)
Partial Differential Equations II
Elements of the Modern Theory. Equations with Constant Coefficients
1995. VII, 263 pp. 5 figs. ISBN 3-540-52001-5

Volume 32: **Yu.V. Egorov, M.A. Shubin** (Eds.)
Partial Differential Equations III
The Cauchy Problem. Qualitative Theory of Partial Differential Equations
1991. VII, 197 pp. ISBN 3-540-52003-1

Volume 33: **Yu.V. Egorov, M.A. Shubin** (Eds.)
Partial Differential Equations IV
Microlocal Analysis and Hyperbolic Equations
1993. VII, 241 pp. 6 figs. ISBN 3-540-53363-X

Volume 63: **Yu.V. Egorov, M.A. Shubin** (Eds.)
Partial Differential Equations VI
Elliptic and Parabolic Operators
1994. VII, 325 pp. 5 figs. ISBN 3-540-54678-2

Volume 64: **M.A. Shubin** (Ed.)
Partial Differential Equations VII
Spectral Theory of Differential Operators
1994. V, 272 pp. ISBN 3-540-54677-4

DURHAM UNIVERSITY LIBRARY
3 0104 00819017 9